Chorea

Federico E. Micheli • Peter A. LeWitt

Editors

Chorea

Causes and Management

 Springer

Editors
Federico E. Micheli, MD, PhD
Parkinson Disease and Movement Disorders
Program and Neurology Department
Hospital de Clínicas, José de San Martín
Buenos Aires
Argentina

Peter A. LeWitt, MD
Parkinson's Disease and Movement
Disorders Program
Henry Ford West Bloomfield Hospital
West Bloomfield
Michigan
USA

ISBN 978-1-4471-6454-8   ISBN 978-1-4471-6455-5   (eBook)
DOI 10.1007/978-1-4471-6455-5
Springer London Heidelberg New York Dordrecht

Library of Congress Control Number: 2014943524

Springer is part of Springer Science+Business Media (www.springer.com)

To my family whose emotional and intellectual support has served as an inspiration to make this book possible. To my patients, from whom I learn every single day.

Federico E. Micheli

Dedicated, in appreciation of their love and support, to Jan, Eli, and Tessa, and to my parents Bernard and Celeste.

Peter A. LeWitt

Prologue

Within the broad categories of neurologic diseases, the abnormal movement *chorea* (from Greek, *choros*, and from the Latin, *choreus*, both meaning dance) belongs among the conditions referred to as movement disorders. The field of movement disorders consists of both hypokinetic and hyperkinetic conditions. The hypokinetic group comprises those disorders that manifest slowness and decreased amplitude of movement, of which parkinsonism is by far the major type and the prime example and has early cardinal signs of bradykinesia, rigidity, and tremor-at-rest. In contrast, the hyperkinetic group has a much larger range of disorders, represented by different motor phenomenologies. The big six in this group – in alphabetical order – are the ataxias, choreas, dystonias, myoclonias, tics and tremors. The first step in diagnosing these hyperkinetic conditions is the precise awareness of the phenomenology so that the correct type of hyperkinesia can be identified. Then the clinical workup searching for a specific etiology of the abnormal movements is undertaken; this can include genetic testing, neuroimaging, investigations for neuroinflammation, hepatic and other metabolic disorders, and other possible etiologies. Depending on the etiology, the appropriate treatment is then applied [5].

Drs. Micheli and LeWitt have embarked on an ambitious project, putting together in a single monograph the complex field of the choreas. Other than tremor (mentioned in the Bible), chorea was the earliest hyperkinesia to be described [1]. Thomas Sydenham, an English physician known as the English Hippocrates because of his keen observations and sound clinical judgment, gave the fullest description of the phenomenology of chorea in children with the condition now referred to as Sydenham chorea in 1686 [13], more than 100 years earlier than James Parkinson [12] described the condition now bearing his name. Sydenham did not use the term *chorea*, but called the condition he was describing as *St. Vitus' Dance*. This was a misapplied term, resulting in confusion, for St. Vitus' Dance was initially used in reference to the dancing mania that occurred in waves of epidemics in the middle ages (fourteenth and fifteenth centuries). As interpreted by Paracelsus, these dancing manias were likely psychogenic in etiology, and today we would describe them as a form of mass hysteria. (Note: Paracelsus' born name was Philippus Aureolus Theophrastus Bombastus von Hohenheim. The English translation of his writing about the dancing mania is given in monographs by Temkin and colleagues [14] and Ehrenwald [4]).

According to Barbeau [1], Bruyn [3], Hayden [8], and Harper [7], Paracelsus was the first to use the term chorea (in the sixteenth century) when he discussed the dancing mania (St. Vitus' Dance). Paracelsus called this *chorea lasciva* (chorea from voluptuous desires). He offered two other types of chorea, namely *chorea imaginativa aestimative* (caused by imagination) and *chorea naturalis coatta* (arising from an organic cause). But Paracelsus provided no detailed description of the choreic movements we recognize today; that credit goes to Sydenham. The chorea described by Sydenham was subsequently referred to as *chorea minor*, and the dancing mania as *chorea major* [8].

Following Sydenham's publication, it was likely that patients with a variety of abnormal involuntary movements were subsequently being called chorea, as emphasized by William Osler in his treatise, *On Chorea and Choreiform Affections* [11]. Attempts to classify all the movement disorders was rather recent in medical history. Baumann [2] attempted a classification based on neuroanatomical localization. Lakke et al. [10] advanced the field by classifying the conditions based on phenomenology of the movements, posture and muscle tone (see history of classification in Fahn [5], which also provides a modern classification). An advance in classifying the choreas was the growing recognition of hereditary forms of chorea. A number of such reports were published prior to the paper of George Huntington in 1872 [9], but it was Huntington's precise and brief description that made a major impact. What made his description stand out, besides the hereditary aspect and the abnormal movements, were his inclusion of features of inappropriate behavior that could lead to insanity and a relentlessly progressive course of the illness.

The abnormal movements of chorea are involuntary, irregular, purposeless, non-rhythmic, abrupt, rapid, and unsustained that seem to flow from one body part to another (from Fahn et al. [6]). Choreic movements are unpredictable in timing, direction, and distribution (i.e., random). Distinguishing chorea from dystonic movements, tics and myoclonus is not always easy, and reports in the literature of unusual cases of chorea need to be viewed with caution, for they could have simply been misdiagnosed. Choreic movements can be of small amplitude involving just the digits of the fingers or toes to flinging and flailing limb movements of the entire limb; the latter condition is usually called ballism, which is simply chorea of extremely large amplitude movements. The phenomenology of choreic movements can differ slightly from one disorder to another. For example, the chorea of Sydenham disease and of withdrawal emergent syndrome (seen in children who were suddenly withdrawn from chronic treatment with dopamine receptor blocking agents) have a restless appearance with fidgetiness, in contrast to the more brief and abrupt involuntary movements seen in Huntington disease and the slightly slower abnormal movements of choreoathetosis seen in hyperosmotic hyperglycemic non-ketotic syndrome that can occur in patients with diabetes mellitus with extremely high blood sugar levels. When chorea is somewhat slow, it thereby merges into the mobile spasms of athetosis and the twisting movements of dystonia; this combination is usually referred to as choreoathetosis.

This monograph edited by Drs. Micheli and LeWitt provides the details and revealing discussions to explain the various disorders that can produce chorea. These chapters will provide rewarding reading to both the expert movement disorder specialist and to others in the health profession with less subspecialization. There is much valuable information in this book.

References

1. Barbeau A. The understanding of involuntary movements: an historical approach. J Nerv Ment Dis. 1958;127(6):469–89.
2. Baumann J. The classification of the diseases of the extrapyramidal system. Acta Neurol Scand Suppl. 1963;39(S4):102–7.
3. Bruyn GW. Huntington's chorea: historical, clinical and laboratory synopsis. In: Vinken PJ, Bruyn GW, editors. Handbook of clinical neurology. Vol. 6, Amsterdam: North-Holland Publ Co; 1968. p. 300–1.
4. Ehrenwald J. Chapter 14. Paracelsus, the magician who turned scientist. In: Ehrenwald J, editor. The history of psychotherapy: from healing magic to encounter. Northvale: Jason Aronson Inc.; 1991. p. 199–200. Available in Google Books.
5. Fahn S. Classification of movement disorders. Mov Disord. 2011;26(6):947–57.
6. Fahn S, Jankovic J. Hallett M. Chapter 1. Principles and practice of movement disorders. 2nd ed. Edinburgh: Saunders Elsevier; 2011.
7. Harper P, Morris M. Chapter 1. Introduction: a historical background. In: Harper PS, editor. Huntington's disease. London: W.B. Saunders Ltd.; 1991. p. 9.
8. Hayden MR. Huntington's Chorea. Berlin: Springer; 1981.
9. Huntington G. On chorea. Medical and surgical reporter. 1872;26:320–1.
10. Lakke JPWF, editor. Classification of extrapyramidal disorders. J Neurol Sci. 1981;51: 311–27.
11. Osler W. On chorea and choreiform affections. Philadelphia: Blakison; 1894.
12. Parkinson J. An essay on the shaking palsy. London: Sherwood, Neely, and Jones; 1817.
13. Sydenham T. Schedula Monitoria de Novae Febris Ingressu. London: G. Kettilby; 1686.
14. Temkin CL, Rosen C, Zilboorg G, Sigerist HE. Theophrastus von Hohenheim called Paracelsus. Baltimore: Johns Hopkins Press; 1941.

New York, NY, USA   Stanley Fahn, MD

Contents

Contributors

Roger L. Albin, MD Department of Neurology, University of Michigan, Ann Arbor, MI, USA

Neurology Service and GRECC, VAAAHS, Ann Arbor, MI, USA

Vanderci Borges Movement Disorders Unit, Departments of Neurology and Neurosurgery, Universidade Federal de São Paulo, São Paulo, Brazil

Francisco Cardoso, MD, PhD Movement Disorders Unit – Neurology Service, Department of Internal Medicine, The Federal University of Minas Gerais, Belo Horizonte, MG, Brazil

Vanessa Cavallera Department of Pediatric Neurology, Fondazione IRCCS Istituto Neurologico Carlo Besta, Milan, Italy

Department of Pediatric Neuroscience, Istituto Neurologico Carlo Besta, Milan, Italy

Luisa Chiapparini Fondazione IRCCS Istituto Neurologico Carlo Besta, Milan, Italy

Department of Neuroradiology, Istituto Neurologico Carlo Besta, Milan, Italy

Emilio Fernández-Alvarez Neuropediatric Department, Hospital Sant Joan de Deu-Barcelona, Lugo, Spain

Héctor Alberto González-Usigli Department of Neurology, Instituto Mexicano del Seguro Social, Guadalajara, Mexico

Felix Gövert Department of Neurology, University of Kiel, Kiel, Germany

Juan Carlos Giugni, MD Department of Neurology, University of Florida Center for Movement Disorders and Neurorestoration, Gainesville, FL, USA

Mônica Santoro Haddad Department of Neurology, Hospital das Clinicas da Faculdade de Medicina da Universidade São Paulo, São Paulo, Brazil

Andreas Hermann Division for Neurodegenerative Diseases,
Department of Neurology, Dresden University of Technology, Dresden, Germany

DZNE, German Centre for Neurodegenerative Diseases, Research Site Dresden,
Dresden, Germany

Hans H. Jung Department of Neurology, University Hospital Zürich,
Zürich, Switzerland

Peter A. LeWitt, MD Department of Neurology, Henry Ford Hospital –
West Bloomfield, West Bloomfield, MI, USA

Department of Neurology, Wayne State University School of Medicine,
Detroit, MI, USA

Raul Martinez-Fernandez Movement Disorder Unit, Department of Psychiatry
and Neurology, CHU de Grenoble, Joseph Fourier University, Grenoble, France

INSERM, Unit 836, Institut des Neurosciences, Grenoble, France

Daniel Martínez-Ramírez, MD Department of Neurology, University of Florida
Center for Movement Disorders and Neurorestoration, Gainesville, FL, USA

Simon Mead MRC Prion Unit, Department of Neurodegenerative Disease,
UCL Institute of Neurology, University College London, London, UK

Federico E. Micheli, MD, PhD Parkinson's Disease and Movement Disorders
Program, Neurology Department, University of Buenos Aires, Ciudad Autónoma
de Buenos Aires, Buenos Aires, Argentina

Neurology Department, Hospital de Clínicas José de San Martín,
University of Buenos Aires, Ciudad Autónoma de Buenos Aires, Buenos Aires,
Argentina

Hiroaki Miyajima First Department of Medicine, Hamamatsu University School
of Medicine, Hamamatsu, Japan

Elena Moro Movement Disorder Unit, Department of Psychiatry and Neurology,
CHU de Grenoble, Joseph Fourier University, Grenoble, France

INSERM, Unit 836, Institut des Neurosciences, Grenoble, France

Renato Puppi Munhoz Movement Disorders Unit, Neurology Service,
Internal Medicine Department, Hospital de Clínicas, Federal University of Paraná,
Curitiba, PR, Brazil

Department of Neurology, University of Toronto, Toronto, ON, Canada

Ignacio Muñoz-Sanjuan CHDI Management/CHDI Foundation Inc.,
Los Angeles, CA, USA

Nardocci Nardo Department of Pediatric Neurology,
Fondazione IRCCS Istituto Neurologico Carlo Besta, Milan, Italy

Department of Pediatric Neuroscience, Istituto Neurologico Carlo Besta,
Milan, Italy

Michael Orth, MD, PhD Department of Neurology and European Huntington's
Disease Network, Ulm University Hospital, Ulm, Germany

Claudia Perandones National Center for Medical Genetics, National Agency
of Laboratories and Health Institutes of Argentina (ANLIS) Dr. Carlos G.
Malbrán, Buenos Aires, Argentina

Marie-Claire Porter The National Prion Clinic, The National Hospital
for Neurology and Neurosurgery, University College London Hospitals NHS
Trust, London, UK

Ramon L. Rodríguez-Cruz, MD Department of Neurology, University of
Florida Center for Movement Disorders and Neurorestoration,
Gainesville, FL, USA

Susanne A. Schneider Department of Neurology, Christian-Albrechts-University
Kiel, University-Hospital-Schleswig-Holstein, Kiel, Germany

Roberta Arb Saba Movement Disorders Unit, Department of Neurology and
Neurosurgery, Universidade Federal de São Paulo, São Paulo, Brazil

Jon Snider, MD Department of Neurology, University of Michigan, Ann Arbor,
MI, USA

Hélio A. Ghizoni Teive Movement Disorders Unit, Neurology Service,
Internal Medicine Department, Hospital de Clínicas, Federal University of Paraná,
Curitiba, PR, Brazil

Giovanna Zorzi Department of Pediatric Neurology,
Fondazione IRCCS Istituto Neurologico Carlo Besta, Milan, Italy

Department of Pediatric Neuroscience, Istituto Neurologico Carlo Besta,
Milan, Italy

Carlos Zúñiga-Ramírez Movement Disorders and Neurodegenerative
Diseases Unit, Hospital Civil de Guadalajara "Fray Antonio Alcalde",
Guadalajara, Mexico

Chapter 1
A Clinician's Approach to Chorea

Peter A. LeWitt

Abstract Chorea is a distinctive movement disorder with a long history of recognition by physicians but little understanding as to its causes. It has a variety of etiologies, including acquired and genetic causes. The workup of a patient with chorea often depends on the "company" that this movement disorder keeps. Various forms of laboratory testing can enhance a clinical diagnosis.

Keywords Chorea • Dyskinesia • Clinical diagnosis • Differential diagnosis • History of chorea

A Clinician's Approach to Chorea

For centuries, the distinctive movement disorder termed *chorea* has held a particular mystique for physicians, adding to many the challenges it presents. The current monograph is just the latest in a long series of medical expositions dealing with this disorder. In 1810, the first text devoted exclusively to chorea, *Traité de la Chorée, ou Danse de St. Guy*, was published by Dr. Etienne Michel Bouteille. For its preface, the 78-year-old author wrote that:

> ...everything is extraordinary in this disease, its name ridiculous, its symptoms peculiar, its character equivocal, its cause unknown, its treatment problematic. Serious authors doubted its existence, others believe it simulated, some considered it supernatural [5].

P.A. LeWitt, MD
Department of Neurology, Henry Ford Hospital – West Bloomfield,
6777 West Maple Road, West Bloomfield, MI 48322, USA

Department of Neurology, Wayne State University School of Medicine, Detroit, MI, USA
e-mail: plewitt1@hfhs.org

F.E. Micheli, P.A. LeWitt (eds.), *Chorea*,
DOI 10.1007/978-1-4471-6455-5_1, © Springer-Verlag London 2014

Fortunately, the modern view of this disorder has become more enlightened, and "serious authors" recognize chorea for what it is. The frustration voiced in Bouteille's comments remains pertinent today, however. As a clinical syndrome with diverse origins (see Table 1.1), chorea can sometimes be a difficult diagnosis to make, and its phenomenology still can't be fully explained by contemporary medical science. Successful management of this problem is not always achievable by current medications or neurosurgical intervention. Nonetheless, recent years have brought considerable progress to understanding of chorea. The chapters in the current monograph reflect this progress, representing the work of an international group of experts who have diligently updated the breadth of experience and current knowledge pertaining to chorea.

Over the past two centuries, several publications have provided important milestones for better understanding of chorea. One of the most enlightening texts was the 1894 monograph by William Osler, M.D. His *On Chorea and Choreiform Affections* was written in the grand style of erudite medical exposition that was also to be found in the contemporaneous writings of the British neurologist William Gowers M.D. (to whom Osler dedicated this publication). Osler provided an encyclopedic view of chorea, bolstered by his extensive understanding and clinical experience with the varieties of involuntary movements. This text was written in an era when Sydenham's chorea, chorea during pregnancy, Huntington disease (HD), and even canine chorea had become well-recognized entities. Even today, Osler's words come alive to illustrate chorea's impact on the patient. Describing the features of HD, he characterized the clinical picture as:

> …Though irregular, involuntary and arrhythmic, the movements and progressive form differ in one important particular from those of Sydenham's chorea, namely, in the absence of the brusque, quick, jerking character. At first indeed the condition is one rather of muscular instability or inquietude, and when the patient is at rest there are irregular movements of the muscles of the hand or arms, which perhaps scarcely alter the position of the limb, or slight, slow contractions past over the muscles of the face. There may be no movements of the hand or arms when in repose, but any attempt at grasping an object may be associated with large, irregular, sweeping movements… ([5], pp. 106–7).

While his analysis of brain pathology was unable to discern the disease process, Osler nonetheless assembled and interpreted as much information as was possible in an era just at the cusp of the medical scientific revolution. He correctly viewed the clinical approach to chorea as in need of a broad differential diagnosis among a variety of movement disorders. If Osler were with us today, the editors of this text would certainly welcome his input!

As in earlier times, the current clinical approach to chorea requires the physician to be a careful observer and well versed in what defines and differentiates the other movement disorders. Since chorea shares some similarities with other hyperkinetic disorders, it needs to be clearly distinguished from them. A precise definition of chorea is not easy to pin down. However, the following concepts about chorea have repeatedly appeared in its descriptions:

> Irregular and brief, jerky and stereotyped involuntary movements that tend to flow from one body part to another in random and sometimes continuous sequences.

Table 1.1 Disorders with chorea

Hereditary disorders
Huntington disease
Huntington disease-like disorders (HD-1, HD-2, HD-3)
Benign hereditary chorea (types 1 and 2)
Dentatorubral-pallidoluysian atrophy
Neuroacanthocytosis
Wilson disease
Spinocerebellar ataxia (types 1, 3, 17, 27)
Pantothenate kinase-associated neurodegeneration
Neuroferritinopathy
Ataxia telangiectasia
Ataxia with oculomotor apraxia
Aceruloplasminemia
Benign hereditary chorea
Infantile neuroaxonal dystrophy
Paroxysmal kinesigenic dyskinesia
Friedreich ataxia
Intoxications
Carbon monoxide
Manganese
Severe hypoxia
Toluene
Mercury
Thallium
Ethanol
Drug induced
Tardive dyskinesia
Withdrawal emergent syndrome
Dopaminergic drugs (in parkinsonism)
Anticholinergics (in parkinsonism)
Amphetamines
Cocaine
Antidepressants
Anticonvulsants
Estrogens
Aminophylline
Vascular disorders
Subthalamic nucleus infarct
Moyamoya disease
Metabolic and other endogenous disorders
Leigh disease
Porphyria
Glucose transporter type 1 deficiency
Hypoglycemia
Hyperglycemia (nonketotic)
Hyperthyroidism
Hypoparathyroidism
Hypocalcemia

(continued)

Table 1.1 (continued)

Hypomagnesemia
Hyper- and hyponatremia
Uremia
Ketosis
Vitamin B12 deficiency
Thiamine deficiency
Lesch-Nyhan syndrome
Glutaric aciduria
GM1 gangliosidosis
Propionic acidemia
Galactosemia
GM1 and GM2 gangliosidosis
Various mitochondrial disorders
Immunological and hematological disorders
Acquired immunodeficiency disease (AIDS)
Systemic lupus erythematosus
Acute disseminated encephalomyelitis
Henoch-Schönlein purpura
Antiphospholipid syndrome
Polycythemia vera
Multiple sclerosis
Infectious and postinfectious disorders
Sydenham's chorea
Poststreptococcal acute disseminated encephalomyelitis
Measles encephalitis
Creutzfeldt-Jakob disease
Encephalitis lethargica
Mycoplasma
Lyme disease
Behçet disease
Anatomical sites for lesions producing chorea
Subthalamic nucleus
Putamen
Globus pallidus externa
Caudate nucleus (usually with other locations as well)
Thalamic degeneration
Diffuse cerebrovascular disease
Other acquired causes
Sarcoidosis
Kernicterus
Multiple infarcts of the white matter and basal ganglia
Hepatocerebral degeneration
Paraneoplastic syndromes
Chorea gravidarum
Postpartum chorea
Psychogenic

Although this 24-word description may be useful as a point of departure in understanding chorea, it doesn't fully engage the full range of its manifestations, nor does it embody all of its characteristic features (such as its temporary worsening during mental concentration or simultaneous volitional motor actions). The best definition is a clinician's personal recollection of a choreatic patient and the distinctive impairments of motor control that these individuals experience. In the past, when video examples of chorea were not available for medical education, words capturing of this movement disorder and its impact on everyday life had to suffice. For this purpose, there is nothing better than Sydenham's 1688 account of chorea to paint a striking portrait of the disorder:

> ... first, it shews itself by a certain lameness, or rather instability of one of the legs, which the patient drags after him like a fool; afterwards, it appears in the hand of the same side; which, he that is affected with this disease, can by no means keep in the same posture for one moment, if it be brought to the breast or other part, but it will be distorted to another position or place by a certain convulsion, let the patient to what he can. If a cup drink be put in his hand, he represents a thousand gestures, like jugglers, before he can bring it to his mouth; for whereas he cannot carry it to his mouth, in a right-line, his hand being drawn hither and thither by the convulsion, he turns it often about for some time, till at length, happily reaching his lips, he flings it suddenly into his mouth and drinks it greedily, as if the poor wretch designed only to make sport. ([10], p. 704)

The reader can appreciate the fact that contemporary clinical notes are lacking in the poetry and precise characterizations that were once the norm of medical writing.

Today, clinicians encountering "chorea" need to be mindful that commonly it is not found in isolation. Enhancing one's diagnostic capabilities regarding chorea can come from analyzing the company it keeps. For example, characteristic choreic movements can be seen in the same patient (even the same limb) that is also manifesting dystonic posturing. For this, the term "choreoathetosis" has been used, reflecting the intermingling of slow, torsional movements ("athetosis") with chorea. An arm afflicted with choreoathetosis might show features of slow, repetitive pronation and supination in the forearm together with fingers engaged in the piano-playing-like movements typical of chorea. In such an instance, applying the more precise description "choreoathetosis" rather than "chorea" might help to focus diagnostic considerations on an acquired disorder like cerebral palsy (for which this pattern of movement disorder is quite characteristic). "Ballism," another pattern of hyperkinetic movement disorder, should also be considered in the spectrum of chorea. Sudden, thrusting involuntary movements of an acutely ballistic limb can evolve, over weeks, to the gentler and flowing movements of chorea. The same brain lesion (typically an infarct) in the subthalamic nucleus can lead either to ballism, to chorea, or to a combination of the two. Tremor is also a hyperkinetic movement disorder that, like chorea, can arise from basal ganglia structures. The two conditions should not be difficult to distinguish: chorea never possesses the constrained regularity of to-and-fro movement that tremor exhibits.

Sometimes it isn't an easy task to distinguish among the various hyperkinetic movement disorders. Even movement disorder specialists need to hone their clinical skills at differentiating the various hyperkinetic conditions. Video teaching files that illustrate different movement disorders are readily available from the journal *Movement Disorders* and others. Viewing typical and not-so-typical cases of hyperkinetic movement disorders is a valuable exercise for the practicing clinician, especially since words alone and the foggy recollections of past medical education generally need better guideposts for sorting out the shakes, jerks, twitches, and posturings of patients encountered in everyday practice.

A long list of medical conditions can be associated with chorea. Table 1.1 provides the categorization of different etiologies that can serve as a starting point in establishing a diagnosis. New entities associated with chorea are occasionally added to the medical literature. Among the major categories listed are chronic and hereditary disorders as well as focal lesions, systemic afflictions, and various chemical exposures that lead to acquired dyskinesias. Perusing this list may be useful for helping the reader to maintain an open mind as to the broad differential diagnosis possible for a patient affected with chorea. The key information for correctly determining the disease category sometimes needs knowledge as to the duration of symptomatology, associated neurological deficits, and family medical history (if available). Many of these conditions are being recognized as genetic in origin. An increasing list of genetic testing options has become available. However, not every new choreic patient needs evaluation for biomarkers of HD, dentatorubral-pallidoluysian atrophy (DRPLA) or the various forms of spinocerebellar atrophy (SCA). One of the major threats for reliance on genetic testing for diagnosis is the incompleteness of screens for these disorders and some of the others. Wilson disease, for example, has hundreds of mutations in the human population and so a single panel of genetic tests may be insufficient for achieving diagnosis. HD presents another diagnostic quandary. While an excess ($\geq$37) of cytosine-adenine-guanine (CAG) exon repeats generally provides strong diagnostic evidence, rare cases of HD are known to manifest CAG repeats in the mid-30s borderland zone or even in seemingly normal territory (<30 repeats) [4].

Diagnosis of chorea (or dyskinesia, a term that is virtually interchangeable in the medical literature) is often enhanced by what arises from the patient's medical history. As should be obvious, knowledge that a child has whose pharyngitis was followed by carditis or other rheumatic manifestations should ring diagnostic bells for considering chorea (even though clinicians should also recognize that some children have tics that readily fit the definition of chorea provided above). Few clinicians today regularly consult Wilson's 1940 textbook of *Neurology*, but the insight he provides as to the history of a child with Sydenham's chorea still provides useful diagnostic clues:

> Someone has said that a choreic child is punished thrice ere his condition is recognized – once for general fidgetiness, once for breaking crockery, and once for making faces at his grandmother. [10]

One of the diagnostic challenges for establishing the duration of a choreic disorder is that patients may lack awareness that they are unable to hold still or that normal movements are embellished. The main complaints of someone with chorea can be something different: weakness, incoordination, or clumsiness. Stumbling leading to falls or near-falls and speech that lacks proper volume or precise enunciation can also be prominent in choreic individuals. It is also important to keep in mind that many patients with disorders manifesting chorea also have other simultaneous motor impairments. For example, ataxia can be seen in concert with chorea in the SCAs [7]. A patient with evolving hemichorea might have previously manifested hemiballism coming on acutely due to a subthalamic nucleus infarct [2]. A patient with chorea induced by one of the various anticonvulsants may also exhibit an action tremor from the same drug. Looking through the conditions in Table 1.1, perhaps only half on the list have manifestations of chorea seen in isolation from other movement disorders.

Examining the patient with a family history that might be HD has a special challenge for the clinician. In its earliest stages, the choreic features of this disorder can be quite minor or intermittent. Observations during a motor task like handwriting or tapping thumb to each of the fingers may bring out involuntary movements not seen with the patient at rest or conversing. The subtle features of dyskinesia sometimes can be seen in the tongue of the half-opened mouth, with the tongue half-extended and held in that position for up to 30 s. Testing for the "milkmaid's grip" (holding the examiner's finger in the clenched hand of the patient) also permits detection of inconstant muscular contractions. Extending the hands forward with spread fingers provides another way to examine for choreic movements. It is also important to keep in mind that chorea in the HD patient usually keeps company with other neurological impairments, especially impaired rapid alternating movements and slowed saccadic pursuit of lateral gaze (which can be the earliest feature of HD, long before onset of other motor impairments) [6]. Finally, clinicians should realize that some persons with HD never manifest chorea. Proven cases of HD have revealed presentations of different movement disorders – parkinsonism, dystonia, or spinocerebellar ataxia [3] – instead of chorea. The availability of gene testing and high-detail MR imaging of the brain might seem to obviate the clinician's need to make the diagnosis by careful examination. However, some gene-positive patients may wish to ascertain if there are any observed clues that they have left the presymptomatic realm.

Another approach to the choreic patient has to do with the neurodiagnostic tools that can augment the clinical examination. CT and MRI pictures of the brain can show the loss of caudate nucleus bulk in HD and DRPLA as well as cerebellar and brainstem atrophy of the SCAs. 2-Deoxyglucose PET and 99mTc-hexamethylpropyleneamine oxime SPECT scans (each imaging neuronal metabolism) can also add additional information. In an analysis of cases in which focal lesions of the caudate nucleus were associated with neurological outcomes, only a small fraction exhibited chorea [1]. Hence, the anatomical substrate for chorea is not readily discerned for most cases. Lesions in the subthalamic nucleus (STN) offer another target for correlating neuroimages with the development of chorea. After initially experiencing hemiballism from the damage to the STN, some patients gradually lose the violent flinging movements and evolve chorea in its place. EEG may also add information to the diagnosis; recordings in HD commonly show relatively low voltage and little rhythmic activity [9]. For other choreic disorders, EEG and other electrophysiological investigations tend not to be informative.

Anatomical locations where chorea arises in the brain are inadequately understood. The most robust connection between anatomy and symptomatology can be found at the subthalamic nucleus, where hemiballism and hemichorea can originate from acute causes (e.g., infarcts) or from more gradual lesions (such as an arteriovenous malformations or a tumor). Damage to other structures of the basal ganglia is less likely to result in chorea. The putamen and globus pallidus externa are sites where lesions have been associated with choreic movements [1]. Even though the HD caudate nucleus loses a major portion of its bulk through neuronal dropout, the prominent chorea in this disorder is not necessarily a consequence of this change. HD patients with little or no chorea also show caudate atrophy. The more likely neural substrate for involuntary movements in choreic disorder is the realm of neurochemical pathology or altered neuronal connections. In this regard, the similarity of levodopa-induced dyskinesias (LID) in Parkinson's disease may be a relevant model. Though either the presynaptic deficiency in dopaminergic innervation or postsynaptic receptors are strongly suspected to be the basis for LID to evolve, other possible mechanisms are also recognized. Recent findings implicate cortical resonant states may be involved in the pathophysiology of LID [8]. Other research has highlighted postsynaptic changes (downstream from the dopamine receptors in the direct and indirect striatal outflow pathways). Movement disorders with similarities to LID (like chorea) may also arise from dysfunction in neural networks composed of interacting components at several levels of cortical and subcortical function.

Dopaminergic neurotransmission is suspected to be a key factor in chorea, since the most effective means of pharmacological control can be achieved by drugs that deplete or block these pathways. In HD, the important role of dopaminergic neurotransmission in generating chorea was demonstrated in experiments carried out in patients at risk for the disorder but presymptomatic. During acute challenges of levodopa administration, choreic movements were unmasked [11].

In summary, the finding of chorea offers a clinical vantage point of high value for differentiating among the various spinocerebellar atrophies or for suspecting HD when other clues are lacking. For a variety of systemic disorders or other neurological conditions, chorea can also be a prominent but sometimes nonspecific finding. Rarely, chorea can be a feigned or functional movement disorder [12], joining a long list of other psychogenic movement disorders. The clinician's approach to the patient with chorea, therefore, should be with an open mind as to a multiplicity of diagnostic possibilities.

References

1. Bhatia KP, Marsden CD. The behavioural and motor consequences of focal lesions of the basal ganglia in man. Brain. 1994;117:859–76.
2. Dewey Jr RB, Jankovic J. Hemiballism-hemichorea: clinical and pharmacologic findings in 21 patients. Arch Neurol. 1989;46:862–7.
3. Dong Y, Sun YM, Liu ZJ, Ni W, Shi SS, Wu ZY. Chinese patients with Huntington's disease initially presenting with spinocerebellar ataxia. Clin Genet. 2013;83:380–3.
4. Ha AD, Jankovic J. Exploring the correlates of intermediate CAG repeats in Huntington disease. Postgrad Med. 2011;123:116–21.
5. Osler W. On chorea and choreiform affections. London: H.K. Lewis; 1894.
6. Penney Jr JB, Young AB, Shoulson I, et al. Huntington's disease in Venezuela: 7 years of follow-up on symptomatic and asymptomatic individuals. Mov Disord. 1990;5:93–9.
7. Perlman SL. Spinocerebellar degenerations. Handb Clin Neurol. 2011;100:113–40.
8. Richter U, Halje P, Petersson P. Mechanisms underlying cortical resonant states: implications for levodopa-induced dyskinesia. Rev Neurosci. 2013;24:415–29.
9. Scott DF, Heathfield KWG, Toone B, Margerison JH. The EEG in Huntington's chorea: a clinical and neuropathological study. J Neurol Neurosurg Psychiatry. 1972;35:97–102.
10. Wilson SAK, Bruce AN. Neurology. 2nd ed. Baltimore: Williams and Wilkins Company; 1955.
11. Klawans HL, Goetz CE, Paulson GW, Barbeau A. Levodopa and presymptomatic detection of Huntington's disease-eight-year follow-up. N Engl J Med. 1980;302:1090.
12. Fekete R, Jankovic J. Psychogenic chorea associated with family history of Huntington disease. Mov Disord. 2010;25:503–4.

Chapter 2
Benign Hereditary Chorea

Carlos Zúñiga-Ramírez and Héctor Alberto González-Usigli

Abstract Benign hereditary chorea (BHC) is a rare autosomal dominant condition, characterized by nonprogressive chorea with neither significant intellectual impairment nor caudate atrophy. It was first described almost 50 years ago. Genetic testing is available since the last decade, leading physicians to classify adequately patients and separate them from other similar conditions, mainly Huntington disease and myoclonus-dystonia, which have a different prognosis and treatment. Associated clinical features have become broader every day, and molecular testing confirms accurately the diagnosis. Different grades of affection in the thyroid gland, lung, and brain are associated with mutations on the TITF-1 gene at chromosome 14q. Nowadays, less than 50 families have been reported and reviewed with typical or atypical symptoms. Since BHC is a nonprogressive condition and some patients respond favorably to some drugs, it is important to characterize it correctly in order to offer the best treatment options and give the right prognosis to the affected subject and relatives.

Keywords Benign hereditary chorea • Brain-lung-thyroid syndrome • Autosomal dominant • NKX2.1 • Basal ganglia development

Introduction

In 1967 two independent publications in the USA reported a new familial nonprogressive choreiform disorder. The first description was done in an African American family from Mississippi [1, 2]. In 1976 Burns et al. [3] described a similar syndrome

C. Zúñiga-Ramírez (✉)
Movement Disorders and Neurodegenerative Diseases Unit,
Hospital Civil de Guadalajara "Fray Antonio Alcalde", Guadalajara, Mexico
e-mail: c.zuniga.ramirez@gmail.com

H.A. González-Usigli
Department of Neurology, Instituto Mexicano del Seguro Social, Guadalajara, Mexico

F.E. Micheli, P.A. LeWitt (eds.), *Chorea*,
DOI 10.1007/978-1-4471-6455-5_2, © Springer-Verlag London 2014

of a benign nonprogressive chorea with childhood onset in 3 siblings of the same family and some of their relatives with an autosomal dominant mode of inheritance with a high penetrance but with variable expressivity on the phenotypic spectrum and a nonprogressive course, differentiating this entity from Huntington disease. Later in 1981, Gillian Sleigh and Richard Lindenbaum [4] described three affected families in which kindred presented with motor milestones delay associated with athetoid and choreic movements that initially were misinterpreted as cerebral palsy. Interestingly, some of their parents were affected from a similar condition since childhood [4]. Since the first description, more than 40 families have been included in other reports, with a growing amount of characteristics that have been described gradually during several years. Nevertheless, many of these original families have been later on found to have other conditions, such as Huntington disease or myoclonus-dystonia [5]. It was not until the last decade when evidence of a genetic abnormality located at chromosome 14q was found, and benign hereditary chorea (BHC) separated adequately from similar disorders [6]. Up to now, there are more than 30 different heterozygous mutations on the thyroid transcription factor-1 (TIFT-1), member of the Nk homeobox gene family. The same mutations had been reported as being responsible for a different clinical entity, the lung-brain-thyroid syndrome, as well as a wide spectrum of clinical features of BHC [7, 8] (see Fig. 2.1).

BHC patients have a normal life expectancy and show nonprogressive mental deterioration, although some individuals fail to finish school and have memory deficits [9]. It usually manifests before the age of 5 years, running a stationary or only a slightly progressive course. Intelligence may be normal or borderline, and mental deterioration found in Huntington disease is not evident. There are only rare and subtle magnetic resonance imaging changes, with normal caudate volume.

A differential diagnosis is broad for chorea during childhood and includes cerebral palsy, Sydenham chorea, drug-induced chorea, inherited metabolic diseases, ataxia-telangiectasia, paroxysmal movement disorders, reflex epilepsies, and familial myoclonus-dystonia, among others. Thus, in the past, before a genetic test was available, BHC was proposed to be a syndrome, not a disease, possibly related to undiagnosed conditions, and this disorder was considered an exclusion diagnosis [5]. Fortunately, genetic testing is now available and allows to differentiate BHC from similar conditions that imply a different prognosis and treatment.

Even though BHC is an unusual disorder, its identification is essential, since some children have suffered mockery that has produced the secondary effects of anxiety, stammering, behavioral disturbances, lack of self-confidence, and increased sensitivity resulting in underachievement. As prognosis is usually good, the reassurance of the parents is very important, and psychotherapy may help patients and family to embrace this condition more successfully reinforcing the patient's self-esteem and therefore decreasing the risk of scholar and work failure that has been reported in the some cases. The discussion of the diagnosis and prognosis is likely to alleviate parental anxiety. Informed genetic counseling is also important for the index family and close relatives [4].

Fig. 2.1 Clinical spectrum of TITF-1 mutations

Even though this disease has a high penetrance with a high expressivity of symptoms, its lack of progression and disabling neuropsychiatric symptoms, the improvement of most cases during adulthood, and the preserved intellect allow us to consider it as a benign disorder. The last page for the scientific knowledge about BHC waits to be written, since we are still learning and discovering previously unknown amazing aspects of this disorder. It is therefore why we are trying to summarize updated data with a very comprehensive approach regarding this condition.

Epidemiology

BHC, previously known as essential chorea, is an autosomal dominant disorder that in theory may affect equally men and women due to its mode of transmission; however, more women than men have been reported in the described families [10]. This disorder is characterized by choreiform movements but with many distinct characteristics from Huntington disease.

Although the epidemiological data is scarce regarding the actual prevalence of TITF-1 and epsilon-sarcoglycan-1 (SGCE) gene mutations of the myoclonus-dystonia syndrome (which is the most important differential diagnosis), some German studies have estimated its prevalence about 0.2 per million for TITF-1 and 2 per million for SGCE mutation carriers. In another study from Wales, the prevalence has been projected to be 1:500,000 based on the number of affected individuals [10]. Its onset is during childhood, mainly before the fifth year, and it is much less prevalent than Huntington disease. Up to 2007, a total of 42 families have been reported and typical or atypical symptoms described [11, 12]. The age at onset of classic BHC ranges through most of childhood and is roughly divided into three age groups: early infancy, around 1 year, and during late childhood and adolescence, the most common being at 1 year of age [11]. However, the earlier onset of symptoms suggests an associated developmental disturbance. Symptom severity peaks around the second decade of life, and thereafter, progression is subtle or negligible [6]. Possible racial differences had been previously described regarding the mode of presentation of the disease related to different mutations and expressivity variations. At the UK, in addition to chorea, either focal or segmental dystonia has been associated. The penetrance of BHC was estimated to be 100 % in males and 75 % in females; however, the actual penetrance seems to be higher after the exclusion of individuals without truly BHC and the inclusion of patients mildly affected once the gene mutation was found [11]. Mutations in this gene have been found in five independent BHC families of Dutch, North American, Welsh, Italian, Portuguese, and Canadian origin as well as in patients with a combination of choreoathetosis, hypothyroidism, and pulmonary symptoms of different origins [13].

Classification

As mentioned above, de Vries and his colleagues [6] performed a genome-wide association linkage study to find that the responsible gene was localized on chromosome 14q. Thirty-three relatives from a four-generation Dutch family with BHC were included in this original description [6]. Since then, other specific mutations have been associated to BHC. This disorder coexists with thyroid and lung diseases where new mutations have been located at the 14q13.3 (NKX2-1, TITF-1, or T/EBP). The classical triad (chorea-hypothyroidism-respiratory distress or repetitive respiratory tract infections) is not always present, and severity of movement disorders, respiratory symptoms, and thyroid features varies largely [7, 14]. Only less than 50 % of affected subjects show the complete triad of symptoms; meanwhile, isolated BHC corresponds only to 21 % of the sample [15]. Variability in the mode of inheritance, either AD or de novo mutations, and multiple phenotypes associated to different mutations in the TITF-1 are now recognized. In general terms, the disease can be divided into a "classical" form in which early-onset chorea is the main feature and the "brain-thyroid-lung" syndrome where the presence of other neuro-logical abnormalities along with hypothyroidism and either respiratory distress or recurrent pulmonary infections coexists.

Clinical Manifestations

Chorea is characterized by a random flow of rapid abnormal movements, unpredict-ably involving many different parts of the body at different times. In BHC, the abnormal movements usually do not subside completely, but they abate consider-ably by early adulthood; the same is true regarding co-joining ataxia. In children with BHC, the most characteristic symptom is the motor delay associated to choreic movements, gaining independent gait by the second to third year of life [16]. The movements are worsened by anxiety but not by voluntary movements, exercise, startle, caffeine, or alcohol. Some reports have depicted educational and behavioral difficulties. In 1984, Leli and colleagues [17] reported cognitive findings from sub-jects with BHC: while the authors found normal intellectual quotient (IQ) in affected subjects comparatively with general population, they observed a lower IQ when compared to unaffected members of their family. Lower verbal intelligence and deficits in verbal abstract concept formation were found as well among affected individuals. Recently, a case series reported up to 71 % of learning difficulties, 15 % of mental retardation, and 15 % of borderline IQ. Moreover, 25 % of subjects pre-sented with ADHD. All subjects assessed in that report had NKX2-1 mutations as confirmed by molecular tests [15]. Phenotype may vary, since important penetrance and expressivity differences were reported; however, no anticipation phenomenon has been ever described [18].

Atypical features previously reported include psychosis, dystonia, mental retar-dation, facial apraxia, epilepsy, hypospadias, pyramidal tract symptoms, and optic nerve glioma. Some individuals have been reported with intention tremor. The dys-tonic features associated with BHC are mild compared to dystonic syndromes. The abnormal postures tend to involve more frequently the upper than the lower half of the body, but severity in the lower limbs varies. Bradykinesia can occur, but it is con-sidered as an atypical feature. Even mild dysarthria has been documented but severe dysarthria is very uncommon. Rarely, cognitive impairment can be present and is usually related with other conditions. Falls are common during early childhood, but disappear over time according to motor maturation. Some families had presented other physical abnormalities, such as congenital deafness, hypospadias, psychosis, epilepsy, and optic nerve glioma, among other signs and symptoms of the disease [8].

From 150 studied subjects with congenital hypothyroidism (CH), only five of them showed choreic movements with or without respiratory distress. All subjects with chorea were found to have a novel mutation on the NKX2-1 gene [19]. In a recent review of 46 cases of brain-thyroid-lung syndrome, 50 % of them had the complete triad, 30 % manifested with hypothyroidism and neurological involve-ment, 13 % isolated chorea, and 4 % with only lung involvement [10]. Table 2.1 shows the different features associated to TITF-1 mutations.

Table 2.1 Reported clinical features of TITF-1 mutations

Study	Reference	No. of individuals	Clinical features	Additional info
Armstrong et al. 2011	[20]	One	Neonatal respiratory distress, delayed motor development and walking, thyroid disease, chorea, superimposed generalized myoclonic jerks, four limbs dystonia	
Asmus et al. 2005	[21]	Six	Premature birth, pneumonia, apnea, respiratory distress at birth, small thyroid gland or partial agenesis, hypothyroidism, delayed motor development, hypotonia, muscular atrophy in legs, choreoathetoid movements, gait difficulties, cerebellar symptoms, dysarthria, pre-B-cell acute lymphocytic leukemia	Some responsiveness to levodopa treatment was found
Asmus et al. 2007	[8]	Two	Chorea, gait problems, jerks during sleep, delayed motor milestones, arm and cervical dystonia	
Asmus et al. 2009	[22]	One	Lower limb dystonia, chorea, ataxic gait, limb jerks	
Breedveld et al. 2002	[9]	One-hundred twenty-six	Delayed walking, ataxic gait, chorea, slow saccadic eye movements, pyramidal signs, lower IQ, myoclonic jerks of head and neck, dystonia, tremor, dysarthria, spasmodic dysphonia, tics	
Carre et al. 2009	[14]	Six	Choreoathetosis, respiratory distress, recurrent pulmonary infections, ataxia, slow saccades, psychomotor delay, severe mental retardation, premature birth, congenital hypothyroidism, hypotonia, delayed walking and speech	Corpus callosum agenesia, athyreosis, hemiagenesis of thyroid gland, thyroid hypoplasia
Devos et al. 2006	[23]	Three	Chorea, asthma, primary hypothyroidism, hypoparathyroidism, osteoporosis, hypodontia, xerostomia, xerophthalmia, cough, dyspnea, interstitial pneumonitis, malabsorption syndrome with diarrhea, respiratory distress, gait disturbances, persistent TSH elevation, hypopallesthesia, pyramidal signs, hypotonia, cerebellar syndrome	Deletions in NKX2-1, PAX9, NKX2-8, and SLC25A21

Do Carmo et al. Costa 2005	[13]	Two	Speech and walk delay, learning difficulties, chorea, stuttering, congenital strabismus, spasticity of lower limbs, pes cavus, hammertoes, tics, balance abnormalities, short stature, obesity, inappropriate laughter, cognitive impairment, explosive speech, areflexia, lower limb hypoesthesia	Enlargement of the fourth ventricle, cerebellar hyperintensities, cortical atrophy, cerebellar atrophy, hyperintensities in both pallidum on MRI
Doyle et al. 2004	[24]	Four	Neonatal respiratory distress, hypothyroidism, global development delay, ataxia, choreoathetosis, dysarthria, gait and balance disturbances, pneumonia, stammering, fatigue, weight gain	
Fernández et al. 2001	[12]	Eight	Chorea, hyperactive reflexes, limb jerks, paroxysmal extension spasms of neck, memory loss, orolingual quivering, hypothyroidism, hypertension, upper limb tremor	Lung cancer
Ferrara et al. 2008	[25]	Three	Pyelectasis, megabladder, respiratory distress, congenital hypothyroidism, patent foramen ovale, vesicourethral reflux, delay in motor development, borderline IQ, meningitis, chorea, low academic performance	
Ferrara et al. 2012	[26]	Ten	Delayed motor milestones, chorea, hypothyroidism, hemithyroid, infant respiratory distress syndrome, learning difficulties, generalized anxiety disorder, exertional dyspnea, stuttering, seizures, leg dystonia, panic disorder, distal limb weakness, psychosis, depression, self-injurious behavior, whispering speech, stammering speech, ataxia of gait	

(continued)

Table 2.1 (continued)

Study	Reference	No. of individuals	Clinical features	Additional info
Glik et al. 2008	[27]	Four	Premature birth, choreoathetosis, psychosis, hypothyroidism, short stature, obesity, slurred and slow speech, downward slow saccades, intention tremor, lower limb dystonia, webbed neck, asthma	Lung carcinoma
Guala et al. 2008	[28]	Two	Chorea, oligodontia, hypothyroidism	Mutations in PAX9, choreic movements nonresponsive to any therapy
Gras et al. 2012	[15]	Twenty-eight	Hypotonia, delayed motor milestones, BHC, myoclonus, dystonia, tics, gait impairment, dysarthria, ataxia, ADHD, hypothyroidism, asthma and other respiratory disorders, learning difficulties, low IQ, mental retardation, brain–lung–thyroid syndrome	New mutations at the NKX2-1 gene, choreic movements responsive to levodopa and tetrabenazine
Krude et al. 2002	[19]	Five	Choreoathetosis, hypotonia, congenital hypothyroidism, thyroid hypoplasia or agenesia, respiratory distress, pulmonary infections	Small-sized pallidum, lack of division of the medial and lateral aspects of the pallidum, cystic mass near to the pituitary gland
Mahajnah et al. 2007	[29]	Three	Chorea, ataxia, gait instability	SPECT with decreased uptake of Tc99 at the right striatum and thalamus
Maquet et al. 2009	[30]	One	Acute respiratory failure with pulmonary hypertension soon after birth. Congenital hypothyroidism	Novel heterozygous TITF1/NKX2.1 mutation (I207F). This mutation was not found in neither parent
Nakamura et al. 2012	[31]	Three	Severe generalized chorea, delayed motor development, subnormal intelligence, congenital hypothyroidism, bronchial asthma, history of pulmonary infection	Novel nonsense mutation in the TITF-1 gene (p.Y98X)
Nettore et al. 2013	[32]	Five	Early motor delay, moderate generalized choreiform movements, cerebellar ataxia, intermittent respiratory insufficiency, short stature, cervical dystonia, mild cognitive deficit	Subclinical hypothyroidism, manifest hypothyroidism, novel NKX2-1 mutation described

Reference		Number	Clinical features	Additional findings
Ngan et al. 2009	[33]	Four	Papillary thyroid carcinoma, multinodular goiter	Novel heterozygous NKX2-1 germ line mutation
Pohlenz et al. 2002	[34]	One	Congenital hypothyroidism, neonatal respiratory distress, chorea, hypotonia, staggering gait, delayed motor milestones	
Provenzano et al. 2008	[35]	One	Delayed motor milestones, wide-based gait, arm dystonia, generalized chorea, dysarthria, oculomotor apraxia, possible learning impairment	Novel mutation R178X
Salvado et al. 2013	[10]	Three	Chorea, myoclonus, ataxia, motor and language development delay, trunk titubation during sedestation, low IQ, respiratory distress, congenital hypothyroidism	Bilobulated thyroid gland, ectopic thyroid tissue, dopa-responsive chorea
Salvatore et al. 2010	[36]	Three	Mild motor delay, childhood-onset chorea, subtle cognitive impairment, congenital hypothyroidism, neonatal respiratory distress, postpartum psychosis, jerky abdominal movements, apraxia, attention and memory deficit	Severe bilateral vesicourethral reflux, pyelectasis, megabladder, patent foramen ovale, mild ventricular enlargement, empty sella on MRI, hypometabolism of basal ganglia and cortex on FDG-PET, slight improvement with tetrabenazine
Sempere et al. 2013	[37]	Five	Chorea, delayed motor development, gait difficulties, axial and leg dystonia, poor scholar performance, attention and memory deficit, apraxia	
Teissier et al. 2012	[38]	Four	Ataxia, hypotonia, gait disturbances, psychomotor delay, congenital hypothyroidism, neonatal respiratory distress syndrome, chronic obstructive pulmonary disease, hexadactyly, arched palate	Thyroid dysgenesis
De Vries et al. 2000	[6]	Nineteen	Chorea, gait ataxia, psychosis, spastic-ataxic diplegia, brisk reflexes	
Willemsen et al. 2005	[7]	One	Pulmonary alveolar proteinosis, primary hypothyroidism, delayed developmental milestones, severe hypotonia, chorea	Aggressive lung carcinoma at short age, emphysema, interstitial fibrosis, first description of the "brain-lung-thyroid syndrome"

Pathophysiology

Thyroid transcription factor-1 or TITF-1, also known as NKX2-1, T/EBP, or TTF, is expressed in the thyroid gland, lungs, and some areas of the brain, namely, the basal ganglia and forebrain. The NKX2-1 homeobox gene results to be an important part for the forebrain and basal ganglia development. It seems that this gene helps to differentiate the telencephalon and diencephalon; besides this, it regulates their size and possibly their function as well, although the latter is still a matter of debate [39]. TITF-1 is essential for the development of the basal ganglia. NKX2-1 or TITF-1 is expressed by day 32 at the mRNA level in the prosencephalon, which will give rise to the hypothalamus. At the seventh week it is found within the thyrocyte precursors and at the eleventh week at the lung epithelium [40]. The external part of the globus pallidus is in charge of regulating the activity from different neurons to the thalamus, dysfunction of the first structure leads to a loss of inhibition of excitatory stimulus from the thalamus to the cortex, with the subsequent appearance of choreic movements. Experimental animal models show that mutations in the NKX2-1 gene result in abnormal dopaminergic nigrostriatal tracts, as well as a failure of neurons to migrate from the medial ganglionic eminence (a pallidal precursor) into the striatum and cortex [41, 42]. Besides this, a PET study in humans showed a 70 % decrease in raclopride-binding levels at the striatum among people with BHC as compared to controls [43]. The authors concluded that this might represent an impairment of the projection neurons at the indirect pathway that could give rise to choreic movements. Previously, a fluorodeoxyglucose PET study showed a hypometabolism of caudate nucleus among subjects with BHC compared to controls [44]. Magnetic resonance has depicted a decreased volume of both caudate nucleus and right putamen in two subjects with BHC. Spectroscopy of these subjects showed an increased myoinositol/creatine ratio and a decreased choline/creatine ratio at the striatum level, as well as a reduced N-acetyl aspartate/creatine ratio in the right striatum of both subjects as compared to controls [45]. Apart from this study, SPECT performed on three subjects with BHC showed a reduced uptake of technetium-99 (Tc99) at both the right striatum and thalamus of all studied subjects [29]. All these findings could be traduced in a loss of striatal cells, mainly cholinergic, leading to a loss of inhibition of the thalamus. Moreover, one postmortem study showed a reduction in size of choline acetyltransferase-staining interneurons, calbindin-positive cells, and calretinin-positive cells at the caudate level, as well as calretinin- and parvalbumin-positive cells at the putamen. A decrease of Met-enkephalin immunoreactive fibers in the ventral caudate nucleus and a decrease of substance P immunostaining at the striatal interneurons were also seen. The authors concluded that TIFT-1 mediates the migration of interneurons from the medial ganglionic eminence to the lateral aspect of it and later to the cortex. TIFT-1 mutations seem to impair this migration, with secondary dysfunction and possible malformations at the forebrain/basal ganglia level [46]. This case has been published previously [47], describing only mild fronto-temporo-parietal atrophy and nonspecific astrocytosis and hyperplasia at the pallidum, thalamus, hippocampus, and periaqueductal gray matter.

In 2002, Breedveld and colleagues [18] described four families from different ethnicity that presented with different mutations of the TITF-1 gene on chromosome 14 associated to BHC. It is believed that the loss of the functional TITF-1 protein could result in an arrested development of the basal ganglia, reduced formation of forebrain neurons, as well as limited migration of GABAergic cells to the cortex, with consequent dysfunction of the thalamus inhibition and further appearance of choreic movements [18]. Apparently, TITF-1 haploinsufficiency interferes not only with GABAergic cells migration but also with dopaminergic cells, as demonstrated in animal models [41].

Approximately, more than 30 heterozygous mutations of the NKX2-1 homeobox gene have been described (see Table 2.2). Clinical variability could be explained due to the type, size, location of the mutation, and affected domain of each protein.

Table 2.2 Mutations found at the TITF-1 gene	
	c.255_256insG
	c.257dupA
	c.278_306del
	p.Tyr114*
	c.373+1_373+4del
	c.373+1G>A
	c.374-2A>C
	c.374-2A>G
	c.374-2A>T
	c.399delC
	p.Ser145Ter
	c.470_479delinsGCG
	p.Arg165Trp
	p.Ser169*
	p.Glu175*
	p.Leu176Val
	p.Arg178*
	p.Tyr185Asp
	p.Ser187*
	p.Leu194Arg
	c.582_583insGG
	c.786_787del
	p.Pro202Leu
	p.Val205Phe
	p.Ile207Phe
	p.Trp208Leu
	p.Gln210Pro
	p.Arg213Ser
	p.Arg213Pro
	p.Cys214*
	p.Gln249*
	c.818delG
	c.825delC
	c.859dupC

Dysfunction of a particular protein generating gain or loss of a specific function could be critical for the development of a specific tissue [13]. Alternatively, haploinsufficiency and a negative disruption of the TITF-1-containing transcriptional complexes have been proposed as the cause of the disease [8, 19, 21, 34, 47]. It seems that NKX2-1 mutations truncate the beginning of the protein homeodomain, leading to haploinsufficiency. With this mutated protein, binding to DNA and activation of specific genes are no longer possible. According to recent studies, approximately 93 % of subjects with TIFT-1 mutations will manifest neurological features, even in the absence of thyroid or lung disease [14]. Besides chorea, hypothyroidism, and respiratory failure, TITF-1 mutations have been associated to thyroid and lung carcinoma as well [43, 48–50].

Diagnosis

Chorea evolving during the first year of life is the main feature of BHC. Even so, other neurological signs and symptoms can be present during diagnosis, making it difficult to distinguish from other diseases. The main concern is differentiating BHC from myoclonus-dystonia, since the former could manifest myoclonus and dystonia as part of the clinical picture [15, 20, 51]. Early-onset hypotonia, associated lung and/or thyroid disease, and lack of responsiveness of abnormal movements to alcohol point to BHC diagnosis instead of myoclonus-dystonia syndrome related to epsilon-sarcoglycan-1 gene mutations [8]. Nonetheless, alcohol has been reported to reduce at least 50 % of the presence of choreic movements in one subject [8]. Myoclonus-dystonia is an inherited condition due to mutations at the epsilon-sarcoglycan-1 gene (DYT11). Other mutations at different loci can produce the same picture. This is the case of DYT15, where the gene defect is still unknown (18p11). Besides this, DYT1, DYT5, and vitamin E deficiency, among others, can reproduce the same clinical features. Sometimes, diagnosis poses a high level of difficulty, since BHC and myoclonus-dystonia can be clinically identical. In these cases, molecular testing is the only way to separate both entities [52].

Recently two Japanese families were described with adult-onset autosomal dominant chorea with caudate atrophy, responsive to haloperidol and with a benign evolution. Linkage analysis was performed among the 21 members of these families (7 affected and 14 unaffected), mapping to chromosome 8q21.3-q23.3. Besides choreic movements, dysarthria, hypotonia, and mild dementing features in some were found. The mean age of onset was 53 years. The term "benign hereditary chorea type 2" was coined for this entity [53].

Huntington disease is an autosomal dominant disorder featured by chorea, behavioral symptoms, and cognitive decline. Besides this, other movement disorders such as parkinsonism, dystonia, myoclonus, tics, and ataxia can arise. Although it is catalogued as an adult-onset disease, about 10 % of the cases can clinically manifest the disease before the age of 20. Most of these subjects will show a rigid-akinetic state as the main picture (Westphal variant). Dementia and other neuropsychiatric disturbances lead to differentiate Huntington disease from BHC; besides, caudate and corticosubcortical global atrophy is only seen in the former [54]. Although the latter statement is the rule for most of the cases, some intermediate cases, according to the number of triplet repeats, may show an atypical and more benign picture [55, 56]. Genetic testing for Huntington mutations is currently available worldwide [57]. Some years ago, a new entity clinically identical to Huntington disease but with a different gene mutation was named Huntington disease-like 2 (HDL-2). The initial signs of the disease are evident after age 20; however, a case of an 11-year-old child with tourettism was previously reported. Near 10 % of the affected subjects have acanthocytes detected by blood smears; this disorder is due to an expanded trinucleotide (CTG/CAG) repeat of the junctophilin-3 gene on chromosome 16q24.3, with a subsequent loss of function of the latter. Anticipation phenomena is seen just as in other polyglutamine diseases. Neuropsychiatric features along with chorea, parkinsonism, dystonia, and brisk tendon reflexes are the dominating features of the disease. Neuromuscular affection is absent; the same is true for seizures and elevation of creatine phosphokinase. Imaging depicts caudate and cortical atrophy as the main finding. Subjects under suspicion of HDL-2 invariably possess an African ancestry [58–61].

Dentatorubral-pallidoluysian atrophy (DRPLA) is a rare autosomal dominant disease that was previously thought to be endemic to Asia, with variable age of onset. Progressive myoclonic epilepsy is the main feature in those who develop the disease before the age of 20, but a combination of different movement disorders, including chorea, plus cognitive changes and neuropsychiatric symptoms may be seen among subjects. Imaging procedures show brainstem and cerebellar atrophy; moreover, white matter involvement could also be seen. Although the disease can be expressed even during the first year of life, the presence of other signs and symptoms, such as epilepsy and marked cognitive decline, and structural changes at the posterior fossa make the diagnosis of BHC unlikely. DRPLA is a polyglutamine disease: CAG expansion in the CTG-37B gene at chromosome 12p13.31 is the responsible for this disease. Genetic testing should be done when necessary [62–64].

Spinocerebellar ataxias (SCA) are a group of diseases that can manifest almost all types of movement disorders. SCA-17 and SCA-3 are the ones in which chorea is a common feature, but other SCAs like types 1, 2, 14, and 27 can present chorea as well [65].

Previously, a recessive inherited form of early-onset chorea was described. The three reported cases only showed childhood-onset chorea; there were no other neurological or extraneurological signs besides the movement disorder. Since genetic testing was not available at that time, it is possible that another type of chorea, different from BHC, was described [66].

Chorea-acanthocytosis is a rare autosomal recessive disease that manifests itself during teenage and young adulthood years. Although autosomal dominant forms of the disease had been previously described, the current accepted mechanism of transmission is recessive [67]. Besides chorea, orofacial dyskinesia, a typical "feeding dystonia" with tongue protrusion and secondary dysphagia, tongue and lip biting, grimacing, involuntary vocalizations, and dysarthria coexist. Besides this, neuro-psychiatric features such as depression, obsessive-compulsive disorder, tics, touret-tism, and a "schizophrenia-like" disorder are present commonly. Subjects affected with chorea-acanthocytosis have a "rubberman" gait, with truncal spasms and flex-ion, and gait unstability as well. Polyneuropathy, myopathy, parkinsonism, dysto-nia, seizures, cognitive decline, elevated creatine phosphokinase, and atrophy of caudate nucleus in magnetic resonance imaging are other clinical clues that can lead to the right diagnosis [60, 68, 69].

Ataxia-telangiectasia (AT) is an autosomal recessive disease characterized by early-onset ataxia, oculocutaneous telangiectasias, endocrinopathy, immunodefi-ciency, respiratory disease, radiosensitivity, and high risk for malignancies. The clinical spectrum of this disease is divided further into a classical and a variant form. Besides ataxia, chorea is present in more than 70 % of the cases, either classic or variant forms. Chorea is an early manifestation of the disease, even as the initial sign, and diagnosis can be easily missed [70]. AT has a broad spectrum of signs and symptoms; the presence of chorea, dystonia, tremor, and other movement disorders is common, and responsiveness to levodopa in familial cervical dystonia has been published recently [71]. Moreover, hyperkinetic movements such as chorea, dysto-nia, and myoclonus are well-described features of AT; they seem to have an impor-tant benefit over symptom control with amantadine use [72]. Furthermore, a recessive form of ataxia without telangiectasia had been previously described in 2 siblings, and a diagnosis of benign hereditary chorea was incorrectly made [73].

Wilson disease comprises an autosomal recessive disease in which a deficient ceruloplasmin is not able to remove copper from different organs, including the brain. The defect gene is the ATP7B located on chromosome 13. Onset can be seen at any age; reported cases depict the beginning of the disease from 3 to 80 years. Diverse movement disorders can be diagnosed as part of the picture; nonetheless, chorea is only seen in up to 10 % of the affected subjects. Coexistence of hepatopa-thy, neuropsychiatric symptoms, and Kayser-Fleischer rings seen at the slit-lamp examination make the diagnosis [74–77].

Sydenham chorea is an autoimmune disorder triggered by group A β-hemolytic streptococcal infection. Besides choreic movements around the face and extremi-ties, neuropsychiatric disorders (depression, attention deficit disorder, obsessive-compulsive disorder, behavioral problems), carditis, and rheumatic fever can be present with the disease. It is the most common cause of acute chorea among chil-dren worldwide. Presence of antistreptolysin antibodies, prior respiratory infection, hemichorea, cardiac abnormalities, signs suggestive of rheumatic fever, and lack of familial history can point to this disease [78, 79].

Treatment

BHC is self-limited most of the time. Chorea tends to lessen over time and move-ments do not seem to bother most of the affected subjects. However, depending on the gravity of chorea and subject's perception of disability due to the movement disorder, some medications should be tried in order to offer some control over the symptoms. Levodopa-responsive cases have been published. Dosages used in BHC are the same as in Parkinson's disease [26]. Tetrabenazine seems to be of benefit in treating symptomatic chorea, with doses from 0.5 to 4.5 mg/kg/day among children, until a maximum daily dosage of 100–150 mg is reached. In adults, good results were obtained with doses of 37.5 mg/day [15]. Other reports with tetrabenazine had shown subtle improvement [36]. Methylphenidate has been reported as useful controlling choreic movements [23, 26]. Diazepam seems to lessen chorea and ataxia through its GABAergic effects over cortical interneurons as well as cerebellar GABA-dependent pathways. Other anti-choreic medications should be tried when the abovementioned drugs do not seem to work: carbamazepine, valproic acid, levetiracetam, gabapentin, pregabalin, amantadine, clozapine, and other neuroleptics. Thyroid and lung diseases must be looked for. A specialized medical team capable to treat neurological, endocrinological, and pneumological issues must conform. Besides this, according to the main clinical features, other specialists in cancer, mental health, physical therapy, and genetics, among others, must be taken into account, in order to provide a better quality of life to the affected individual and the rest of the family.

Conclusions

BHC is a rare autosomal dominant disease that affects subjects during their first 5 years of life. Apart from choreic movements, thyroid and lung diseases can coexist, and different abnormalities at diverse levels could be evident. Although it is considered to be a "benign" disease, with improvement of symptoms over time, some features can produce a considerable worsening of the quality of life or even life-threatening situations. Other neurological diseases should be included in differential diagnosis, mainly myoclonus-dystonia syndrome. Treatment options are scarce, mainly due to the rarity of the disease. Genetic counseling must be looked for those affected families when necessary.

References

1. Haerer AF, Currier RD, Jackson JF. Hereditary nonprogressive chorea of early onset. N Engl J Med. 1967;276:1220–4.
2. Pincus JH, Chutorian A. Familial benign chorea with intention a clinical entity. J Pediatr. 1967;70:724–9.
3. Burns J, Neuhäuser G, Tomasi L. Benign hereditary non-progressive chorea of early onset. Clinical genetics of the syndrome and report of a new family. Neuropadiatrie. 1976;7:431–8.
4. Sleigh G, Lindenbaum RH. Benign (non-paroxysmal) familial chorea. Paediatric perspectives. Arch Dis Child. 1981;56:616–21.
5. Schrag A, Quinn NP, Bhatia KP, Marsden CD. Benign hereditary chorea—entity or syndrome? Mov Disord. 2000;15:280–8.

6. De Vries BBA, Arts WFM, Breedveld GJ, Hoogeboom JJM, Niermeijer MF, Heutink P. Benign hereditary chorea of early onset maps to chromosome 14q. Am J Hum Genet. 2000;66:136–42.
7. Willemsen MA, Breedveld GJ, Wouda M, Otten BJ, Yntema JL, Lammens M, et al. Brain-Thyroid-Lung syndrome: a patient with a severe multi-system disorder due to a de novo mutation in the thyroid transcription factor 1 gene. Eur J Pediatr. 2005;164:28–30.
8. Asmus F, Devlin A, Munz M, Zimprich A, Gasser T, Chinnery PF. Clinical differentiation of genetically proven benign hereditary chorea and myoclonus-dystonia. Mov Disord. 2007;22:2104–9.
9. Breedveld GJ, Percy AK, MacDonald ME, de Vries BBA, Yapijakis C, Dure LS, et al. Clinical and genetic heterogeneity in benign hereditary chorea. Neurology. 2002;59:579–84.
10. Salvado M, Boronat-Guerrero S, Hernández-Vara J, Álvarez-Sabin J. Chorea due to TITF1/NKX2-1 mutation: phenotypical description and therapeutic response in a family. Rev Neurol. 2013;56:515–20.
11. Kleiner-Fisman G, Lang AE. Benign hereditary chorea revisited: a journey to understanding. Mov Disord. 2007;22:2297–305.
12. Fernández M, Raskind W, Matsushita M, Wolff J, Lipe H, Bird T. Hereditary benign chorea, clinical and genetic features of a distinct disease. Neurology. 2001;57:106–10.
13. Do Carmo Costa M, Costa C, Silva AP, Evangelista P, Santos L, Ferro A, et al. Nonsense mutation in TITF1 in a Portuguese family with benign hereditary chorea. Neurogenetics. 2005;6:209–15.
14. Carre A, Szinnai G, Castanet M, Sura-Trueba S, Tron E, Broutin-L'Hermite I, et al. Five new TTF1/NKX2.1 mutations in brain lung–thyroid syndrome: rescue by PAX8 synergism in one case. Hum Mol Genet. 2009;18:2266–76.
15. Gras D, Jonard L, Roze E, Chantot-Bastaraud S, Koht J, Motte J, et al. Benign hereditary chorea: phenotype, prognosis, therapeutic outcome and long term follow-up in a large series with new mutations in the TITF-1/NKX2-1 gene. J Neurol Neurosurg Psychiatry. 2012;83:956–62.
16. Bird TD, Carlson CB, Hall JG. Familial essential ('benign') chorea. J Med Genet. 1976;13:357–62.
17. Leli DA, Furlow Jr TW, Falgout JC. Benign familial chorea: an association with intellectual impairment. J Neurol Neurosurg Psychiatry. 1984;47:471–4.
18. Breedveld GJ, van Dongen JWF, Danesino C, Guala A, Percy AK, Dure LS, et al. Mutations in TITF-1 are associated with benign hereditary chorea. Hum Mol Genet. 2002;11:971–9.
19. Krude H, Schütz B, Bierbermann H, von Moers A, Schnabel D, Neitzel H, et al. Choreoathetosis, hypothyroidism, and pulmonary alterations due to NKX2.1 haploinsufficiency. J Clin Invest. 2002;109:475–80.
20. Armstrong MJ, Shah BB, Chen B, Angel MJ, Lang AE. Expanding the phenomenology of benign hereditary chorea: evolution from chorea to myoclonus and dystonia. Mov Disord. 2011;26:2296–7.
21. Asmus F, Horber V, Pohlenz J, Schwabe D, Zimprich A, Munz M, et al. A novel TITF-1 mutation causes benign hereditary chorea with response to levodopa. Neurology. 2005;64:1952–4.
22. Asmus F, Langseth A, Doherthy E, Nestor T, Munz M, Gasser T, et al. 'Jerky" dystonia in children: spectrum of phenotypes and genetic testing. Mov Disord. 2009;24:702–9.
23. Devos D, Vuillaume I, de Becdelievre A, de Martinville B, Dhaenens CM, Cuvellier JC, et al. New syndromic form of benign hereditary chorea is associated with a deletion of TITF-1 and PAX-9 contiguous genes. Mov Disord. 2006;21:2237–40.
24. Doyle DA, González I, Thomas B, Scavina M. Autosomal dominant transmission of congenital hypothyroidism, neonatal respiratory distress, and ataxia caused by a mutation of NKX2-1. J Pediatr. 2004;145:190–3.
25. Ferrara AM, Da Michele G, Salvatore E, Di Maio L, Zampella E, Capuano S, et al. A novel NKX2.1 mutation in a family with hypothyroidism and benign hereditary chorea. Thyroid. 2008;18:1005–9.
26. Ferrara JM, Adam OR, Kirwin SM, Houghton DJ, Shepherd C, Vinette KMB, et al. Brain-lung-thyroid disease: clinical features of a kindred with a novel thyroid transcription factor 1 mutation. J Child Neurol. 2012;27:68–73.

27. Glik A, Vuillaume I, Devos D, Inzelberg R. Psychosis, short-stature in benign hereditary chorea: a novel thyroid transcription factor-1 mutation. Mov Disord. 2008;23:1744–7.
28. Guala A, Falco V, Breedveld G, De Filippi P, Danesino C. Deletion of PAX9 and oligodontia: a third family and review of the literature. Int J Paediatr Dent. 2008;18:441–5.
29. Mahajnah M, Inbar D, Steinmetz A, Heutink P, Breedveld GJ, Straussberg R. Benign hereditary chorea: clinical, neuroimaging, and genetic findings. J Child Neurol. 2007;22:1231–4.
30. Maquet E, Costagliola S, Parma J, Christophe-Hobertus C, Oligny LL, Fournet JC, et al. Lethal respiratory failure and mild primary hypothyroidism in a term girl with a de novo heterozygous mutation in the TITF1/NKX2.1 gene. J Clin Endocrinol Metab. 2009;94: 197–203.
31. Nakamura K, Sekijima Y, Nagamatsu K, Yoshida K, Ikeda S. A novel nonsense mutation in the TITF-1 gene in a Japanese family with benign hereditary chorea. J Neurol Sci. 2012;313:189–92.
32. Nettore IC, Mirra P, Ferrara AM, Sibilio A, Pagliara V, Kay SC, et al. Identification and functional characterization of a novel mutation in the NKX2-1 gene: comparison with the data in the literature. Thyroid. 2013;23:675–82.
33. Ngan ESW, Lang BHH, Liu T, Shum CK, So MT, Lau DK, et al. A germline mutation (A339V) in thyroid transcription factor-1 (TITF-1/NKX2.1) in patients with multinodular goiter and papillary thyroid carcinoma. J Natl Cancer Inst. 2009;101:162–75.
34. Pohlenz J, Dumitrescu A, Zundeol D, Martiné U, Schönberger W, Koo E, et al. Partial deficiency of thyroid transcription factor 1 produces predominantly neurological defects in humans and mice. J Clin Invest. 2002;109:469–73.
35. Provenzano C, Veneziano L, Appleton R, Frontali M, Civitareale D. Functional characterization of a novel mutation in TITF-1 in a patient with benign hereditary chorea. J Neurol Sci. 2008;264:56–62.
36. Salvatore E, Di Maio L, Filla A, Ferrara AM, Rinaldi C, Saccà F, et al. Benign hereditary chorea: clinical and neuroimaging features in an Italian family. Mov Disord. 2010;25:1491–6.
37. Sempere AP, Aparicio S, Mola S, Pérez-Tur J. Benign hereditary chorea: clinical features and long-term follow-up in a Spanish family. Parkinsonism Relat Disord. 2013;19:394–6.
38. Teissier R, Guillot L, Carré A, Morandini M, Stuckens C, Ythier H, et al. Multiplex ligation-dependent probe amplification improves the detection rate of NKX2.1 mutations in patients affected by brain-lung-thyroid syndrome. Horm Res Paediatr. 2012;77:146–51.
39. Van den Akker WMR, Brox A, Puelles L, Durston AJ, Medina L. Comparative functional analysis provides evidence for a crucial role for the homeobox gene Nkx2.1/Titf-1 in forebrain evolution. J Comp Neurol. 2008;506:211–23.
40. Ghaffari M, Zeng X, Whitsett JA, Yan C. Nuclear localization domain of thyroid transcription factor-1 in respiratory epithelial cells. Biochem J. 1997;328:757–61.
41. Kawano H, Horie M, Honma S, Kawamura K, Takeuchi K, Kimura S. Aberrant trajectory of ascending dopaminergic pathway in mice lacking Nkx2.1. Exp Neurol. 2003;182:103–12.
42. Sussel L, Marin O, Kimura S, Rubenstein JL. Loss of Nkx2.1 homeobox gene function results in a ventral to dorsal molecular respecification within the basal telencephalon: evidence for a transformation of the pallidum into the striatum. Development. 1999;126:3359–70.
43. Konishi T, Kono S, Fujimoto M, Terada T, Matsushita K, Ouchi Y, et al. Benign hereditary chorea: dopaminergic brain imaging in patients with a novel intronic NKX2.1 gene mutation. J Neurol. 2013;260:207–13.
44. Suchowersky O, Hayden MR, Martin WRW, Stoessl AJ, Hildebrand AM, Pate BD. Cerebral metabolism of glucose in benign hereditary chorea. Mov Disord. 1986;1:33–44.
45. Maccabelli G, Pichiecchio A, Guala A, Ponzio M, Palesi F, Maranzana Rt D, et al. Advanced magnetic resonance imaging in benign hereditary chorea: study of two familial cases. Mov Disord. 2010;25:2670–4.
46. Kleiner-Fisman G, Calingasan NY, Putt M, Chen J, Beal MF, Lang AE. Alterations of striatal neurons in benign hereditary chorea. Mov Disord. 2005;20:1353–7.
47. Kleiner-Fisman G, Rogaeva E, Halliday W, Houle S, Kawarai T, Sato C, et al. Benign hereditary chorea: clinical, genetic and pathological findings. Ann Neurol. 2003;54:244–7.

48. Cantara S, Capuano S, Formichi C, Pisu M, Capezzone M, Pacini F. Lack of germline A339V mutation in thyroid transcription factor-1 (TITF-1/NKX2.1) gene in familial papillary thyroid cancer. Thyroid Res. 2010;3:1–4.
49. Kwei KA, Kim YH, Girard L, Kao J, Pacyna-Gengelbach M, Salari K, et al. Genomic profiling identifies TITF1 as a lineage- specific oncogene amplified in lung cancer. Oncogene. 2008;27:3635–40.
50. Tang X, Kadara H, Behrens C, Liu DD, Xiao Y, Rice D, et al. Abnormalities of the TITF-1 lineage-specific oncogene in NSCLC: implications in lung cancer pathogenesis and prognosis. Clin Cancer Res. 2011;17:2434–43.
51. Kurlan R, Behr J, Shoulson I. Hereditary myoclonus and chorea: the spectrum of hereditary nonprogressive hyperkinetic movement disorders. Mov Disord. 1987;2:301–6.
52. Nardocci N. Myoclonus-dystonia syndrome. Handb Clin Neurol. 2011;100:563–75.
53. Shimohata T, Hara K, Sanpei K, Nunomura J, Maeda T, Kawachi I, et al. Novel locus for benign hereditary chorea with adult onset maps to chromosome 8q21.3-q23.3. Brain. 2007;130:2302–9.
54. Bordelon YM. Clinical neurogenetics Huntington disease. Neurol Clin. 2013;31:1085–94.
55. MacMillan JC, Morrison PJ, Nervin NC, Shaw DJ, Harper PS, Quarrell OW, et al. Identification of an expanded CAG repeat in the Huntington's disease gene (IT15) in a family reported to have benign hereditary chorea. J Med Genet. 1993;30:1012–3.
56. Britton JW, Uitti RJ, Ahlskog JE, Robinson RG, Kremer B, Hayden MR. Hereditary late-onset chorea without significant dementia: genetic evidence for substantial phenotypic variation in Huntington's disease. Neurology. 1995;45:443–7.
57. Kremer B, Goldberg P, Andrew SE, Theilmann J, Telenius H, Zeisler J, et al. A worldwide study of the Huntington's disease mutation, the sensitivity and specificity of measuring CAG repeats. N Engl J Med. 1994;330:1401–6.
58. Margolis RL, O'Hearn E, Rosenblatt A, Willour V, Holmes SE, Franz ML, et al. A disorder similar to Huntington's disease is associated with a novel CAG repeat expansion. Ann Neurol. 2001;50:373–80.
59. Bardien S, Abrahams F, Soodyall H, van der Merwe L, Greenberg J, Brink T, et al. A South African mixed ancestry family with Huntington disease-like 2: clinical and genetic features. Mov Disord. 2007;22:2083–9.
60. Walker RH, Jung HH, Danek A. Neuroacanthocytosis. Handb Clin Neurol. 2011;100: 141–51.
61. Schneider SA, Bhatia KP. Huntington's disease look-alikes. Handb Clin Neurol. 2011;100:101–12.
62. Wardle M, Morris HR, Robertson NP. Clinical and genetic characteristics of non-Asian dentatorubral-pallidoluysian atrophy: a systematic review. Mov Disord. 2009;24:1636–40.
63. Rajput A. Dentatorubral pallidoluysian atrophy. Handb Clin Neurol. 2011;100:153–9.
64. Tsuji S. Dentatorubral-pallidoluysian atrophy. Handb Clin Neurol. 2012;103:587–94.
65. Van Gaalen J, Giunti P, van de Warrenburg BP. Movement disorders in spinocerebellar ataxias. Mov Disord. 2011;26:792–800.
66. Nutting PA, Cole BR, Schimke RN. Benign, recessively choreo-athetosis of early onset. J Med Genet. 1969;6:408–10.
67. Ichiba M, Nakamura M, Kusumoto A, Mizuno E, Kurano Y, Matsuda M, et al. Clinical and molecular genetic assessment of a chorea-acanthocytosis pedigree. J Neurol Sci. 2007;263:124–32.
68. Jung HH, Danek A, Walker RH. Neuroacanthocytosis syndromes. Orphanet J Rare Dis. 2011;6:68–77.
69. Sokolov E, Schneider SA, Bain PG. Chorea-acanthocytosis. Pract Neurol. 2012;12:40–3.
70. Verhagen MMM, Abdo WF, Willemsen MAAP, Hogervorst FB, Smeets DF, Hiel JA, et al. Clinical spectrum of ataxia-telangiectasia in adulthood. Neurology. 2009;73:430–7.
71. Charlesworth G, Mohire MD, Schneider SA, Stamelou M, Wood NW, Bhatia KP. Ataxia telangiectasia presenting as dopa-responsive cervical dystonia. Neurology. 2013;81:1148–51.

72. Nissenkorn A, Hassin-Baer S, Lerman SF, Levi YB, Tzadok M, Ben-Zeev B. Movement disorder in ataxia telangiectasia: treatment with amantadine sulphate. J Child Neurol. 2013;28:155–60.
73. Klein C, Wenning GK, Quinn NP, Marsden CD. Ataxia without telangiectasia masquerading as benign hereditary chorea. Mov Disord. 1996;11:217–20.
74. Wilson DC, Phillips MJ, Cox DW, Roberts EA. Severe hepatic Wilson's disease in preschool-aged children. J Pediatr. 2000;137:719–22.
75. Czlonkowska A, Rodo M, Gromadzka G. Late onset Wilson's disease: therapeutic implications. Mov Disord. 2008;23:896–8.
76. Mihaylova V, Todorov T, Helev H, Kotsev I, Angelova L, Kosseva O, et al. Neurological symptoms, genotype-phenotype correlations and ethnic-specific differences in Bulgarian patients with Wilson disease. Neurologist. 2012;18:184–9.
77. Seo JK. Diagnosis of Wilson disease in young children: molecular genetic testing and a paradigm shift from the laboratory diagnosis. Pediatr Gastroenterol Hepatol Nutr. 2012;15:197–209.
78. Pavone P, Parano E, Rizzo R, Trifiletti RR. Topical review: autoimmune neuropsychiatric disorders associated with streptococcal infection: Sydenham chorea, PANDAS, and PANDAS variants. J Child Neurol. 2006;21:727–36.
79. Cardoso F. Sydenham's chorea. Handb Clin Neurol. 2011;100:221–9.

Suggested Reading

Inzelberg R, Weinberger M, Gak E. Benign hereditary chorea: an update. Parkinsonism Relat Disord. 2011;17:301–7.

Chapter 3
Chorea-Acanthocytosis

Andreas Hermann

Abstract Neuroacanthocytosis (NA) syndromes are a group of rare disorders displaying neurodegeneration and misshaped spiky red blood cells (acanthocytes). NA syndromes include chorea-acanthocytosis (ChAc), McLeod syndrome (MLS), Huntington's disease-like 2, and pantothenate kinase-associated neurodegeneration (PKAN) with ChAc as the prototype of this disease family. ChAc is caused by loss-of-function mutations within the gene *VPS13A* encoding for a protein of unknown function named chorein. This leads to movement disorders most often showing involuntary movements as the main symptom and thus being a differential to Huntington's disease. Additional symptoms are cognitive decline, psychiatric disturbances, peripheral neuropathy, epilepsy, parkinsonism, and blood cell acanthocytosis together with elevated creatine kinase and liver enzymes. Interestingly, the non-neurological changes do not translate in relevant clinical symptoms in the respective system. There is no cure for the disease, and treatment is purely symptomatic. Recent data hints towards chorein being involved in different kinase pathways and actin cytoskeleton disturbances. This book chapter tries to shed light into the clinical phenotype, treatment, neuroimaging, neuropathology, and the current state of pathophysiology of ChAc.

Keywords Neuroacanthocytosis syndromes • Elevated creatine kinase • Cortical actin depolymerization • Psychosis and akathisia • Involuntary movements

A. Hermann
Division for Neurodegenerative Diseases, Department of Neurology,
Dresden University of Technology, Dresden 01307, Germany

DZNE, German Centre for Neurodegenerative Diseases,
Research Site Dresden, Dresden, Germany
e-mail: andreas.hermann@uniklinikum-dresden.de

F.E. Micheli, P.A. LeWitt (eds.), *Chorea*,
DOI 10.1007/978-1-4471-6455-5_3, © Springer-Verlag London 2014

Introduction and Classification

The first reports of neuroacanthocytosis (NA) syndromes appeared in the late 1970s by Levine and Critchley [1–3]. NA syndromes are rare diseases affecting only one person in three million inhabitants and have been defined by neurological abnormalities in combination with misshaped acanthocytic red blood cells (Fig. 3.1). NA syndromes are characterized by a wide variety of dysfunctions of the nervous system including epileptic seizures and basal ganglia-related movement disorder symptoms, dementia, psychosis, and speech/swallowing difficulties leading to a markedly reduced life span [4–6]. So far, only fairly effective symptomatic treatment is available for NA syndromes. No curative treatment approach is available. NA syndromes can be classified as follows [7]:

I. Core NA syndromes

   (a) Chorea-acanthocytosis (ChAc, OMIM #200150)
   (b) McLeod syndrome (MLS; OMIM #300842)

II. Degenerative disorders where acanthocytosis is occasionally seen

   (a) Pantothenate kinase-associated neurodegeneration (PKAN; OMIM #234200)
   (b) Huntington's disease-like 2 (HDL 2; OMIM #606438)

III. Disorders with reduced blood lipoproteins and acanthocytosis

   (a) Bassen-Kornzweig syndrome (OMIM #200100)
   (b) Hypobetalipoproteinemia

ChAc is caused by mutations within the gene *VPS13A* encoding for the protein named chorein. Pathological studies revealed that basal ganglia, mainly the striatum and substantia nigra, are selectively affected [5, 8]. The striatum suffers from a pronounced atrophy that correlates with extensive striatal neuronal loss and reactive astrogliosis. Clinical symptoms comprise a progressive movement disorder with involuntary movements, oromandibular dyskinesias, and sometimes parkinsonism. Epilepsy is another often noted phenotype. Creatine kinase is elevated in serum, and blood acanthocytosis of 7–50 % is seen in almost all patients. Until now, rather anything is known about the pathophysiology of the disease, and thus the treatment is still mainly symptomatic.

Epidemiology

There are no real epidemiology data on the prevalence or incidence of ChAc. Approximately 1,000 patients are estimated worldwide. ChAc is not restricted to a specific ethnical background. However, higher prevalence was found in Japan and within the French-Canadian populations most likely due to a founder effect [6].

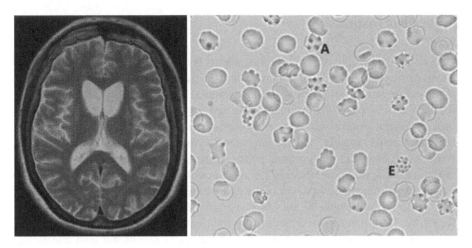

Fig. 3.1 (*Left*) T2-weighted cMRI from a female ChAc patient. Please note caudate degeneration with subsequent widening of anterior horns of the lateral ventricles. Additionally, slight bilateral hyperintensity of putamen is seen (Image kindly provided by Prof. Dr. R. v. Kummer, Dresden, Germany). (*Right*) Blood smear diluted 1:1 in isotonic NaCl with 10E/ml heparin showing significant acanthocytosis (protocol according to Storch et al. [9]). *A* acanthocyte, *E* echinocyte

Clinical Manifestations

Mean age of onset of ChAc is in the early 30s, although it has been reported appearing from the first to the seventh decade [5, 10]. However, this might change in future since the clear differentiation of neuroacanthocytosis syndromes with genetic testing is available only recently. Life expectancy is dramatically reduced with most of the patients dying within approximately 10 years after disease onset.

ChAc is a classical differential diagnosis to Huntington's disease (HD). This especially holds true for cases without autosomal dominant inheritance. However, there are some clear differences enabling to identify ChAc patients.

Full-blown ChAc is characterized by a movement disorder with chorea, dystonia and tics, cognitive impairment, dysarthria, neuropathy, and epilepsy. Up to 7–50 % of red blood cells will appear acanthocytic. Severe trunk spasms and head banging may be hard to treat and dangerous for the patients. Creatine kinase and liver enzymes are elevated [5, 6]. However, not all of these symptoms have to appear, and they might become visible at different stages in the course of the disease. Additionally, different phenotypes appear within the same family [11], and there is no genotype-phenotype correlation yet reported [5, 6, 8].

Neuropathy/Myopathy

In large case series, the most frequent phenotype of ChAc patients is a peripheral neuropathy [8] present in almost all patients. Myopathic changes are rare and most often secondary to chronic denervation [12].

Movement Disorder

Chorea is seen in more than half of the patients typically involving the limbs and to a lesser extent the face [5, 11]. Many patients suffer from limb dystonia often leading to severe gait disturbances and facial dystonia. Involuntary movements affecting the face are composed of oromandibular and lingual dystonia and/or choreiform orofaciolingual dyskinesias. These involuntary movements include movements of the eyes, mouth, tongue, and pharynx and larynx with involuntary vocalizations ("tics") including sucking, grunting noises, perseveration of words, humming, sighing, bruxism, spitting, or belching [13, 14], possibly misdirecting the diagnosis to Tourette syndrome in early stages of the disease [13]. These involuntary movements may lead to lip and tongue biting which is often mentioned as being pathognomonic for ChAc. The often reported feeding dystonia with tongue protrusion pushing out the food of the mouth [15] is seldom seen (one out of five of my own genetically proven ChAc patients) [11]. Head drop and axial extension have been reported in later stages of the disease [16]. Slurred speech followed by progressive dysarthria is the typical course of the disease. Some patients become mute and dependent on computer-based speech aids [5].

Dysphagia is a major problem of ChAc patients, and the oral phase rather than the pharyngeal or esophageal phase of swallowing is impaired [5]. This leads to severe weight loss, often requiring special alimentation or tube feeding [5, 6, 17].

Parkinsonism is seen in approximately one-third of ChAc patients [5, 6, 11]. This might be the initial manifestation of the disease; however, it is often seen in later stages of the disease, replacing the chorea ending by an akinetic-rigid syndrome [11]. This has to be taken into account for therapeutical interventions (see below).

Interestingly, involuntary movements of ChAc patients might be different from HD patients possibly due to differences in the pathophysiology of these diseases. In an early study of ChAc in the early 1980s, Shibasaki and colleagues compared three ChAc patients with three HD patients. In ChAc, the patients were able to suppress the choreic movements (by calculation, instruction to stop them, or muscle contraction), whereas HD patients were not. Using back averaging, they could track a negative cortical potential prior to the movements only in ChAc patients which resembled the "Bereitschaftspotential" of voluntary movements [18].

Behavioral Changes

Some ChAc patients were diagnosed in psychiatric departments presenting as schizophrenia or Tourette syndrome [7, 13, 19]. However, when screening schizophrenic patients or patients with Tourette syndrome without additional neurological deficits for mutations of *VPS13A* or *XK*, they are rare [20]. However, this highlights that one should be alert in psychiatric patients who develop movement disorders, as they not only represent cases of tardive dyskinesia due to neuroleptic treatment but also can have an underlying neurodegenerative disorder.

Neuropsychiatric symptoms are however often part of ChAc. They include depression, executive disturbances, emotional instability, distractibility, obsessive-compulsive disorder, and schizophrenia-like psychosis (for a review, see [7]). In sum, approximately two-thirds of patients suffer from behavioral disturbances [21]. Some authors have interpreted the lip and tongue biting as a kind of autoaggression along the obsessive-compulsive disease spectrum [22].

Cognitive Dysfunction

Hardie and colleagues reported 11 out of 19 NA cases having evidence for generalized intellectual deterioration often together with signs of frontal lobe dysfunction and problems with attention and planning tasks [11]. In a subsequent study, they reported 50 % of the NA patients (5 of 10) presenting with memory impairment, two of whom had visuoperceptual deficits [23]. Impairment in frontal lobe executive skills was however the most consistent finding, while none of the patients showed language deficits [23]. This would mean a frontal but not temporal lobe dysfunction as seen in patients with the behavioral variant of frontotemporal lobar degeneration.

Epilepsy

Epileptic seizures are common in ChAc (approx. 30–40 %) [11, 24]. Epilepsy might occur during the course of the disease but also as the initial manifestation sometimes even 10 years before the movement disorders appears [25]. It is important to realize that elevated creatine kinase is not always due to seizures but might also be a hint for underlying neuroacanthocytosis. There are no systematic studies on the classification of seizures in ChAc, but generalized [26] as well as focal epileptic seizures [27] have been reported. Most of these patients suffered from both focal and generalized seizures, which might be primary generalized or more often secondary after focal aura [25, 27–30]. Auras consisted of déjà vu, hallucinations, palpitations, fear, and somatosensory and abdominal auras [25, 28]. Most of the reported focal epileptic syndromes were localized to the temporal regions, suggesting a link between ChAc and familial temporal lobe epilepsy (FTLE). However, the semiology of these temporal lobe seizures in ChAc did often not clearly fit in the classification of either mesial or lateral TLE [31, 32] and may represent another example of pseudotemporal epilepsy [33]. It is still not clear why epilepsy develops in ChAc. Is this a primary phenotype suggesting that pathophysiological changes in ion channels or neurotransmitter release/turnover lead to unstable membrane properties? Or is it a correlate of secondary cellular changes? The latter is however difficult to explain in cases in which epilepsy is the initial symptom of the disease many years prior to onset of the movement disorder [25]. Furthermore, there is yet no neuropathological report with abnormalities of the temporal pole/hippocampus (see below). However, some studies report hippocampal sclerosis in ChAc patients suffering from temporal lobe epilepsy [28, 29], and one report shows interictal hypometabolism and ictal hypermetabolism in the left mesial temporal lobe of a ChAc patient with temporal lobe epilepsy [28]. It remains enigmatic whether these changes are related to the underlying neurodegenerative disease or due to recurrent seizures [28, 29].

The take-home message is that within patients presenting as familial temporal lobe epilepsy, one should have neuroacanthocytosis syndromes in mind especially if increased values of liver enzymes or creatine kinases are seen [28].

Blood/Spleen/Liver

Even though over 80 % of ChAc patients are showing up to 50 % acanthocytes, there are no reports about significant hemolysis or splenomegaly in these patients. None of the 19 patients reported by Hardie and colleagues had history of splenectomy, hemolysis, or liver disease [11]. Twenty-two percent of the patients ($n=20$) reported by Rampoldi showed splenomegaly, while 11 % showed hepatomegaly [24]. However, 100 % (20/20) of the patients showed reduced serum haptoglobin which is a sensitive marker for hemolysis [24].

The degree of blood cell acanthocytosis does not predict disease severity [24].

Pathophysiology

Chorein is a protein of yet unknown function. Analyzing the *VPS13A* sequence, Rampoldi and colleagues could not identify possible patterns, domains, or structural features providing indirect hints for possible functions [10]. The gene was named *VPS13A* to acknowledge the high homology to the yeast vps13/soi1. The latter was described to be important for the cycling of *trans*-Golgi network particles between the *trans*-Golgi network and the prevacuolar compartment [34]. The last corresponds to the multivesicular body/late endosome in animal cells. VPS13A belongs to a family of four proteins VPS13A-D. Mutations in VPS13B (COH1) are known to cause Cohen syndrome, a rare autosomal recessive disorder. COH1 is a Golgi matrix protein, and patient fibroblasts showed disrupted Golgi organization [35].

Tipc gene (the only other possible chorein homologue) mutants in *Dictyostelium discoideum* showed aberrant cell-sorting behavior which might hint towards a role in the cytoskeleton or cell transport mechanisms [36].

There is yet only one mouse model reported carrying an exon 60–61 deletion of the *VPS13A* gene. This led to a rather complete loss of chorein expression. The mice were viable and showed no differences in survival. Upon week 70, the homozygous ChAc mice developed some locomotor changes including reduced stride length in the footprint test and shorter latency to fall from rotarod. Neuropathology revealed gliosis of the striatum and increased apoptotic cells in the striatum [37]. Further neuropathological analysis including neuronal subtype characterization, enhanced neurochemical analysis, or the search for possible neuronal inclusions is still lacking. The mice failed to show involuntary movements or clear behavioral changes. However, red blood cell acanthocytosis was obvious [37]. In a subsequent study, GABA$_A$ receptor γ2-subunit and gephyrin were found to be increased in the striatum of chorein-deficient mice [38]. Chorein was found to be expressed in many brain regions, and subcellular analysis of brain lysates showed high level of chorein in the microsomal and synaptosomal fraction [39]. Interestingly, male homozygotes were unfertile which is different to human ChAc patients.

The first insides in possible functions of chorein were recently gained in red blood cells (RBC) of ChAc patients. Disturbances in membrane fluidity were suggested as membrane skeletal network was compact in some areas but not in others [40]. Lipid composition of RBC membrane seems to be rather normal. Increased phosphorylation of erythrocyte membrane proteins was reported, mainly the anion transporter band 3 and spectrin β-subunit. In a recent study, De Franceschi and colleagues identified the kinase Lyn being responsible at least for the band 3 phosphorylation [41]. Increased ion efflux associated with decreased intracellular potassium concentration was shown in ChAc erythrocytes [42]. In a very recent study, we could show that ChAc red blood cells harbor a reduced response in drug-induced endovesiculation, lysophosphatidic acid-induced phosphatidylserine exposure, and calcium uptake [43]. Interestingly, this was also seen in RBC samples from PKAN patients with acanthocytosis in contrast to RBC samples from PKAN patients without acanthocytosis. This suggests that these changes in functional membrane properties arise from the acanthocytic cell shape [43].

Band 3 is an important structural and transport membrane molecule and is expressed in cellular membranes including the cell membrane, nuclear membrane, mitochondrial membrane, and Golgi membrane. Band 3 proteins functions are ankyrin binding, anion transport, and generation of senescent age antigen. The latter is an aging antigen terminating the life of cells. Differences have been found between young and old mice indicating a role in aging [44], and antibrain antibodies were identified in ChAc patients [45]. Additionally, anti-band 3 immunoreactivity is seen in amyloid plaque of Alzheimer's disease patients [44, 46].

We recently identified chorein as regulator of cortical actin cytoskeleton of red blood cells. Loss of chorein resulted in actin depolymerization which was promoted by decreased PI3K p85 phosphorylation, decreased Rac1 activity, and PAK1 phosphorylation [47]. Additionally, chorein silencing in K562 cells resulted in decreased Bad phosphorylation and increased cell death [47]. Same changes in patient thrombocytes resulted in decreased platelet degranulation and impaired platelet aggregation [48]. Whether chorein loss of function also leads to changes of neuronal actin cytoskeleton or vesicle degranulation is not known.

At which timepoint of the disease acanthocytes appear during the cause of disease is still enigmatic. Are these already abundant in early life years before symptom onset? This is unfortunately also not known from the animal model. This would be of interest when concerning red blood cell modeling using human-induced pluripotent stem cells. Furthermore, it is not known why only up to 50 % of RBC become acanthocytic even though all cells carry the genetic mutation. Additionally, are the changes described above only seen in the acanthocytic RBC or in all RBC independent from their shape?

Pathology

There are many reports on postmortem examination of single ChAc cases, most of them prior to the availability of the molecular diagnostic techniques for ChAc. Systematic studies including many samples from genetically proven ChAc patients are yet lacking.

Beyond doubt however is the hallmark pathology of atrophy of the caudate nucleus and to a lesser extent the putamen. To my knowledge, there are 16 cases described in the literature all reporting severe caudate nucleus atrophy [11, 26, 49–56]. The putamen also was reported to be slightly less severely affected, whereas the pallidum and the substantia nigra were moderately affected.

Stereological countings and 3D reconstruction nicely showed the extent of overall volume reduction within the basal ganglia (Fig. 3.2).

Spared regions included the cerebral cortex [11, 26, 54], cerebellum [11, 54], subthalamic nucleus [54], and pons and medulla [11, 54]. The locus coeruleus appeared normal [11]. This differs from Huntington's disease or Friedreich's ataxia (see below).

Some authors describe neuronal loss in the anterior spinal horn [54], which have been reported as amyotrophic chorea-acanthocytosis. This was however not seen by other reports [11, 52].

Marked cell loss together with severe gliosis and extraneuronal pigment was found in the substantia nigra of the one reported autopsied patient by Hardie and

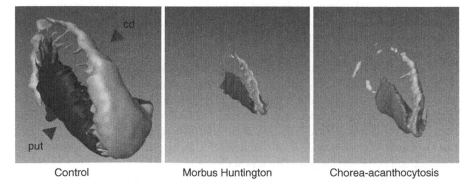

| Control | Morbus Huntington | Chorea-acanthocytosis |

Fig. 3.2 Showing the pattern of neurodegeneration within the striatum measured stereologically in postmortem human tissue (Reprinted with permission from Bader et al. [49])

colleagues [11]. Cell loss was almost complete in the pars reticulata, while the pars compacta was less affected [11]. Clinically, the patient also suffered from parkinsonism. Rinne and colleagues reported a comparative analysis of the substantia nigra (SN) of ChAc and Parkinson's disease patients. They found a significant reduction of SN neurons in two ChAc patients with clinical parkinsonism, whereas the third patient without clinical parkinsonism had SN neuron cell numbers at the lower limit of the control range. The ventrolateral region of the SN was more affected, while neuroacanthocytosis patients seemed to have a more widespread neuronal loss compared to Parkinson's disease patients [57]. No Lewy bodies were found within the substantia nigra or cortex [11]. In contrast to that, Sato and colleagues found no histopathological abnormalities within the SN [55].

There was no evidence for neuritic plaques, neurofibrillary tangles or other neuronal inclusions within the cortex including the hippocampus and the temporal lobe [11]. This is, however, to my best knowledge only reported in one single patient. Autopsy of two siblings with clinical features of ChAc was reported showing iron deposition and spheroid bodies in the atrophic striatum [58]. Thus, whether ChAc belongs to neurodegenerative diseases with protein aggregations and deposits remains elusive and clearly needs further investigation in a cohort of genetically proven ChAc.

Since all mentioned reports on neuropathology are not very recent reports, we do lack state-of-the-art immunocytochemistry studies. Thus, the major cell type affected within the striatum can only be speculated on by old studies. The major neuronal cell type of the striatum (approx. 90 % of all neurons) is the medium-sized spiny neurons (MSNs). These are the GABAergic output neurons of the striatum. Large-sized neurons are the cholinergic aspiny interneurons. Additionally, other medium-sized interneurons exist in the striatum, some of them expressing neuropeptide Y (for details, see also [59]). Two different subpopulations of MSN are known, the enkephalin- and D2-receptor-positive MSN projecting to the external part of the globus pallidus (the so-called indirect pathway) and the substance P- and D1-receptor-positive MSN projecting to the internal part of the globus pallidus and the SN pars reticulata (the so-called direct pathway) (for a review, see [59]).

Sato and colleagues report a reduction of small neurons to 1 % compared to the control within the caudate and to 20 % compared to the control in the putamen. A number of large neurons were spared [55]. Similar observations were made by Rinne and Iwata [26, 54]. Thus, it is very likely that the MSN neurons are the major affected nerve cells within the striatum of ChAc patients. Whether MSN neurons of the direct or the indirect pathway posses a different vulnerability towards neurodegeneration in ChAc patients as it was reported for Huntington's disease patients (see below) is not known yet.

Neurochemical studies were performed in some autopsies. Sato and colleagues reported a decrease of substance P immunoreactivity within the striatum, while normal reactivity against choline acetyltransferase (CHAT) and glutamic acid decarboxylase (GAD) was noted. In contrast, a decrease of GAD activity and substance P level was detected in the substantia nigra [55]. Bird and colleagues also noted normal GAD and CHAT activity in the cortex, caudate nucleus, and putamen [50].

Why there is no decrease of GAD activity (a key enzyme of GABA synthesis) while there is such a huge neurodegeneration of mainly GABAergic neurons remains enigmatic. A reduction of substance P level within the striatum and substantia nigra was also reported by de Yebenes and colleagues [51]. They also report that dopamine and its metabolites are depleted in most brain areas most notably within the striatum [51] in line with reports showing a neurodegeneration within the substantia nigra [11, 57]. They also reported an elevation of norepinephrine within the putamen and the pallidum [51].

Together, central nervous system pathology of ChAc is by far not understood. The rather selective affection of the caudate nucleus and putamen and to a lesser extent of the substantia nigra seems to be the only validated data. However, whether the neuropathology also occurs in a gradient as in Huntington's disease is as elusive as the differential vulnerability of the neuronal subpopulations within the striatum. Furthermore, a detailed analysis of neurochemical changes is just as needed as a detailed and state-of-the-art investigation for aggregates or inclusions.

A long discussion whether the elevation of creatine kinase is a sign for myopathy is nowadays clearly answered by "no." Peripheral neuropathy including nerve and muscle specimens was reported in many ChAc patients, and all authors report of neurogenic changes rather than signs of a separate myopathy [12, 53, 55, 56, 60, 61]. Nevertheless, some authors report signs of myopathy such as central nucleation and fiber splitting. However, these were finally seen as secondary effects due to chronic denervation [61]. Malandrini and colleagues recorded a loss of myelinated fibers which were accentuated distally with cytoskeletal changes in the axoplasm typically seen in distal neuropathies [53]. This fits well to the clinical data of reduced tendon reflexes and neurogenic changes in eletroneurographical and myographical investigations [5, 11]. This is interestingly different in McLeod syndrome, in which myopathic changes in muscle biopsy were found in up to 80 % of patients [4, 10].

Cardiomyopathy is not seen in ChAc, whereas it is a common feature of McLeod syndrome [6, 10].

Lessons from Huntington's Disease

In Huntington's disease the course of pathology is caudo-rostral, dorsoventral, and medio-lateral [62]. The extent of striatal degeneration was rated by Vonsattel and colleagues giving five stages of degeneration (grades 0–4) [63, 64]. The tail and body of the caudate is more affected than the head of the caudate nucleus (for a review, see [65]). The MSNs are the main affected neurons. The MSNs of the indirect pathway, the enkephalin and D2-receptor-expressing MSN projecting to the globus pallidus externus, are more vulnerable and affected before the D1-positive, substance P, and dynorphin-expressing MSNs.

Additional data suggests that the different involvement of the striosomal neurons and the matrix neurons depends on the clinical phenotype (or vice versa). Pronounced mood symptoms in HD patients are associated with profound degeneration in the striosomes. In contrast, motor symptoms correlate with matrix neuronal degeneration. In addition, different cortical structures are degenerated; patients suffering mainly from mood symptoms show neuronal loss in the limbic cingulated cortex, while patients mainly affected by motor symptoms show neuronal loss in the primary motor cortex but no cell loss in the limbic cortex [66] (for a review, see [65]).

Aggregates are well known in the neuropathology of HD. Both neuronal intranuclear inclusions and neuronal extranuclear inclusions are present. They mostly contain ubiquitinated huntingtin. These are found prior to symptom onset and are more abundant throughout the cortex than the striatum (for a review, see [65]).

In summary, there is no doubt that ChAc patients suffer from striatal atrophy and MSN loss accompanied by reactive gliosis. But we still need to know which MSN cell type is preferentially affected, if there is a striosome-matrix compartmental degeneration similar to that present in HD, and if there are aggregates/tangles as in other neurodegenerative diseases.

Imaging

Structural brain imaging using computed tomography or magnetic resonance imaging revealed widening of the anterior horn of the lateral ventricles due to caudate atrophy in clinical diagnosed ChAc [11, 67, 68]. Recent studies including genetically confirmed ChAc patients reproduced this finding [69–71]. These results were similar to those seen in HD patients [67, 68] and not helpful in the differential diagnosis of these diseases. Additionally, there are hints of abnormal hyperintensity in T2-weighted images within the striatum (Fig. 3.1 and [72]). Newer studies using voxel-based morphometry revealed selective atrophy only of the caudate nucleus but no significant global brain atrophy measured by brain parenchymal fraction method [69] (Fig. 3.3), similar to the neuropathology. This might be useful to distinguish ChAc from HD where gray matter loss was visualized by VBM in the insular cortex and the premotor and sensorimotor cortices [73]. However, two ChAc patients (not genetically proven) were recently reported to have cerebellar atrophy [74].

There are only few reports on functional imaging studies on ChAc patients. [18F]-2-fluoro-2-deoxyglucose PET (FDG-PET) clearly showed marked reduction of glucose utilization in the caudate and putamen [72, 75, 76]. This was already obvious in patients without structural striatal atrophy in MRI [75]. Müller-Vahl and colleagues interestingly found an increase in glucose uptake throughout the cerebral cortex [72], which was however not seen in the study of Hosokawa et al. [76]. FDG-PET is similar to that seen in HD patients [76, 77]. However, in HD hypometabolism is seen already in very early stages of the disease in the bilateral frontal, temporal, and parietal cortices [78] which is different to the ChAc patients but fits to the neuropathology data (see above).

Serotonin transporter (SERT) imaging using ^{123}I-2-β-carbomethoxy-3-β-(4-iodophenyl)-tropane (^{123}I-β-CIT) and single-photon emission computed

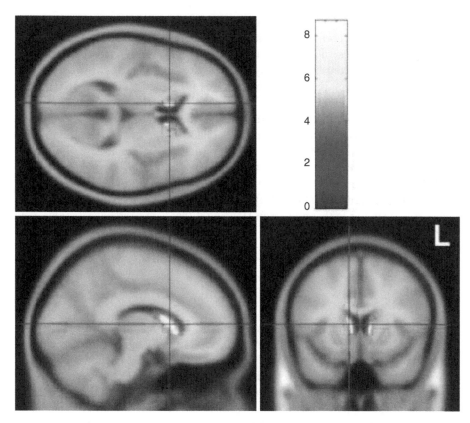

Fig. 3.3 Showing overlay of areas of significant gray matter reduction of the group analysis on the SPM standard brain template, demonstrating alterations in the caudate nucleus bilaterally (*blue crosshair* indicating the maximum). Z-score is indexed by the *color bar* (Reprinted with permission from Henkel et al. [69])

tomography (SPECT) showed normal values in the hypothalamus and midbrain in two ChAc patients [72]. Striatal presynaptic dopamine transporters visualized by 123I-β-CIT did not show significant changes compared to 20 healthy controls [72]. It is not clear whether the patients were suffering also from a parkinsonian syndrome but were reported to have a hyperkinetic and dystonic phenotype [72]. In a larger study including six patients suffering from neuroacanthocytosis, 16 suffering from idiopathic Parkinson's disease (PD) and 30 controls, Brooks and colleagues investigated the pre- and postsynaptic dopaminergic system using both F-DOPA-PET and raclopride-PET, respectively. Thereby, they could show normal presynaptic F-DOPA uptake in the caudate and anterior putamen but reduced uptake in the posterior putamen which was similar in PD patients [79]. Raclopride-PET showed a more than 50 % reduction of D2-receptor-binding sites in the caudate and putamen of NA patients. Even though no genetic analysis or Western blot of chorein could be performed in the early 1990s, the clinical syndrome (chorea, epilepsy, and orofacial dyskinesias) is likely to be ChAc [79]. 99mTc HMPAO SPECT was normal in two

NA patients [80] but showed marked reduction in frontal lobe in one patient suffering from ChAc with frontal lobe dysfunction [81].

Two proton magnetic resonance spectroscopy (^{1}H-MRS) published studies in NA/ChAc patients each of them investigating two patients by Molina and colleagues report a marked increase of myoinositol (Ino) in the basal ganglia as compared to healthy controls, accompanied by an increase of choline but without differences in the N-acetyl-aspartate (NAA) [80]. In a more recent study, investigating ChAc patients proven by the absence of chorein in the Western blot, the authors report an increase in myoinositol (Ino) without changes in choline in one patient and a decrease of Ino in the other. In the first patient, a decrease of NAA was also observed [82]. Increased Ino and choline could point towards a gliosis which would be in agreement with the neuropathology. However, why NAA as a marker for neuronal integrity is not changed remains elusive.

In summary, caudate and putamen atrophy and hypometabolism in ChAc seem to be proven and in accordance with most neuropathological studies. All other described phenotypes including possible cerebral cortex atrophy vs hypermetabolism or normal values of DAT and SERT imaging are very premature and lack sufficient number of patients. The same is true for H-MRS studies. However, the latter technique could be an interesting tool to study basic pathophysiology since inositol signal could also give insights on membrane phenotypes.

Diagnosis (Including Differential Diagnosis)

Full-blown ChAc is maybe easy to diagnose since neuropathy and creatine elevation together with chorea and epilepsy are a condition fairly different from classical Huntington's or Wilson's disease. However, in earlier stages of the disease when only some of the above-mentioned symptoms appear, it might be more difficult. Therefore, the overall goal must be to have ChAc in mind while thinking of the diagnosis of a patient with involuntary movements with additional symptoms. In principle, CK, AST and ALT elevations are more helpful for diagnosis than the often reported pathognomonic facial dyskinesia with lip biting.

In principle, a wide range of diseases have to be considered in differential diagnosis including hyperkinetic/hypokinetic movement disorders, epileptic syndromes, and mitochondrial disorders (see also Table 3.1).

Huntington's Disease

Huntington's disease is the main differential diagnosis if chorea is the leading symptom. Age at onset could be the same; however, creatine kinase or liver enzyme elevation is not usually seen in HD. Additionally, neuropathy hints towards ChAc, and neuropathology is much more widespread in HD which might be seen already

Table 3.1 Differential diagnosis

Main symptom/ syndrome	Differential	Red flags (pointing towards ChAc)
Chorea	Huntington	Neuropathy
		Epilepsy
		CK↑, ALT↑, AST↑
	Wilson	Epilepsy
		CK↑, ALT↑, AST↑
Dystonia	Wilson	Epilepsy
		CK↑, ALT↑, AST↑
	PKAN	No iron deposits in cMRI
		Epilepsy
		CK↑, ALT↑, AST↑
Epilepsy	Familial epileptic syndromes	Involuntary movements (not always drug induced)
	Sporadic epileptic syndromes	CK↑(not always due to seizures)
		AST, ALT↑ (not always drug induced)
Psychiatric disturbances	Tourette syndrome	Involuntary movements (not always drug induced)
	Obsessive-compulsive disorder	Parkinsonism (not always neuroleptic induced)

in neuroimaging (see above). Epilepsy and acanthocytes are additionally normally not seen in HD. Inheritance is autosomal dominant with high penetrance and anticipation in HD.

Wilson's Disease

Wilson's disease (WD) is always a differential when neuropsychiatric symptoms and elevated liver enzymes are present. Whereas in ChAc liver enzyme elevation remains mainly asymptomatic, untreated patients suffering from WD are likely to develop liver failure with the need of transplantation. The movement disorder of WD includes chorea, tremor, dystonia, and hypokinetic-rigid syndromes which are not to distinguish from ChAc. Neuropathy and epilepsy are rare in WD. Diagnosis is made by reduced ceruloplasmin concentration in serum and increased copper excretion in urine. Inheritance is autosomal recessive with mutations in the *ATP7B* gene.

McLeod Syndrome

McLeod syndrome is a ChAc multisystem disorder. There is significant overlap to ChAc including the movement disorder, neuropsychiatric disturbances, acanthocytosis, and creatine kinase elevation [4]. Some clinicians called McLeod syndrome "ChAc of old men" [10]. Nevertheless, there are some differences. Neuropathy is often not that prominent in McLeod, whereas McLeod patients show myopathic changes in neuropathology and myography. Compensated hemolysis, splenomegaly, and hepatomegaly are more common in McLeod than ChAc. Hemolysis most likely results from reduced Kell blood antigen expression which is to be considered when transfusion needs to be done in McLeod patients [83]. There are several reports of severe transfusion reactions [6]. Cardiomyopathy is a common feature of McLeod (>60 % of patients) and needs to be monitored on a regular basis. Inheritance is X-linked, mutations are found in the *XK* gene, and heterozygous females only rarely show symptoms.

Pantothenate Kinase-Associated Neurodegeneration (PKAN)

Pantothenate kinase-associated neurodegeneration (PKAN) normally affects young children with onset before the age of 6 years. However, there are some atypical forms also appearing in adolescence or early adulthood. PKAN belongs to the group of neurodegeneration with brain iron accumulation (NBIA syndromes) which are characterized by increased iron deposits within the basal ganglia. Thus, normally cMRI is the crucial diagnostic instrument showing the "eye of the tiger" sign. Clinically, the patients most often suffer from progressive generalized dystonia but also have features of hypokinesia and rigidity (especially in the atypical forms). Pigmentary retinopathy is common in the typical young onset form, while acanthocytosis is only seen in approximately 10 % of PKAN patients [84]. HARP syndrome is allelic with PKAN (hypobetalipoproteinemia, acanthocytosis, retinitis pigmentosa, and pallidal degeneration) [5]. Inheritance is autosomal recessive, and mutations are found in the *PANK2* gene.

Huntington's Disease-Like 2 (HDL2)

Huntington's disease-like 2 (HDL2) is a rare disorder up to now only reported in individuals of African ancestry. It shows a similar clinical picture as ChAc or HD and manifests in midlife. More than 41 CTG trinucleotide repeats in the *JPH3* gene are considered diagnostic. Acanthocytosis is only found in few patients [85].

Hypolipoproteinemia (HLP) and Abetalipoproteinemia (ALP)

Acanthocytosis was a long-known feature which could occur together with spinocerebellar degeneration and neuropathy in ALP (Bassen-Kornzweig syndrome). Patients suffer from steatorrhoea and pigmentary retinopathy and normally lack basal ganglia movement disorders but have significant vitamin E deficiency. ABL is inherited in an autosomal recessive manner; HBL can be symptomatic in heterozygote or homozygote patients [5].

Lesch-Nyhan Syndrome

Severely affected patients of this X-linked disorder often show self-mutilations with lip and finger biting. Additional symptoms are cognitive and behavior changes as well as hyperuricemia. Disease onset is usually within the first year and thus unlikely to be misdiagnosed as ChAc. Enzyme activity of hypoxanthine-guanine phosphoribosyltransferase (HPRT) is significantly reduced.

Other rare disorders including dentatorubral-pallidoluysian atrophy, benign hereditary chorea, and infantile neuroaxonal dystrophy could also mimic ChAc.

Testing Strategy

Testing for Acanthocytosis

It is important to use special methods to determine blood cell acanthocytosis (Fig. 3.4). If these are not used, it is likely that one fail to detect acanthocytosis [2, 86]. A rather simple but highly specific test was established by Storch and colleagues. Using a 1:1 dilution of fresh blood with isotonic NaCl supplemented with 10E/ml heparin and an incubation time of 30–60 min in room temperature leads to unmask acanthocytosis with high specificity (0.98). A normal range of <6.3 % of total erythrocytes is recommended to search for significant acanthocytosis [9].

Western Blot

Chorein detection by Western blot analysis in ChAc patients normally shows significant reduction if not totally lost of full-length chorein [87]. In contrast, normal levels of chorein are seen in McLeod syndrome or Huntington's disease. It is however noted that in some *VPS13A* mutations there might be normal chorein levels and thus genetic testing is required. The Western blot can be performed on a research basis; contact is Prof. Dr. Adrian Danek (http://www.euro-hd.net/html/na/network/docs/chorein-wb-info.pdf).

Genetic Testing

Since 2002, the gene mutated in ChAc is known [10, 88]. The protein was named chorein, and initially the gene was also reported as *CHAC* gene [10]. Due to the sequence homology to vesicular sorting protein 13 (vps13) in *Saccharomyces cerevisiae*, the chorein gene was named *VPS13A*. *VPS13A* is located on chromosome 9q21 and is organized in 73 exons spanning about 250 kb of the genome. The transcript possess a full-length sequence of 11,262 bp leading a protein of 3,174 amino

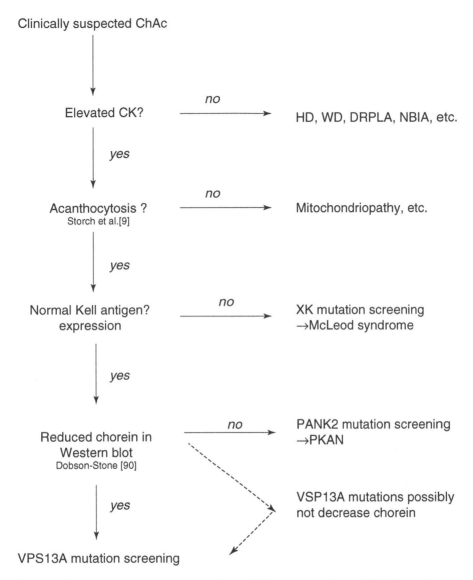

Fig. 3.4 Diagnostic pathway in suspected ChAc. *HD* Huntington's disease, *WD* Wilson's disease, *NBIA* neurodegeneration with brain iron accumulation, *DRPLA* dentato-rubro-pallidolysial degeneration, *PKAN* pantothenate kinase-associated neurodegeneration

acids (transcript variant A). Additionally, a splicing variant was identified lacking exons 70–30 encoding a protein of 3,095 amino acids (transcript variant B) [10]. Other splicing variants have also been described [89]. The protein is about 360 kDa. Mutations were found throughout the gene with no hot spot causing the disease [90]. This however makes genetic testing difficult and expensive. For details of genetics, please also refer to [5]. To the best of our knowledge, the disease appears autosomal recessive. Carriers (heterozygotes) are asymptomatic [5].

Treatment (Including the Management of Complications)

There is up to now no causative treatment for ChAc. All efforts are thus purely symptomatic.

Chorea

Chorea is treated as in Huntington's disease. Tetrabenazine or tiapride is the common anti-choreic treatment. Trihexyphenidyl can be given when dystonia is the leading symptom; however, anticholinergic effects must be kept in mind. Of note, treatment of involuntary movements should only be initiated if really necessary since the used neuroleptics may worsen epilepsy as well as induce hypokinesia/parkinsonism [91]. Furthermore, involuntary movements often decrease in the course of the disease, and patients develop a hypokinetic-rigid syndrome.

Parkinsonism

It is important to change the therapy in these stages by removal of anti-choreic drugs to dopaminergic drugs. Levodopa or amantadine was proven helpful in some patients.

Single case reports exist about deep brain stimulations of ChAc patients with different target regions. Currently, a prospective study is performed assessing the safety and tolerability using bilateral pallidal DBS in ChAc patients within the EMINA consortium (http://www.klinikum.uni-muenchen.de/Klinik-und-Poliklinik-fuer-Neurologie/de/Klinik/Neurologische_Poliklinik/Kognitive_Neurologie/Forschung/emina/index.html).

Epilepsy

There are no systematic reports on the best treatment of epilepsy in ChAc. The course of the epilepsy in ChAc is however often difficult, and most patients reported received multidrug therapy in order to control seizures [25, 29]. In our own not published patients, we fortunately saw a reduction of seizures in the course of the disease enabling us to reduce the multidrug therapy in later stages. Even though often appearing as temporal lobe epilepsy with drug resistance, there are yet no reports about epilepsy surgery, which is more considered as inappropriate in such a fatal neurodegenerative disease [25].

Valproic acid is often not recommended in diseases which might involve dysfunction of the mitochondrias; it is the antiepileptic drug (AED) with the best known mitochondrial toxicity. However, also other AEDs may affect mitochondrial function as carbamazepine, oxcarbazepine, phenytoin, phenobarbital, zonisamide, gabapentin, topiramate, and vigabatrin (for a review, see [92]). However, there are reports of ChAc patients treated with valproic acid without any obvious problems [25]. On the other hand, lamotrigine and carbamazepine were reported to worsen involuntary movements in ChAc patients [25]. This is in line with reports of other non-ChAc movement disorders [93, 94] and argues against the use of lamotrigine and carbamazepine in ChAc patients as first-line treatments.

Oromandibular Dyskinesia

Lip biting and tongue protrusions are hard to treat. Botulinum toxin treatment was reported to be helpful [95]; however, one should be careful not to deteriorate swallowing problems. Pallidal bilateral deep brain stimulation was beneficial in two cases [95]. Mechanical devices were used to protect from mouth injuries [96].

Physiotherapy is mandatory to preserve motor function as long as possible, and logotherapy to improve dysarthria and dysphagia. However, tube feeding is often not avoidable in later stages of the disease [5].

Psychiatric Disturbances

Psychiatric disturbances are treated symptomatically in a way used in primary psychiatric illness. Depression should be treated with modern antidepressants as serotonin reuptake inhibitors (SSRIs). Tricyclic antidepressants – due to their anticholinergic properties – may impair cognitive function or worsen confusion. However, the latter could help to treat hypersalivation. Data on electroconvulsive therapy however is limited and showed more side effects than amelioration of psychiatric disturbances [97–99]. Psychosis could be difficult to treat since neuroleptic drugs may induce or worsen parkinsonism and epilepsy. Atypical neuroleptics such as clozapine, olanzapine, or quetiapine are the treatment of choice (for a review, see [7]). Long treatment with high doses of SSRI is often needed in obsessive-compulsive disorder [7].

Cognitive Impairment

There exist no data on specific treatment of cognitive decline. Since the pathomechanism is not understood so far, classical treatment as for Alzheimer's disease cannot be recommended yet. Ergotherapy and cognitive training however should be done as in other entities of cognitive impairment.

Of note, since the pathophysiology of ChAc is not understood, many (co-)treatments are still to be done by trial and error. A report about an oral contraceptive-induced exacerbation of chorea in chorea-acanthocytosis is exemplarily noted [100].

Conclusions

Chorea-acanthocytosis is a rare differential diagnosis of Huntington's disease. It should be considered when besides chorea/dystonia other symptoms are obvious such as epilepsy, elevated liver enzymes and creatine kinases, and acanthocytosis. Neuroimaging shows caudate atrophy and reduced glucose uptake in the striatum. Neuropathological examination revealed severe neuronal cell loss and astrogliosis as well as microglial and oligodendroglial proliferation. Additionally, neuronal loss in the substantia nigra was shown. Pathophysiology is currently not understood; however, the first insights gained in red blood cells of patients hint towards chorein being involved in regulation of cortical actin depolymerization mediated via PI3K-RAC-PAK as well as Lyn kinase-mediated phosphorylation of red cell membrane proteins. There is currently no cure for the disease; all therapeutical options concentrate on symptomatic treatment. Further studies unraveling the pathophysiology are urgently needed that can help to understand the disease mechanisms and thereby possibly develop better treatments.

Acknowledgment The author is supported by the Advocacy for Neuroacanthocytosis Patients and by the Federal Ministry of Education and Research (BMBF) under the frame of E-Rare-2, the ERA-Net for Research on Rare Diseases.

References

1. Critchley EM, Clark DB, Wikler A. Acanthocytosis and neurological disorder without betalipoproteinemia. Arch Neurol. 1968;18:134–40.
2. Estes JW, Morley TJ, Levine IM, Emerson CP. A new hereditary acanthocytosis syndrome. Am J Med. 1967;42:868–81.
3. Levine IM, Estes JW, Looney JM. Hereditary neurological disease with acanthocytosis. A new syndrome. Arch Neurol. 1968;19:403–9.
4. Danek A, Rubio JP, Rampoldi L, Ho M, Dobson-Stone C, Tison F, et al. McLeod neuroacanthocytosis: genotype and phenotype. Ann Neurol. 2001;50:755–64.
5. Dobson-Stone C, Rampoldi L, Bader B, Velayos Baeza A, Walker RH, Danek A, et al. Chorea-acanthocytosis. In: Pagon RA, Adam MP, Ardinger HH, Bird TD, Dolan CR, Fong CT, Stephens K, editors. GeneReviews® [Internet]. University of Washington: Seattle; 1993–2014.
6. Jung HH, Danek A, Walker RH. Neuroacanthocytosis syndromes. Orphanet J Rare Dis. 2011;6:68.
7. Walterfang M, Evans A, Looi JC, Jung HH, Danek A, Walker RH, et al. The neuropsychiatry of neuroacanthocytosis syndromes. Neurosci Biobehav Rev. 2011;35:1275–83.
8. Danek A, Dobson-Stone C Velayos Baeza A, Monaco AP. The phenotype of chorea-acanthocytosis: a review of 106 patients with VPS13A mutations. In: 2nd International Neuroacanthocytosis Symposium. Movement disorders. Wiley Journals. vol. 20. 2005. p. 1678.
9. Storch A, Kornhass M, Schwarz J. Testing for acanthocytosis A prospective reader-blinded study in movement disorder patients. J Neurol. 2005;252:84–90.
10. Rampoldi L, Danek A, Monaco AP. Clinical features and molecular bases of neuroacanthocytosis. J Mol Med. 2002;80:475–91.

11. Hardie RJ, Pullon HW, Harding AE, Owen JS, Pires M, Daniels GL, et al. Neuroacanthocytosis. A clinical, haematological and pathological study of 19 cases. Brain. 1991;114(Pt 1A):13–49A.

12. Ohnishi A, Sato Y, Nagara H, Sakai T, Iwashita H, Kuroiwa Y, et al. Neurogenic muscular atrophy and low density of large myelinated fibres of sural nerve in chorea-acanthocytosis. J Neurol Neurosurg Psychiatry. 1981;44:645–8.

13. Saiki S, Hirose G, Sakai K, Matsunari I, Higashi K, Saiki M, et al. Chorea-acanthocytosis associated with Tourettism. Mov Disord. 2004;19:833–6.

14. Sibon I, Ghorayeb I, Arne P, Tison F. Distressing belching and neuroacanthocytosis. Mov Disord. 2004;19:856–9.

15. Bader B, Walker RH, Vogel M, Prosiegel M, McIntosh J, Danek A. Tongue protrusion and feeding dystonia: a hallmark of chorea-acanthocytosis. Mov Disord. 2010;25:127–9.

16. Schneider SA, Lang AE, Moro E, Bader B, Danek A, Bhatia KP. Characteristic head drops and axial extension in advanced chorea-acanthocytosis. Mov Disord. 2010;25:1487–91.

17. Yamamoto T, Hirose G, Shimazaki K, Takado S, Kosoegawa H, Saeki M. Movement disorders of familial neuroacanthocytosis syndrome. Arch Neurol. 1982;39:298–301.

18. Shibasaki H, Sakai T, Nishimura H, Sato Y, Goto I, Kuroiwa Y. Involuntary movements in chorea-acanthocytosis: a comparison with Huntington's chorea. Ann Neurol. 1982;12:311–4.

19. Yamada H, Ohji T, Sakurai S, Yamaguchi E, Uchimura N, Morita K, et al. Chorea-acanthocytosis presenting with schizophrenia symptoms as first symptoms. Psychiatry Clin Neurosci. 2009;63:253–4.

20. Shimo H, Nakamura M, Tomiyasu A, Ichiba M, Ueno S, Sano A. Comprehensive analysis of the genes responsible for neuroacanthocytosis in mood disorder and schizophrenia. Neurosci Res. 2011;69:196–202.

21. Danek A, Sheesley L, Tierney M, Uttner I, Grafman J. Cognitive and neuropsychiatric findings in McLeod syndrome and in chorea-acanthocytosis. In: Danek A, editor. Neuroacanthocytosis syndromes. Dordrecht: Springer; 2004. p. 95–115.

22. Walker RH, Liu Q, Ichiba M, Muroya S, Nakamura M, Sano A, et al. Self-mutilation in chorea-acanthocytosis: manifestation of movement disorder or psychopathology? Mov Disord. 2006;21:2268–9.

23. Kartsounis LD, Hardie RJ. The pattern of cognitive impairments in neuroacanthocytosis. A frontosubcortical dementia. Arch Neurol. 1996;53:77–80.

24. Rampoldi L, Dobson-Stone C, Rubio JP, Danek A, Chalmers RM, Wood NW, et al. A conserved sorting-associated protein is mutant in chorea-acanthocytosis. Nat Genet. 2001; 28:119–20.

25. Al-Asmi A, Jansen AC, Badhwar A, Dubeau F, Tampieri D, Shustik C, et al. Familial temporal lobe epilepsy as a presenting feature of choreoacanthocytosis. Epilepsia. 2005;46:1256–63.

26. Iwata M, Fuse S, Sakuta M, Toyokura Y. Neuropathological study of chorea-acanthocytosis. Jpn J Med. 1984;23:118–22.

27. Tiftikcioglu BI, Dericioglu N, Saygi S. Focal seizures originating from the left temporal lobe in a case with chorea-acanthocytosis. Clin EEG Neurosci. 2006;37:46–9.

28. Bader B, Vollmar C, Ackl N, Ebert A, la Fougere C, Noachtar S, et al. Bilateral temporal lobe epilepsy confirmed with intracranial EEG in chorea-acanthocytosis. Seizure. 2011;20:340–2.

29. Scheid R, Bader B, Ott DV, Merkenschlager A, Danek A. Development of mesial temporal lobe epilepsy in chorea-acanthocytosis. Neurology. 2009;73:1419–22.

30. Schwartz MS, Monro PS, Leigh PN. Epilepsy as the presenting feature of neuroacanthocytosis in siblings. J Neurol. 1992;239:261–2.

31. Cendes F, Lopes-Cendes I, Andermann E, Andermann F. Familial temporal lobe epilepsy: a clinically heterogeneous syndrome. Neurology. 1998;50:554–7.

32. Vadlamudi L, Scheffer IE, Berkovic SF. Genetics of temporal lobe epilepsy. J Neurol Neurosurg Psychiatry. 2003;74:1359–61.

33. Andermann F. Pseudotemporal vs neocortical temporal epilepsy: things aren't always where they seem to be. Neurology. 2003;61:732–3.

34. Brickner JH, Fuller RS. SOI1 encodes a novel, conserved protein that promotes TGN-endosomal cycling of Kex2p and other membrane proteins by modulating the function of two TGN localization signals. J Cell Biol. 1997;139:23–36.
35. Seifert W, Kuhnisch J, Maritzen T, Horn D, Haucke V, Hennies HC. Cohen syndrome-associated protein, COH1, is a novel, giant Golgi matrix protein required for Golgi integrity. J Biol Chem. 2011;286:37665–75.
36. Stege JT, Laub MT, Loomis WF. Tip genes act in parallel pathways of early Dictyostelium development. Dev Genet. 1999;25:64–77.
37. Tomemori Y, Ichiba M, Kusumoto A, Mizuno E, Sato D, Muroya S, et al. A gene-targeted mouse model for chorea-acanthocytosis. J Neurochem. 2005;92:759–66.
38. Kurano Y, Nakamura M, Ichiba M, Matsuda M, Mizuno E, Kato M, et al. Chorein deficiency leads to upregulation of gephyrin and GABA(A) receptor. Biochem Biophys Res Commun. 2006;351:438–42.
39. Kurano Y, Nakamura M, Ichiba M, Matsuda M, Mizuno E, Kato M, Agemura A, et al. In vivo distribution and localization of chorein. Biochem Biophys Res Commun. 2007;353:431–5.
40. Terada N, Fujii Y, Ueda H, Kato Y, Baba T, Hayashi R, et al. Ultrastructural changes of eryth-rocyte membrane skeletons in chorea-acanthocytosis and McLeod syndrome revealed by the quick-freezing and deep-etching method. Acta Haematol. 1999;101:25–31.
41. De Franceschi L, Tomelleri C, Matte A, Brunati AM, Bovee-Geurts PH, et al. Erythrocyte membrane changes of chorea-acanthocytosis are the result of altered Lyn kinase activity. Blood. 2011;118:5652–63.
42. Clark MR, Aminoff MJ, Chiu DT, Kuypers FA, Friend DS. Red cell deformability and lipid composition in two forms of acanthocytosis: enrichment of acanthocytic populations by den-sity gradient centrifugation. J Lab Clin Med. 1989;113:469–81.
43. Siegl C, Hamminger P, Jank H, Ahting U, Bader B, Danek A, et al. Alterations of red cell membrane properties in neuroacanthocytosis. PLoS One. 2013;8(10):e76715.
44. Kay MM. Band 3 in aging and neurological disease. Ann N Y Acad Sci. 1991;621:179–204.
45. Bosman GJ, Bartholomeus IG, De Grip WJ, Horstink MW. Erythrocyte anion transporter and antibrain immunoreactivity in chorea-acanthocytosis. A contribution to etiology, genetics, and diagnosis. Brain Res Bull. 1994;33:523–8.
46. Kay MM, Goodman J, Lawrence C, Bosman G. Membrane channel protein abnormalities and autoantibodies in neurological disease. Brain Res Bull. 1990;24:105–11.
47. Foller M, Hermann A, Gu S, Alesutan I, Qadri SM, Borst O, et al. Chorein-sensitive polymer-ization of cortical actin and suicidal cell death in chorea-acanthocytosis. FASEB J. 2012;26:1526–34.
48. Schmidt EM, Schmid E, Munzer P, Hermann A, Eyrich AK, Russo A, et al. Chorein sensitivity of cytoskeletal organization and degranulation of platelets. FASEB J. 2013;27:2799–806.
49. Bader B, Arzberger T, Heinsen H, Dobson-Stone C, Kretzschmar HA, Danek A. Neuropathology of chorea-acanthocytosis. In: Walker RH, Saiki S, Danek A, editors. Neuroacanthocytosis syn-dromes II. Berlin: Springer; 2008. p. 187–95.
50. Bird TD, Cederbaum S, Valey RW, Stahl WL. Familial degeneration of the basal ganglia with acan-thocytosis: a clinical, neuropathological, and neurochemical study. Ann Neurol. 1978;3:253–8.
51. de Yebenes JG, Brin MF, Mena MA, De Felipe C, del Rio RM, Bazan E, et al. Neurochemical findings in neuroacanthocytosis. Mov Disord. 1988;3:300–12.
52. Galatioto S, Serra S, Batolo D, Marafioti T. Amyotrophic choreo-acanthocytosis: a neuro-pathological and immunocytochemical study. Ital J Neurol Sci. 1993;14:49–54.
53. Malandrini A, Fabrizi GM, Palmeri S, Ciacci G, Salvadori C, Berti G, et al. Choreo-acanthocytosis like phenotype without acanthocytes: clinicopathological case report. A contribution to the knowledge of the functional pathology of the caudate nucleus. Acta Neuropathol. 1993;86:651–8.
54. Rinne JO, Daniel SE, Scaravilli F, Pires M, Harding AE, Marsden CD. The neuropathological features of neuroacanthocytosis. Mov Disord. 1994;9:297–304.

55. Sato Y, Ohnishi A, Tateishi J, Onizuka Y, Ishimoto S, Iwashita H, et al. An autopsy case of chorea-acanthocytosis. Special reference to the histopathological and biochemical findings of basal ganglia. No To Shinkei. 1984;36:105–11.
56. Vital A, Bouillot S, Burbaud P, Ferrer X, Vital C. Chorea-acanthocytosis: neuropathology of brain and peripheral nerve. Clin Neuropathol. 2002;21:77–81.
57. Rinne JO, Daniel SE, Scaravilli F, Harding AE, Marsden CD. Nigral degeneration in neuroacanthocytosis. Neurology. 1994;44:1629–32.
58. Spencer SE, Walker FO, Moore SA. Chorea-amyotrophy with chronic hemolytic anemia: a variant of chorea-amyotrophy with acanthocytosis. Neurology. 1987;37:645–9.
59. Gerfen CR, Bolam JP. Handbook of basal ganglia structure and function. Amsterdam: Elsevier; 2010. p. 1–28.
60. Alonso ME, Teixeira F, Jimenez G, Escobar A. Chorea-acanthocytosis: report of a family and neuropathological study of two cases. Can J Neurol Sci. 1989;16:426–31.
61. Limos LC, Ohnishi A, Sakai T, Fujii N, Goto I, Kuroiwa Y. "Myopathic" changes in choreaacanthocytosis. Clinical and histopathological studies. J Neurol Sci. 1982;55:49–58.
62. Vonsattel JP, DiFiglia M. Huntington disease. J Neuropathol Exp Neurol. 1998;57:369–84.
63. Vonsattel JP, Myers RH, Stevens TJ, Ferrante RJ, Bird ED, Richardson EPJ. Neuropathological classification of Huntington's disease. J Neuropathol Exp Neurol. 1985;44:559–77.
64. Vonsattel JP, Keller C, Del Pilar Amaya M. Neuropathology of Huntington's disease. Handb Clin Neurol. 2008;89:599–618.
65. Waldvogel HJ, Kim EH, Thu DC, Tippett LJ, Faull RL. New perspectives on the neuropathology in Huntington's disease in the human brain and its relation to symptom variation. J Huntingtons Dis. 2012;1:143–53.
66. Thu DC, Oorschot DE, Tippett LJ, Nana AL, Hogg VM, Synek BJ, et al. Cell loss in the motor and cingulate cortex correlates with symptomatology in Huntington's disease. Brain. 2010;133:1094–110.
67. Kutcher JS, Kahn MJ, Andersson HC, Foundas AL. Neuroacanthocytosis masquerading as Huntington's disease: CT/MRI findings. J Neuroimaging. 1999;9:187–9.
68. Serra S, Xerra A, Scribano E, Meduri M, Di Perri R. Computerized tomography in amyotrophic choreo-acanthocytosis. Neuroradiology. 1987;29:480–2.
69. Henkel K, Danek A, Grafman J, Butman J, Kassubek J. Head of the caudate nucleus is most vulnerable in chorea-acanthocytosis: a voxel-based morphometry study. Mov Disord. 2006;21:1728–31.
70. Huppertz HJ, Kroll-Seger J, Danek A, Weber B, Dorn T, Kassubek J. Automatic striatal volumetry allows for identification of patients with chorea-acanthocytosis at single subject level. J Neural Transm. 2008;115:1393–400.
71. Walterfang M, Looi JC, Styner M, Walker RH, Danek A, Niethammer M, et al. Shape alterations in the striatum in chorea-acanthocytosis. Psychiatry Res. 2011;192:29–36.
72. Muller-Vahl KR, Berding G, Emrich HM, Peschel T. Chorea-acanthocytosis in monozygotic twins: clinical findings and neuropathological changes as detected by diffusion tensor imaging, FDG-PET and (123)I-beta-CIT-SPECT. J Neurol. 2007;254:1081–8.
73. Douaud G, Gaura V, Ribeiro MJ, Lethimonnier F, Maroy R, Verny C, et al. Distribution of grey matter atrophy in Huntington's disease patients: a combined ROI-based and voxel-based morphometric study. Neuroimage. 2006;32:1562–75.
74. Katsube T, Shimono T, Ashikaga R, Hosono M, Kitagaki H, Murakami T. Demonstration of cerebellar atrophy in neuroacanthocytosis of 2 siblings. AJNR Am J Neuroradiol. 2009;30:386–8.
75. Dubinsky RM, Hallett M, Levey R, Di Chiro G. Regional brain glucose metabolism in neuroacanthocytosis. Neurology. 1989;39:1253–5.
76. Hosokawa S, Ichiya Y, Kuwabara Y, Ayabe Z, Mitsuo K, Goto I, et al. Positron emission tomography in cases of chorea with different underlying diseases. J Neurol Neurosurg Psychiatry. 1987;50:1284–7.

77. Antonini A, Leenders KL, Spiegel R, Meier D, Vontobel P, Weigell-Weber M, et al. Striatal glucose metabolism and dopamine D2 receptor binding in asymptomatic gene carriers and patients with Huntington's disease. Brain. 1996;119(Pt 6):2085–95.
78. Shin H, Kim MH, Lee SJ, Lee KH, Kim MJ, Kim JS, Cho JW. Decreased metabolism in the cerebral cortex in early-stage Huntington's disease: a possible biomarker of disease progression? J Clin Neurol. 2013;9:21–5.
79. Brooks DJ, Ibanez V, Playford ED, Sawle GV, Leigh PN, Kocen RS, et al. Presynaptic and postsynaptic striatal dopaminergic function in neuroacanthocytosis: a positron emission tomographic study. Ann Neurol. 1991;30:166–71.
80. Molina JA, Garcia-Segura JM, Benito-Leon J, Jiminez-Jiminez FJ, Martinez V, Viano J, et al. In vivo proton magnetic resonance spectroscopy in patients with neuroacanthocytosis. Parkinsonism Relat Disord. 1998;4:11–5.
81. Delecluse F, Deleval J, Gerard JM, Michotte A, Zegers de Beyl D. Frontal impairment and hypoperfusion in neuroacanthocytosis. Arch Neurol. 1991;48:232–4.
82. Ismailogullari S, Caglayan AO, Bader B, Danek A, Korkmaz S, Sharifov E, et al. Magnetic resonance spectroscopy in two siblings with chorea-acanthocytosis. Mov Disord. 2010;25:2894–7.
83. Russo DC, Oyen R, Powell VI, Perry S, Hitchcock J, Redman CM, Reid ME. First example of anti-Kx in a person with the McLeod phenotype and without chronic granulomatous disease. Transfusion. 2000;40:1371–5.
84. Hayflick SJ, Westaway SK, Levinson B, Zhou B, Johnson MA, Ching KH, et al. Genetic, clinical, and radiographic delineation of Hallervorden-Spatz syndrome. N Engl J Med. 2003;348:33–40.
85. Walker RH, Rasmussen A, Rudnicki D, Holmes SE, Alonso E, Matsuura T, et al. Huntington's disease–like 2 can present as chorea-acanthocytosis. Neurology. 2003;61:1002–4.
86. Feinberg TE, Cianci CD, Morrow JS, Pehta JC, Redman CM, Huima T, et al. Diagnostic tests for choreoacanthocytosis. Neurology. 1991;41:1000–6.
87. Dobson-Stone C, Danek A, Rampoldi L, Hardie RJ, Chalmers RM, Wood NW, et al. Mutational spectrum of the CHAC gene in patients with chorea-acanthocytosis. Eur J Hum Genet. 2002;10:773–81.
88. Ueno S, Maruki Y, Nakamura M, Tomemori Y, Kamae K, Tanabe H, et al. The gene encoding a newly discovered protein, chorein, is mutated in chorea-acanthocytosis. Nat Genet. 2001;28:121–2.
89. Velayos-Baeza A, Vettori A, Copley RR, Dobson-Stone C, Monaco AP. Analysis of the human VPS13 gene family. Genomics. 2004;84:536–49.
90. Dobson-Stone C, Velayos-Baeza A, Filippone LA, Westbury S, Storch A, Erdmann T, et al. Chorein detection for the diagnosis of chorea-acanthocytosis. Ann Neurol. 2004;56:299–302.
91. Borchardt CM, Jensen C, Dean CE, Tori J. Case study: childhood-onset tardive dyskinesia versus choreoacanthocytosis. J Am Acad Child Adolesc Psychiatry. 2000;39:1055–8.
92. Finsterer J, Zarrouk Mahjoub S. Mitochondrial toxicity of antiepileptic drugs and their tolerability in mitochondrial disorders. Expert Opin Drug Metab Toxicol. 2012;8:71–9.
93. Jacome D. Movement disorder induced by carbamazepine. Neurology. 1981;31:1059–60.
94. Sotero de Menezes MA, Rho JM, Murphy P, Cheyette S. Lamotrigine-induced tic disorder: report of five pediatric cases. Epilepsia. 2000;41:862–7.
95. Schneider SA, Aggarwal A, Bhatt M, Dupont E, Tisch S, Limousin P, et al. Severe tongue protrusion dystonia: clinical syndromes and possible treatment. Neurology. 2006;67:940–3.
96. Fontenelle LF, Leite MA. Treatment-resistant self-mutilation, tics, and obsessive-compulsive disorder in neuroacanthocytosis: a mouth guard as a therapeutic approach. J Clin Psychiatry. 2008;69:1186–7.
97. Kennedy R, Mittal D, O'Jile J. Electroconvulsive therapy in movement disorders: an update. J Neuropsychiatry Clin Neurosci. 2003;15:407–21.

98. Rutherford M. Use of electroconvulsive therapy in a patient with chorea neuroacanthocytosis and prominent delusions. J ECT. 2012;28:e5–6.
99. Vazquez MJ, Martinez MC. Electroconvulsive therapy in neuroacanthocytosis or McLeod syndrome. J ECT. 2009;25:72–3.
100. Munhoz RP, Kowacs PA, Soria MG, Ducci RD, Raskin S, Teive HA. Catamenial and oral contraceptive-induced exacerbation of chorea in chorea-acanthocytosis: case report. Mov Disord. 2009;24:2166–7.

Chapter 4
Sydenham's Chorea

Francisco Cardoso

Abstract Sydenham's chorea (SC), a major manifestation of acute rheumatic fever, has been recognized in reports going back at least five centuries. In spite of its declining prevalence (particularly in developed countries), it is the most common explanation for acute onset of chorea in children. It is more common in girls than boys, and its features include chorea, decreased muscle tone, tics, and nonmovement disorder findings such as behavioral abnormalities (emotional lability, obsessions, compulsions, and others) and cognitive changes such as decreased verbal fluency and dysexecutive syndrome. In up to 50 % of patients, SC has a persistent course and can recur during pregnancy or in women with estrogen exposure. Its pathogenesis is not clearly understood, but it is most likely caused by *Streptococcus-induced* antibodies that cross-react with basal ganglia antigens. The diagnosis of SC relies on the presence of typical clinical features and exclusion of alternative causes. The management of patients with SC includes the use of antichoreic agents as well as secondary prophylaxis of *Streptococcus* infection with penicillin.

Keywords Chorea • Sydenham's chorea • Rheumatic fever • Carditis • Anti-basal ganglia antibodies

Introduction and Classification

Sydenham's chorea (SC) is also known as *rheumatic chorea* or *minor chorea*. Thomas Sydenham first described postinfectious choreic movements of children in 1686. However, the term chorea was coined by Paracelsus in the sixteenth century

F. Cardoso, MD, PhD
Movement Disorders Unit – Neurology Service,
Department of Internal Medicine, The Federal University of Minas Gerais,
Av Pasteur 89/1107, Belo Horizonte, MG 30150-290, Brazil
e-mail: cardosofe@terra.com.br

F.E. Micheli, P.A. LeWitt (eds.), *Chorea*,
DOI 10.1007/978-1-4471-6455-5_4, © Springer-Verlag London 2014

when describing the movement disorder seen in people who were afflicted by a "dancing mania" in the Medieval era. Although most cases may have been of psychogenic origin, some were in all likelihood caused by SC [1]. The casual relationship of SC with streptococcal infection was firmly established only midway in the twentieth century [2]. Despite SC's drastic reduction in incidence after World War II (particularly in North America and Western Europe), it remains worldwide the most common cause of chorea in children [3–6]. The aim of this chapter is to provide a review of epidemiology, clinical aspects, diagnosis, pathogenesis, and management of SC.

Epidemiology

A recent review of all inpatient notes seen between 1878 and 1911 by Sir William Gowers at the National Hospital, Queen Square, showed that almost all chorea cases were SC [7]. These figures seem to exemplify the etiology of chorea throughout the world at that time. The incidence of rheumatic fever and SC in the United States and Western Europe has declined substantially since World War II as a result of improved health care, penicillin and other antibiotic availability, and lower virulence of streptococcal strains [8]. This decline in incidence is demonstrated by the finding that, in 1980, the annual age-adjusted incidence rate of initial attacks of rheumatic fever declined from 3.0 in 1970 to 0.5 per 100,000 children in Fairfax County, Virginia, in the United States [9]. Furthermore, Nausieda and colleagues [10] showed that SC accounted for 0.9 % of childhood admissions to hospitals in Chicago before 1940, whereas this number dropped to 0.2 % during the period between 1950 and 1980. Nevertheless, despite its declining incidence, SC remains the most common cause of acute chorea in children. The continuous importance of SC even in developed areas is shown by a study performed in a university hospital pediatrics unit in Pennsylvania, United States, that showed it accounted for 96 % of all patients with chorea seen during 1980–2004 [11]. Studies from Australia also confirm that SC is a relatively common cause of acute chorea in children [12, 13]. A national survey of all cases of rheumatic fever in Australia identified 151 children in a 3-year period; SC was diagnosed in 19 % of them. Not surprisingly, the majority (86.7 %) were in indigenous Australians living in environments without urbanized health care. However, the authors identified ten cases in nonindigenous Australians, of whom three had atypical presentations. The authors concluded that rheumatic fever may be more common than previously thought among children with low health risks [14]. Outbreaks of rheumatic fever associated with chorea occurrence have also been identified in the United States [3, 15]. Variable manifestations may occur among people of differing ethnic backgrounds. Australian aboriginals appear to be a group at particularly high risk of developing rheumatic fever, whether with or without chorea [16]. Rheumatic fever has remained a significant public health problem in developing areas, particularly within the low-income population. That has been the situation in Latin America, Africa, Turkey, and the Indian subcontinent. For instance, at the Movement Disorders Clinic of the Federal University of Minas Gerais, Brazil, SC accounted for 64 % of all patients with chorea. This incidence far exceeded other choreic conditions such as Huntington's disease (HD), among others. The role of socioeconomic influences is nicely illustrated by the situation in Australia, where the overall incidence of rheumatic fever is low. However, in the northern portion of the Northern Territory, an area predominantly inhabited by Aboriginal people, the point prevalence of rheumatic fever in 1995 was 9.6 per 1,000 people aged 5–14 years [17]. The percentage of patients with rheumatic fever who go on to develop SC is variable according to time, geography, and ethnicity. In the author's experience with a tertiary care health unit, chorea occurs in about 26 % of patients with rheumatic fever [18]. In this setting (and despite the lack of formal community-based studies), there has been a gradual decline of SC incidence, suggesting a similar phenomenon is currently taking place in Brazil similar to that which occurred in North America and Western Europe decades ago.

Clinical Manifestations

The usual age for onset of SC is 8–9 years. However, there are reports on patients who developed chorea during the third decade of life. In most series, there is a female preponderance [18]. Typically, patients develop this disease 4–8 weeks after an episode of group A beta-hemolytic streptococcal pharyngitis. It does not occur after streptococcal infection of the skin. The chorea, characterized by a random and continuous flow of contractions, spreads rapidly and usually becomes generalized. Approximately 20 % of patients remain affected with hemichorea [10, 18]. SC patients display motor impersistence, particularly noticeable during tongue protrusion and ocular fixation. The muscle tone is usually decreased; in severe and rare cases (8 % of all patients seen at the author's clinical unit), the decrease of muscle tone can be so pronounced that the patient may become bedridden (chorea paralytica). After careful assessment of adult subjects with SC in remission, we found that 16 (64 %) of the 25 individuals had residual bradykinesia. This finding suggests that there is a nigrostriatal lesion caused by SC and it may explain the proclivity of such patients to develop drug-induced parkinsonism when treated with neuroleptics [19]. There are reports that tics commonly occur in SC [20]. In a cohort of 108 SC patients carefully followed up at the author's clinical unit, we have identified vocalizations in only 8 % of subjects. We have avoided the term "tic" in describing these vocalizations because with them, there has not been a premonitory sign or complex sound. These vocalizations tended to be associated with severe cranial chorea. Taken together, these findings suggest that involuntary sounds produced by a few SC patients result from choreic contractions of the upper respiratory tract muscles rather than sharing an identity with genuine tics [21].

Patients with SC often display clinical features that comprise other neurologic, "nonmotor disorders" and nonneurologic symptoms and signs. There is evidence that many patients with active chorea have eye movements with hypometric saccades and a few of them also show oculogyric crisis. Dysarthria is also common. Gowers recognized that SC patients present with a "disinclination to speak." A case-control study of patients described a pattern of decreased verbal fluency that reflected reduced phonetic output [22]. This result is consistent with dysfunction of the dorsolateral prefrontal-basal ganglia circuit. In a study of adults with a past history of SC, we have explored this observation further, showing that rheumatic chorea can be included among the causes of dysexecutive syndrome [23, 24]. These results were confirmed by another group of investigators [25]. Prosody of speech is also impaired in SC. Patients with rheumatic chorea had various voice alterations including decreased range of fundamental frequency, higher intensity, and decreased speed. These findings are similar to those observed in Parkinson's disease [26–29]. In a survey of 100 patients with rheumatic fever (half of whom also had chorea), we found that migraine is more frequent in SC (21.8 %) than in normal controls (8.1 %, $p=0.02$) [30]. In the older medical literature, there are also references to papilledema, central retinal artery occlusion, and seizures in a few patients with SC.

Attention has also been drawn to behavioral abnormalities in SC. Maia and colleagues [31] investigated behavioral abnormalities in 50 healthy subjects, 50 patients with rheumatic fever without chorea, and 56 patients with SC. The authors found that obsessive-compulsive behavior, obsessive-compulsive disorder, and attention deficit and hyperactivity disorder were more frequent in the SC group (19, 23.2, and 30.4 %) than in healthy controls (11, 4, and 8 %) or in patients with rheumatic fever without chorea (14, 6, and 8 %). In this study, the authors demonstrated that obsessive-compulsive behavior confers little interference in performing activities of daily living. This is similar to what has also been found by other authors [32–34]. Another study compared the phenomenology of obsessions and compulsions of SC patients to subjects diagnosed with tic disorders [35]. The authors demonstrated that the symptoms observed among the SC patients were different from those reported by patients with tic disorders, but were similar to those previously noted among pediatric patients with primary obsessive-compulsive disorder. An investigation comparing healthy controls with patients with rheumatic fever showed that obsessive-compulsive behavior is more commonly seen in patients with SC who have relatives also affected by obsessions and compulsions [36]. This study emphasizes that there is an interplay between genetic factors and environment in the development of behavioral problems in SC. A recent systematic review confirms that all studies of behavioral changes in patients with SC have found an association with obsessive-compulsive disorder [37]. We also confirmed that finding, although rarely, SC may induce psychosis during the acute phase of the illness [38]. We also found that one of our patients with SC developed trichotillomania [39]. Another study by our group suggested that SC is not a cause for nonspecific behavioral problems; in comparing patients and controls, there was no difference on a rating scale of anxiety symptoms [40]. Finally, clinical experience demonstrates that the most commonly seen behavioral problem in acute SC is emotional lability (i.e., patients becoming tearful).

There is evidence indicating that the peripheral nervous system is not targeted in SC [41, 42]. Finally, it is important to consider that SC is a major manifestation of rheumatic fever. Sixty percent to 80 % of patients display cardiac involvement, particularly mitral valve dysfunction, when they are also affected with SC, whereas the association with arthritis is less common (in 30 % of patients); however, in approximately 20 % of patients, chorea has been the sole finding [18]. Cardiac lesions are the main source of serious morbidity in SC. There is also evidence suggesting that SC may be associated with other autoimmune disorders such as Takayasu arteritis [43].

There is now a validated scale to rate SC. The Universidade Federal de Minas Gerais Sydenham Chorea Rating Scale was designed to provide a detailed quantitative description of performance in activities of daily living, behavioral abnormalities, and motor function of patients with SC. It comprises 27 items, and each is scored from 0 (no symptom or sign) to 4 (severe disability or finding) [44].

Diagnosis

The current diagnostic criteria of SC are a modification of the Jones criteria: chorea with acute or subacute onset, and lack of clinical and laboratory evidence of an alternative cause are requirements. The diagnosis is further supported by the presence of additional major or minor manifestations of rheumatic fever [18, 45, 46].

At present, there is no biological or laboratory marker specific to the diagnosis of SC. Children and young adults with chorea should undergo complete neurologic examination and diagnostic testing to distinguish between the various causes of chorea. The aim of the diagnostic workup in patients suspected to have rheumatic chorea is threefold: (1) to identify evidence of recent streptococcal infection or acute phase reaction, (2) to search for cardiac injury associated with rheumatic fever, and (3) to rule out alternative causes. The laboratory assessment of SC includes:

1. Tests of acute phase reactants, such as erythrocyte sedimentation rate, C-reactive protein, and leukocytosis
2. Other blood tests like rheumatoid factor, mucoproteins, and protein electrophoresis
3. Supporting evidence of preceding streptococcal infection (increased antistreptolysin-O, anti-DNAse-B, or other antistreptococcal antibodies)
4. Positive throat culture for group A *Streptococcus*, or recent scarlet fever, which is much less helpful in SC than in other forms of rheumatic fever due to the usually long latency between the infection and onset of the movement disorder. Elevated antistreptolysin-O titer may be found in populations with a high prevalence of streptococcal infection. Furthermore, the antistreptolysin-O titer declines if the interval between infection and rheumatic fever is greater than 2 months. Anti-DNase-B titers, however, may remain elevated up to 1 year after streptococcal pharyngitis. Cardiac evaluation (i.e., Doppler echocardiography) is mandatory because an association between SC and carditis is found in up to 80 % of patients. Serologic studies for systemic lupus erythematosus and primary antiphospholipid antibody syndrome must be ordered to rule out these conditions. Cerebrospinal fluid analysis is usually normal, but it may show a slightly increased lymphocyte count.

In general, neuroimaging will help rule out structural causes such as moyamoya disease and other vascular causes. CT scan of the brain invariably fails to display abnormalities. Similarly, head MRI is often normal, although there are case reports of reversible hyperintensity in the basal ganglia area. In one study, Giedd and colleagues [47] showed increased signal images in just 2 of 24 SC patients, although morphometric techniques revealed mean values for the size of the striatum and pallidum that were larger than in controls. Unfortunately, these findings are of little help on an individual basis because there was an extensive overlap between findings in controls and patients. PET and SPECT imaging may prove to be useful tools in the evaluation, since they can show transient increases in striatal metabolism [48–51]. Barsottini and colleagues showed that six of ten patients with SC had hyperperfusion of the basal ganglia [52]. This contrasts with other choreic disorders, such as Huntington disease that are associated with hypometabolism. Of note, however, one investigation showed hyperperfusion in two patients with SC, whereas the remaining five had hypometabolism [53]. It is possible that the inconsistencies in these studies reflect heterogeneity in the patient populations.

Increasing interest is now directed to autoimmune markers that may be useful for diagnosis. Testing for antineuronal antibodies, however, is not commercially available and is performed only for research purposes. Preliminary evidence, however, suggests that these antibodies are not specific for SC. Similarly, the low sensitivity and specificity of the alloantigen D8/17 renders this test unsuitable for the diagnosis of SC.

Several conditions may present with clinical manifestations resembling SC [54]. The most important differential diagnosis is systemic lupus erythematosus, for which up to 2 % of patients may develop chorea. The majority of subjects with this condition will have other nonneurologic manifestations such as arthritis, pericarditis, and other serositis as well as skin abnormalities. Moreover, the neurologic picture of systemic lupus erythematosus tends to be more complex and may include psychosis, seizures, other movement disorders, and even mental status and level of consciousness changes. Only in rare instances will chorea, with a tendency for spontaneous remissions and recurrences, be an isolated manifestation of systemic lupus erythematosus. The difficulty in distinguishing these two conditions is increased by the finding that up to 20 % of patients with SC display recurrence of the movement disorder. Eventually, patients with systemic lupus erythematosus will develop other clinical features and so will meet diagnostic criteria for this condition [55]. Primary antiphospholipid antibody syndrome is differentiated from SC by the absence of other clinical and laboratory features of rheumatic fever as well as the usual association with repeated abortions, venous thrombosis, other vascular events, and the presence of typical laboratory abnormalities. Encephalitides, either as a result of direct viral invasion or by means of an immune-mediated postinfectious process, can cause chorea. This usually happens, however, in younger children; the clinical picture is more diversified to include seizures, pyramidal signs, impairment of the psychomotor development, and laboratory abnormalities suggestive of the underlying condition. Among the various encephalitides, there has been a growing interest in anti-NMDAR encephalitis, a condition occasionally paraneoplastic but more often seen as a postinfectious phenomenon [56]. Drug-induced choreas are readily distinguished by careful history demonstrating temporal relationship between onset of the movement disorder and exposure to the agent.

Pathophysiology

Taranta and Stollerman established the casual relationship between infection with group A beta-hemolytic streptococci and the occurrence of SC [2]. Based on the assumption of molecular mimicry between streptococcal and central nervous system antigens, it has been proposed that the bacterial infection in genetically predisposed subjects leads to formation of cross-reactive antibodies that disrupt the basal ganglia function. Several studies have demonstrated the presence of such circulating antibodies in 50–90 % of patients with SC [57, 58]. A specific epitope of streptococcal M proteins that cross-reacts with basal ganglia has been identified [59]. In one study, it was demonstrated that all patients with active SC have anti-basal ganglia antibodies demonstrated by ELISA and Western blot. In subjects with persistent SC (i.e., duration of disease greater than 2 years despite best medical treatment), there were positive findings in about 60 % [57]. Another investigation confirmed that antibodies from patients with SC bind to neuronal surface, although this pattern was not found in patients with Tourette's syndrome or meeting criteria for PANDAS [60]. It must be emphasized that the biological value of the anti-basal ganglia antibodies remains to be determined, although there is growing evidence that they may interfere with neuronal function. Kirvan and colleagues demonstrated that IgM of one patient with SC induced expression of calcium-dependent calmodulin in a culture of neuroblastoma cells [61]. Our finding of a linear correlation between the increase of intracellular calcium levels in PC12 cells and anti-basal ganglia antibody titer in the serum from SC patients suggests that these antibodies may have pathogenic properties [62]. However, an in vivo study failed to demonstrate that antibodies from SC patients infused in the basal ganglia of rodents would induce behavioral changes although they were found to bind to an ~50-kDa molecule in the striatum extract [63]. One possible explanation for the finding of this study is that a low titer of the antibodies prevented the occurrence of detectable behavioral manifestations. More recent data lend support, however, to the molecular mimicry hypothesis. First, we have demonstrated that infusion of sera of SC patients in rodents with 6-hydroxydopamine-induced unilateral lesion of the nigrostriatal system caused circling behavior similar to apomorphine [64]. This finding suggests that the circulating antibodies act on dopamine receptors. Another study gives support to this hypothesis. Rats exposed to *Streptococcus* antigens not only developed behavioral abnormalities suggestive of SC but also had IgG that reacted with tubulin and D1 and D2 dopamine receptors and caused elevated calcium-/calmodulin-dependent protein kinase II signaling [65]. There is also evidence that antibodies circulating in SC patients target dopamine receptors [66].

It remains unclear why up to 50 % of patients with SC develop a persistent course of the illness. Also unexplained is why patients with previous history of SC may have recurrence of chorea when pregnant or using hormones [67]. In this subset of patients, the titers of anti-basal ganglia antibodies are low. Taking into account this finding, as well as our observation that serum BDNF levels are high in this group of patients, we have hypothesized that the acute immune process causes structural brain lesions that result in permanent dysfunction of the basal ganglia [68].

Although some investigations suggest that susceptibility to rheumatic chorea is linked to human leukocyte antigen expression [69], there are studies that fail to identify any relationship between SC and human leukocyte antigen (HLA) class I and II alleles [70]. One investigation has shown, however, an association between HLA-DRB1*07 and recurrent streptococcal pharyngitis and rheumatic heart disease [71]. The genetic marker for rheumatic fever and related conditions would be the B-cell alloantigen D8/17 [72]. Despite repeated claims by the group that developed the assay as to its high specificity and sensitivity [73, 74], findings of other authors have not confirmed this. Another suspected genetic risk factor for development of acute rheumatic fever (but not SC) is polymorphism within the promoter region of the tumor necrosis factor-alpha gene [75].

There have also been studies that address the role of immune cellular mechanisms in SC. In an investigation of sera and CSF samples from subjects in our clinic, we found elevation of cytokines that take part in the Th2 (antibody-mediated) response, interleukins 4 (IL-4) and 10 (IL-10), in the serum of acute SC in comparison to persistent SC [76]. We concluded that SC is characterized by a Th2 response. However, as we also found an elevation of interleukin 12 and chemokines CXCL9 and CXCL10 in the sera of acute SC [77], it seems that Th1 (cell-mediated) mechanisms also may be involved in the pathogenesis of this disorder. Another study confirmed that cellular immune mechanisms may be relevant to the pathogenesis of SC because there is a dysfunction of monocytes [68].

Some authors have suggested that streptococcal infection induces vasculitis of medium-sized vessels, leading to neuronal dysfunction. Such vascular lesions could be produced by antiphospholipid antibodies. Although essentially all patients with SC are negative for antiphospholipid antibodies, another study demonstrates many immunologic similarities between primary antiphospholipid antibody syndrome and SC [78]. There has also been a suggestion that cellular immune mechanisms participate in the pathogenesis of *Streptococcus*-related movement disorders. However, most of these findings have not been replicated so far.

In summary, the predominant evidence suggests that the pathogenesis of SC is related to circulating cross-reactive antibodies.

Treatment

There are few controlled studies of symptomatic treatment of SC [79]. The first choice of this author is valproic acid, using an initial dosage of 250 mg per day that can be increased during a 2-week period to 250 mg three times a day. If the response is not satisfactory, dosage can be increased gradually to 1,500 mg per day. As this drug has a rather slow onset of action, it is prudent to wait 2 weeks before concluding that a regimen is ineffective. If the patient fails to respond to this medication, the next option is to prescribe neuroleptics. Neuroleptics can also be prescribed as a first-line treatment in patients who present with chorea paralytica. Risperidone, a relatively potent dopamine D2 receptor blocker, is usually effective in controlling the chorea. The usual initial regimen is 1 mg twice a day. If, by 2 weeks later, the chorea is still troublesome, the dosage can be increased to 2 mg twice a day. Haloperidol and pimozide are also occasionally used in the management of chorea in SC. However, they are not as well tolerated as risperidone. In general, neuroleptics must be used with great caution in patients with this condition. We demonstrated that 5 % of 100 patients with chorea developed extrapyramidal complications, whereas these findings were not observed among patients with tics matched for age and neuroleptics dosage [80]. One study demonstrated that carbamazepine (15 mg/kg per day) is as effective as valproic acid (20–25 mg/kg per day) to induce remission of chorea [81].

Finally, the most important measure in the treatment of patients with SC is a secondary prophylaxis with penicillin or, if there is allergy, with sulfa drugs (in patients up to 21 years of age). If the onset of SC occurs after this age, the recommendation is to maintain prophylaxis indefinitely [5].

Some controversy exists as to the role of immunosuppression in the management of SC. Despite mention in some reports regarding effectiveness of prednisone in suppressing chorea, this drug only is used when there is associated severe carditis. A placebo-controlled study showed that oral prednisone only accelerates the control of chorea; rates of remission and recurrence were not changed by the active treatment [82]. A few reports describe the usefulness of plasma exchange or intravenous immunoglobulin in SC. A recent open-label study of ten children treated with conventional treatment associated with immunoglobulin showed a better outcome than ten patients who just received conventional management [83]. Because of the efficacy of other therapeutic agents described in the previous paragraph, potential complications, and the high cost of the latter treatment modalities, these options are not usually recommended. Intravenous methylprednisolone is reserved for patients with persistent disabling chorea refractory to antichoreic agents. We reported that 25 mg/kg per day in children and 1 g/day in adults of methylprednisolone for 5 days followed by 1 mg/kg per day of prednisone is an effective and well-tolerated treatment for patients with SC refractory to conventional treatment with antichoreic drugs and penicillin [84–86].

Conclusions

SC is a major manifestation of acute rheumatic fever. It remains as the most common of acute chorea in children worldwide despite the decline of incidence. From a clinical point of view, it is an acute onset chorea of children whose features often include nonmovement disorder findings such as behavioral abnormalities (emotional lability, obsessions, compulsions, and others) and dysexecutive syndrome. In a significant proportion of patients, SC has a persistent course. Although its underlying mechanism remains to be fully elucidated, it most likely is caused by *Streptococcus*-induced antibodies that cross-react with basal ganglia antigens. Its diagnosis is based on the recognition of typical clinical features and by ruling out alternative causes. The management of patients with SC includes the use of antichoreic drugs as well as secondary prophylaxis of *Streptococcus* infection with penicillin.

References

1. Krack P. Relicts of dancing mania: the dancing procession of Echternach. Neurology. 1999;53:2169–72.
2. Taranta A, Stollerman GH. The relationship of Sydenham's chorea to infection with group A streptococci. Am J Med. 1956;20:1970.
3. Ryan M, Antony JH, Grattan-Smith PJ. Sydenham chorea: a resurgence of the 1990s? J Paediatr Child Health. 2000;36:95–6.
4. Cardoso F. Chorea gravidarum. Arch Neurol. 2002;59(5):868–70.
5. Cardoso F. Infectious and transmissible movement disorders. In: Jankovic J, Tolosa E, editors. Parkinson's disease and movement disorders. 4th ed. Baltimore: Williams and Wilkins; 2002. p. 930–40.
6. Cardoso F, Seppi K, Mair KJ, Wenning GK, Poewe W. Seminar on choreas. Lancet Neurol. 2006;5(7):589–602.
7. Vale TC, Glass PG, Lees A, Cardoso F. Gowers' queen square case notes on chorea: a 21st century re-appraisal. Eur Neurol. 2013;69(1):48–52.
8. Quinn RW. Comprehensive review of morbidity and mortality trends for rheumatic fever, streptococcal disease, and scarlet fever: the decline of rheumatic fever. Rev Infect Dis. 1989;11:928–53.
9. Schwartz RH, Hepner SI, Ziai M. Incidence of acute rheumatic fever. A suburban community hospital experience during the 1970s. Clin Pediatr. 1983;22:798–801.
10. Nausieda PA, Grossman BJ, Koller WC, Weiner WJ, Klawans HL. Sydenham's chorea: an update. Neurology. 1980;30:331–4.
11. Zomorrodi A, Wald ER. Sydenham's chorea in western Pennsylvania. Pediatrics. 2006;117:e675–9.
12. Dale RC, Singh H, Troedson C, Pillai S, Gaikiwari S, Kozlowska K. A prospective study of acute movement disorders in children. Dev Med Child Neurol. 2010;52(8):739–48.
13. Smith MT, Lester-Smith D, Zurynski Y, Noonan S, Carapetis JR, Elliott EJ. Persistence of acute rheumatic fever in a tertiary children's hospital. J Paediatr Child Health. 2011;47(4):198–203.
14. Noonan S, Zurynski YA, Currie BJ, et al. A national prospective surveillance study of acute rheumatic fever in Australian children. Pediatr Infect Dis J. 2013;32(1):e26–32.
15. Ayoub EM. Resurgence of rheumatic fever in the United States. The changing picture of a preventable illness. Postgrad Med. 1992;92:133–42.
16. Carapetis JR, Currie BJ. Rheumatic chorea in northern Australia: a clinical and epidemiological study. Arch Dis Child. 1999;80:353–8.
17. Carapetis JR, Wolff DR, Currie BJ. Acute rheumatic fever and rheumatic heart disease in the top end of Australia's Northern Territory. Med J Aust. 1996;164:146–9.
18. Cardoso F, Silva CE, Mota CC. Sydenham's chorea in 50 consecutive patients with rheumatic fever. Mov Disord. 1997;12:701–3.
19. Barreto LB, Maciel RO, Maia DP, Teixeira AL, Cardoso F. Parkinsonian signs and symptoms in adults with a history of Sydenham's chorea. Parkinsonism Relat Disord. 2012;18(5):595–7.
20. Mercadante MT, Campos MC, Marques-Dias MJ, et al. Vocal tics in Sydenham's chorea. J Am Acad Child Adolesc Psychiatry. 1997;36:305–6.
21. Teixeira AL, Cardoso F, Maia DP, et al. Frequency and significance of vocalizations in Sydenham's chorea. Parkinsonism Relat Disord. 2009;15(1):62–3.

22. Cunningham MC, Maia DP, Teixeira Jr AL, Cardoso F. Sydenham's chorea is associated with decreased verbal fluency. Parkinsonism Relat Disord. 2006;12(3):165–7.
23. Cardoso F, Beato R, Siqueira CF, Lima CF. Neuropsychological performance and brain SPECT imaging in adult patients with Sydenham's chorea. Neurology. 2005;64 Suppl 1:A76.
24. Beato R, Maia DP, Teixeira Jr AL, Cardoso F. Executive functioning in adult patients with Sydenham's chorea. Mov Disord. 2010;25(7):853–7.
25. Cavalcanti A, Hilário MO, dos Santos FH, Bolognani SA, Bueno OF, Len CA. Subtle cognitive deficits in adults with a previous history of Sydenham's chorea during childhood. Arthritis Care Res (Hoboken). 2010;62(8):1065–71.
26. Cardoso F, Oliveira PM, Reis CC, et al. Prosody in Sydenham chorea – I: Tessitura. Mov Disord. 2006;21:S359–60.
27. Cardoso F, Oliveira PM, Reis CC, et al. Prosody in Sydenham chorea – II: duration of statements. Mov Disord. 2006;21:S360.
28. Oliveira PM, Cardoso F, Maia DP, Cunningham MC, Teixeira Jr AL, Reis C. Acoustic analysis of prosody in Sydenham's chorea. Arq Neuropsiquiatr. 2010;68(5):744–8.
29. Azevedo LL, Cardoso F, Reis C. Acoustic analysis of prosody in females with Parkinson's disease: comparison with normal controls. Arq Neuropsiquiatr. 2003;61(4):999–1003.
30. Teixeira Jr AL, Meira FC, Maia DP, Cunningham MC, Cardoso F. Migraine headache in patients with Sydenham's chorea. Cephalalgia. 2005;25(7):542–4.
31. Maia DP, Teixeira Jr AL, Quintao Cunningham MC, Cardoso F. Obsessive compulsive behavior, hyperactivity, and attention deficit disorder in Sydenham chorea. Neurology. 2005;64(10):1799–801.
32. Swedo SE, Leonard HL, Garvey M, et al. Pediatric autoimmune neuropsychiatric disorders associated with streptococcal infections: clinical description of the first 50 cases. Am J Psychiatry. 1988;155:264–71.
33. Asbahr FR, Negrao AB, Gentil V, et al. Obsessive-compulsive and related symptoms in children and adolescents with rheumatic fever with and without chorea: a prospective 6-month study. Am J Psychiatry. 1998;155:1122–4.
34. Mercadante MT, Busatto GF, Lombroso PJ, et al. The psychiatric symptoms of rheumatic fever. Am J Psychiatry. 2000;157:2036–8.
35. Asbahr FR, Ramos RT, Costa AN, Sassi RB. Obsessive-compulsive symptoms in adults with history of rheumatic fever, Sydenham's chorea and type I diabetes mellitus: preliminary results. Acta Psychiatr Scand. 2005;111:159–61.
36. Hounie AG, Pauls DL, do Rosario-Campos MC, et al. Obsessive-compulsive spectrum disorders and rheumatic fever: a family study. Biol Psychiatry. 2007;61:266–72.
37. Ben-Pazi H, Jaworowski S, Shalev RS. Cognitive and psychiatric phenotypes of movement disorders in children: a systematic review. Dev Med Child Neurol. 2011;53:1077–84.
38. Teixeira Jr AL, Maia DP, Cardoso F. Psychosis following acute Sydenham's chorea. Eur Child Adolesc Psychiatry. 2007;16(1):67–9.
39. Kummer A, Maia DP, Cardoso F, Teixeira AL. Trichotillomania in acute Sydenham's chorea. Aust N Z J Psychiatry. 2007;41(12):1013–4.
40. Teixeira AL, Athayde GR, Sacramento DR, Maia DP, Cardoso F. Depressive and anxiety symptoms in Sydenham's chorea. Mov Disord. 2007;22(6):905–6.
41. Cardoso F. Tourette syndrome: autoimmune mechanism. In: Fernández-Alvarez E, Arzimanoglou A, Tolosa E, editors. Pediatric movement disorders. Progress in understanding. Montrouge: John Libbey Eurotext; 2005. p. 23–46.
42. Vijayalakshmi IB, Mithravinda J, Deva AN. The role of echocardiography in diagnosing carditis in the setting of acute rheumatic fever. Cardiol Young. 2005;15:583–8.
43. Vale TC, Maciel RO, Maia D, Beato R, Cardoso F. Takayasu's Arteritis in a patient with Sydenham's Chorea: is there an association? Tremor Other Hyperkinet Mov (NY). 2012;2. pii: tre-02-94-542-1.
44. Teixeira Jr AL, Maia DP, Cardoso F. UFMG Sydenham's chorea rating scale (USCRS): reliability and consistency. Mov Disord. 2005;20(5):585–91.
45. Guidelines for diagnosis of rheumatic fever, Jones criteria, 1992 update. Special Writing Group of the Committee of Rheumatic Fever, Endocarditis, and Kawasaki Disease of the

Council on Cardio-Vascular Disease of the Young of the American Heart Association. Guidelines for the diagnosis of rheumatic fever. JAMA. 1992;268:2069–73.

46. Cardoso F, Vargas AP, Oliveira LD, Guerra AA, Amaral SV. Persistent Sydenham's chorea. Mov Disord. 1999;14:805–7.

47. Giedd JN, Rapoport JL, Kruesi MJ, et al. Sydenham's chorea: magnetic resonance imaging of the basal ganglia. Neurology. 1995;45:2199–202.

48. Goldman S, Amrom D, Szliwowski HB, et al. Reversible striatal hypermetabolism in a case of Sydenham's chorea. Mov Disord. 1993;8(3):355–8.

49. Weindl A, Kuwert T, Leenders KL, et al. Increased striatal glucose consumption in Sydenham's chorea. Mov Disord. 1993;8:437–44.

50. Lee PH, Nam HS, Lee KY, Lee BI, Lee JD. Serial brain SPECT images in a case of Sydenham chorea. Arch Neurol. 1999;56:237–40.

51. Paghera B, Caobelli F, Giubbini R, Premi E, Padovani A. Reversible striatal hypermetabolism in a case of rare adult-onset Sydenham chorea on two sequential 18F-FDG PET studies. J Neuroradiol. 2011;38(5):325–6.

52. Barsottini OG, Ferraz HB, Seviliano MM, Barbieri A. Brain SPECT imaging in Sydenham's chorea. Braz J Med Biol Res. 2002;35:431–6.

53. Citak EC, Gukuyener K, Karabacak NI, et al. Functional brain imaging in Sydenham's chorea and streptococcal tic disorders. J Child Neurol. 2004;19:387–90.

54. Cardoso F. Chorea: non-genetic causes. Curr Opin Neurol. 2004;17:433–6.

55. Bakdash T, Goetz CG, Singer HS, Cardoso F. A child with recurrent episodes of involuntary movements. Mov Disord. 1999;14:146–54.

56. Baizabal-Carvallo JF, Stocco A, Muscal E, Jankovic J. The spectrum of movement disorders in children with anti-NMDA receptor encephalitis. Mov Disord. 2013;28(4):543–7.

57. Husby G, Van De Rijn U, Zabriskie JB, Abdin ZH, Williams Jr RC. Antibodies reacting with cytoplasm of subthalamic and caudate nuclei neurons in chorea and acute rheumatic fever. J Exp Med. 1976;144:1094–110.

58. Church AJ, Cardoso F, Dale RC, et al. Anti-basal ganglia antibodies in acute and persistent Sydenham's chorea. Neurology. 2002;59:227–31.

59. Bronze MS, Dale JB. Epitopes of streptococcal M proteins that evoke antibodies that cross-react with human brain. J Immunol. 1993;151:2820–8.

60. Brilot F, Merheb V, Ding A, Murphy T, Dale RC. Antibody binding to neuronal surface in Sydenham chorea, but not in PANDAS or tourette syndrome. Neurology. 2011;76:1508–13.

61. Kirvan CA, Swedo SE, Heuser JS, Cunningham MW. Mimicry and autoantibody-mediated neuronal cell signaling in Sydenham chorea. Nat Med. 2003;9(7):914–20.

62. Teixeira Jr AL, Guimaraes MM, Romano-Silva MA, Cardoso F. Serum from Sydenham's chorea patients modifies intracellular calcium levels in PC12 cells by a complement-independent mechanism. Mov Disord. 2005;20(7):843–5.

63. Ben-Pazi H, Sadan O, Offen D. Striatal microinjection of Sydenham chorea antibodies: using a rat model to examine the dopamine hypothesis. J Mol Neurosci. 2012;46:162–6.

64. Doyle F, Cardoso F, Lopes L, et al. Infusion of Sydenham's chorea antibodies in striatum with up-regulated dopaminergic receptors: a pilot study to investigate the potential of SC antibodies to increase dopaminergic activity. Neurosci Lett. 2012;523(2):186–9.

65. Brimberg L, Benhar I, Mascaro-Blanco A, et al. Behavioral, pharmacological, and immunological abnormalities after streptococcal exposure: a novel rat model of Sydenham chorea and related neuropsychiatric disorders. Neuropsychopharmacology. 2012;37(9):2076–87.

66. Cunningham MW. Streptococcus and rheumatic fever. Curr Opin Rheumatol. 2012;24:408.

67. Maia DP, Fonseca PG, Camargos ST, et al. Pregnancy in patients with Sydenham's Chorea. Parkinsonism Relat Disord. 2012;18(5):458–61.

68. Torres KC, Dutra WO, de Rezende VB, Cardoso F, Gollob KJ, Teixeira AL. Monocyte dysfunction in Sydenham's chorea patients. Hum Immunol. 2010;71(4):351–4.

69. Ayoub EM, Barrett DJ, Maclaren NK, Krischer JP. Association of class II human histocompatibility leukocyte antigens with rheumatic fever. J Clin Invest. 1986;77:2019–26.

70. Donadi EA, Smith AG, Louzada-Junior P, Voltarelli JC, Nepom GT. HLA class I and class II profiles of patients presenting with Sydenham's chorea. J Neurol. 2000;247:122–8.
71. Haydardedeoglu FE, Tutkak H, Kose K, Duzgun N. Genetic susceptibility to rheumatic heart disease and streptococcal pharyngitis: association with HLA-DR alleles. Tissue Antigens. 2006;68:293–6.
72. Feldman BM, Zabriskie JB, Silverman ED, Laxer RM. Diagnostic use of B-cell alloantigen D8/17 in rheumatic chorea. J Pediatr. 1993;123:84–6.
73. Eisen JL, Leonard HL, Swedo SE, et al. The use of antibody D8/17 to identify B cells in adults with obsessive-compulsive disorder. Psychiatry Res. 2001;104:221–5.
74. Harel L, Zeharia A, Kodman Y, et al. Presence of the D8/17 B-cell marker in children with rheumatic fever in Israel. Clin Genet. 2002;61:293–8.
75. Ramasawmy R, Fae KC, Spina G, et al. Association of polymorphisms within the promoter region of the tumor necrosis factor-alpha with clinical outcomes of rheumatic fever. Mol Immunol. 2007;44:1873–8.
76. Church AJ, Dale RC, Cardoso F, et al. CSF and serum immune parameters in Sydenham's chorea: evidence of an autoimmune syndrome? J Neuroimmunol. 2003;136(1–2):149–53.
77. Teixeira Jr AL, Cardoso F, Souza AL, Teixeira MM. Increased serum concentrations of monokine induced by interferon-gamma/CXCL9 and interferon-gamma-inducible protein 10/CXCL-10 in Sydenham's chorea patients. J Neuroimmunol. 2004;150(1–2):157–62.
78. Blank M, Krause I, Magrini L, et al. Overlapping humoral autoimmunity links rheumatic fever and the antiphospholipid syndrome. Rheumatology (Oxford). 2006;45:833–41.
79. Cardoso F. Sydenham's chorea. Curr Treat Options Neurol. 2008;10(3):230–5.
80. Teixeira AL, Cardoso F, Maia DP, Cunningham MC. Sydenham's chorea may be a risk factor for drug induced parkinsonism. J Neurol Neurosurg Psychiatry. 2003;74(9):1350–1.
81. Genel F, Arslanoglu S, Uran N, Saylan B. Sydenham's chorea: clinical findings and comparison of the efficacies of sodium valproate and carbamazepine regimens. Brain Dev. 2002;24:73–6.
82. Paz JA, Silva CA, Marques-Dias MJ. Randomized double-blind study with prednisone in Sydenham's chorea. Pediatr Neurol. 2006;34:264–9.
83. Walker K, Brink A, Lawrenson J, Mathiassen W, Wilmshurst JM. Treatment of Sydenham chorea with intravenous immunoglobulin. J Child Neurol. 2012;27(2):147–55.
84. Cardoso F, Maia D, Cunningham MC, Valenca G. Treatment of Sydenham chorea with corticosteroids. Mov Disord. 2003;18(11):1374–7.
85. Barash J, Margalith D, Matitiau A. Corticosteroid treatment in patients with Sydenham's chorea. Pediatr Neurol. 2005;32:205–7.
86. Teixeira Jr AL, Maia DP, Cardoso F. Treatment of acute Sydenham's chorea with methyl-prednisolone pulse-therapy. Parkinsonism Relat Disord. 2005;11(5):327–30.

Chapter 5
Huntington's Disease: Clinical Phenotypes and Therapeutics

Michael Orth

Abstract Twenty years after the discovery of the causal CAG repeat expansion mutation in the *HTT* gene, Huntington's disease remains an incurable devastating disorder. However, using disease models, and studies in human patients, great progress has been made in understanding the pathophysiology of HD. Research has led to the development of the first gene therapy approaches with promising results in HD model systems. This raises hopes that HD, a monogenetic fully penetrant autosomal dominant disorder, may be a model for novel therapeutics in neurodegenerative disorders. In addition, thanks to the efforts of the HD community – families, clinicians, health professionals, and researchers – standards of care are improving patients' and families' quality of life. HD networks (HSG, EHDN, RLAH), and their observational studies, have further laid the groundwork for conducting clinical trials of high quality on a global stage. This includes collaboration with clinical trial sponsors in designing and conducting clinical trials, preparing and training investigators, and developing the right assessment tools. The time seems right for the clinical trials of the future that hopefully will change our treatment options to relieve the plight of all those affected by HD.

Keywords Predictive testing • Symptomatic treatment • Movement disorder • Behavioural phenotype • Irritability • Apathy • Pathophysiology • REGISTRY • COHORT • ENROLL

Disclosure
The author reports no conflict of interest.

M. Orth, MD, PhD
Department of Neurology and European Huntington's Disease Network,
Ulm University Hospital, Oberer Eselsberg 45/1, Ulm 89081, Germany
e-mail: michael.orth@uni-ulm.de

Introduction

George Huntington, in a family from New England, gave a detailed account of the phenotype of an inherited movement disorder with cognitive impairment and behavioral problems that progresses relentlessly until death [1]. This disorder now bears his name, Huntington's disease (HD). HD is the most common inherited cause of chorea. The genetic mutation causing HD was first mapped to chromosome 4 in 1983 [2] with the gene and its mutation identified as a CAG repeat expansion in the *HTT* gene in 1993 [3]. This enabled the establishment of disease models, most of them in rodents, fruit flies, or worms [4] but recently also in large animals such as sheep or nonhuman primates. These models have contributed greatly to our understanding of the pathophysiology of HD and are suitable to explore therapeutic interventions. New treatment approaches, e.g., gene silencing, are promising; however, 20 years after the identification of the disease-causing gene mutation, there is still no causal treatment. There is rightfully hope that this will change in the foreseeable future; in fact, HD could be a model disease and the first neurodegenerative disease in which novel therapeutic approaches prove successful. However, until then HD treatments remain symptomatic and supportive. This can without doubt improve quality of life of HD patients and their families and carers. Nonetheless, until causal treatment will be available, HD will continue to wreck lives and cause suffering for those affected and their families.

The HD community has made great efforts to advance knowledge and improve standards of care of HD. To this end, networks including clinicians, scientists, and family members have formed in Europe (EHDN; www.euro-hd.net), North America and Australia (Huntington Study Group; www.huntington-study-group.org), and Latin America (Red Latinoamericana de Huntington; www.rlah.net) with the ultimate goal to improve quality of life of those affected by HD. The networks also provide the platforms on which systematic efforts to study and treat HD can build. Good examples are the observational natural history studies of HD, REGISTRY in Europe [5] and COHORT in North America and Australia [6]. These now merge into a global effort called ENROLL-HD that also includes Latin American sites and potentially sites from other regions, e.g., Asia (www.enroll-hd.org). The CHDI Foundation, Inc. (www.chdifoundation.org), a not-for-profit organization, supports in a collaborative way research into HD, with a particular emphasis on developing treatments.

The following chapter will give an overview of the epidemiology, pathophysiology, genetic diagnosis, phenotype, and management of HD. It closes with an outlook towards novel therapeutics that will hopefully change the course of this devastating disease for the better.

Epidemiology

The prevalence of manifest Huntington's disease in North America and Europe is about 10 per 100,000 inhabitants (for a meta-analysis see [7]). More precise estimates can be difficult given the challenges of ascertainment that may result in bias with subsequent over-inflation or underestimation of prevalence rates; for instance, more genetically confirmed patients may attend multidisciplinary specialist interest clinics and research programs so that prevalence estimates in regions with such services can be higher. In addition, because of improved clinical care, patients may survive for longer and, considering the increasing overall life expectancy, more individuals may develop late-onset HD. Thus, the prevalence of HD may be much higher. HD is prevalent worldwide. Compared with North America and Europe, studies in Japan and Africa have shown lower prevalence rates; among the lowest are found in black South Africans with 0.01 in 100,000 [8, 9]. However, in the absence of epidemiological studies, estimating the prevalence in other geographical regions, and other ethnicities, is more difficult.

Notably in Latin America (e.g., Peru, Colombia, Venezuela) HD also occurs in clusters with the Venezuelan cluster around Lake Maracaibo one of several where the prevalence of HD far exceeds that seen in Europe or North America [10]. Given its dominant inheritance for every person with manifest HD, about five persons live at risk of having also inherited the HD gene mutation. This means that in Europe, North America, and Latin America, there are probably about 100,000 individuals with manifest HD with a further about 500,000 individuals at risk. If one includes the genetically unaffected family as impacted by HD, the number of those impacted by HD is even larger. This suggests that HD imposes a substantial burden on health-care systems and societies.

Genetics

HD is an autosomal dominant disorder. The HD gene, *HTT*, resides on the short arm of chromosome 4 [3]. *HTT* contains at its N-terminus a trinucleotide (CAG) repeat in exon 1 that codes for a polyglutamine repeat within the huntingtin protein. Healthy humans have up to 27 CAG repeats; 40 and more CAG repeats invariably (full penetrance) lead to clinical manifestations of HD, while individuals with 36–39 CAG repeats may or may not suffer from HD in their lifetime (reduced penetrance). It is possible that with increasing life expectancy more people in the reduced penetrance range will develop signs of HD so that the lower end of the complete penetrance range may actually be lower. Alleles with 28–35 CAG repeats (intermediate range) also do not cause HD in their carriers; however, these expansions are unstable so that the number of repeats may differ in the subsequent generation. Most commonly, the number of repeats increases, especially when the mutation is inherited from the father suggesting that during spermatogenesis the expanded CAG stretch is more unstable than during oogenesis. This phenomenon is called anticipation; it can explain the occurrence of apparently de novo HD in a family without any other family member affected by HD. In addition, in juvenile HD most commonly the mutation was passed on from an affected father [11, 12]. The risk of a large increase of CAG repeat length depends on the size of the parental CAG repeat. The risk is higher if the parental CAG repeat expansion is already large.

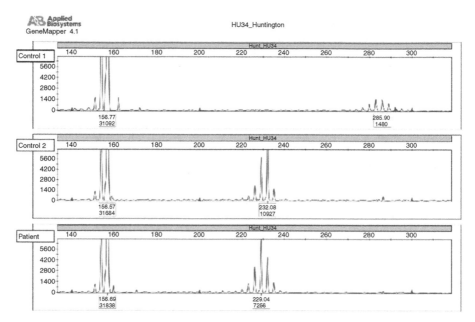

Fig. 5.1 *HTT* genetic test result using PCR capillary electrophoresis. A fragment containing the CAG repeat section within the first exon of the *HTT* gene is amplified by PCR and labeled with a fluorescent dye. The labeled fragments are then separated according to size using capillary electrophoresis. The highest fluorescent peak is called automatically. A standardized marker indicates the size in base pairs (bp, *red triangles*). Because the number of triplets of the amplified fragment flanking the CAG repeat section in exon one is known, it is known that the first peak in the controls is at 18 CAG repeats. One can then calculate the difference in bp between the second peak and the first peak, which is divided by three results in the number of CAG repeats of the second allele. In control 1 (healthy control) there is only one peak. In control 2 (disease control) this is 43 CAG repeats; in the patient the CAG repeat size is 42. Note that for technical reasons there is more than one peak. This is why the result is given as 42 ± 1 CAG repeats

The analytical methods to measure CAG repeat length involve a PCR analysis of the region containing the CAG repeat followed by capillary gel electrophoresis to determine the size of the fragment. This should allow the separation of alleles that differ by one repeat [13, 14] (Fig. 5.1). It is critical that laboratories adhere to published guidelines for genotyping in particular since a considerable proportion of genotyping results are still outside acceptable error limits [15, 16].

Predictive and Diagnostic Genetic Testing

There are two main reasons for requesting *HTT* genetic testing. If the diagnosis of HD is suspected on clinical grounds, most commonly when clinical signs of the typical movement disorder and a family history for HD are present, genetic testing is requested by a clinician to confirm the clinical diagnosis. A diagnostic test may also be requested to rule out HD in a progressive neuropsychiatric disorder of unknown cause. From when the possibility of HD is considered in the clinically affected relative, family members should be prepared for the potential implications to their own risk in case the diagnostic test confirms the diagnosis of HD. If it does, family members should be offered genetic counseling about their own risk of having inherited the *HTT* mutation.

Anyone who is healthy but has a parent with genetically confirmed HD is at risk. Genetic testing in this context is predictive of whether an individual can expect HD to manifest some time later in life. This form of genetic testing is called predictive testing. The decision of an at-risk individual to undergo predictive testing is very personal. As long as there is no causal treatment available, that decision is not objectively right or wrong. Appropriate genetic counseling should follow international guidelines so that the participant can make an informed decision and has the necessary support in coping with the predictive testing procedure and the test result [17, 18]. The genetic test for the *HTT* mutation is deceptively simple. However, the process of undergoing genetic counseling, making a decision about predictive testing and coping with the test result can be very complex. The interested reader is referred to the appropriate literature for in-depth information (e.g., [18]). In brief, it is very important that the person seeking counseling for predictive testing has a clear understanding about what the test results mean for many life decisions. Each participant undergoing genetic counseling has to make an autonomous decision. This requires that the participant is mature and independent enough; this is an important reason why minors should not be tested. In addition, participants should not be coerced to have the test by any third party – such as family members, insurance companies, or employers.

Counseling should consist of three appointments with at least 4 weeks between them. At the first appointment the participant's motivation for seeking advice should be established. This includes taking a careful family history and verifying that indeed there is a confirmed diagnosis of HD in the family. The individual's knowledge of HD and predictive testing should be probed as well as the personal experience with HD in the family and the current life circumstances. This helps in providing the participant and his partner and/or family with information about what is known about the genetics, HD phenotype(s), and disease evolution, as well as the current treatment and management options. It is very important that the participant understands that with very few exceptions (intermediate allele, reduced penetrance range; see above) genetic testing can give a clear answer to the question whether the participant has inherited the mutant allele and will thus in his lifetime develop HD. While there is an association of CAG repeat length with age-at-onset, a substantial percentage of the variability of age-at-onset (about 40 %) cannot be explained by the CAG repeat [10]. This means that in a given participant it is not possible to predict accurately when exactly HD will manifest. It is also not possible to make predictions about how HD will manifest, i.e., what the phenotype of a gene carrier's future HD will be, or how HD will develop once a clinical diagnosis of manifest HD is made. Fortunately, in recent years, information from reliable sources has become available on the Internet about HD and HD research using language and a format that is suitable for laypersons and young people (www.hdbuzz.net; www.hdyo.org; www.predictivetestingforhd.com). At the second appointment, the participant has the opportunity for further counseling. If the decision is to have the test, it is now important to prepare for the disclosure of the test result and the implications this may have for the participant. This includes a detailed assessment of the participant's risk of not coping well with the test result and his support by family and friends, and professionals, e.g., a psychologist. It is important to acknowledge that the predictive test result can negatively impact on mood so it is paramount to carefully assess the participant for depressive symptoms [19]. The participant needs to consider insurance issues since some insurance policies can only be taken out without knowledge of the *HTT* gene status. The laws regulating disclosure of genetic test results may differ between countries. In some countries the implications of the disclosure of *HTT* gene status are such that they may deter people from having the test. This is probably one explanation for the fairly low number of people undergoing predictive testing.

Following the disclosure of the test result, it is very important to arrange for a formal follow-up. This should consist of a telephone call within a few days of the disclosure, another visit to the clinic a few weeks later, and the offer for regular follow-up once a year or on demand to provide a port of call for all questions and concerns the participant may have. This can include counseling regarding the reproductive options, which, e.g., consist of having children without testing, prenatal diagnosis, and preimplantation genetic diagnosis (PIGD) [20, 21]. Such a post-predictive testing follow-up can take place within a specialist HD clinic preferably on a clinic day dedicated to people affected by HD but without signs of manifest HD.

Pathophysiology

Huntingtin (HTT) is a very large soluble acidic protein that consists of 3,144 amino acids. HTT is expressed in every tissue. Its N-terminus contains a polyglutamine stretch and a poly-proline domain; it has nuclear import and export signals and harbors so-called HEAT repeats, about 40 amino acid sequences that are present several times. These repeats are composed of two antiparallel α-helices with helical hairpin configuration [22, 23]. HTT has hundreds of binding partners [24, 25]. In addition, it is extensively modified posttranslationally including phosphorylation, ubiquitination, sumoylation, acetylation, and palmitoylation [26].

HTT is predominantly localized in the cytosol where it can associate with cell membranes such as those of the endoplasmatic reticulum or Golgi [27]. In addition, HTT can shuttle into the nucleus where it may contribute to the regulation of transcriptional activity [27]. Since HTT loss of function is lethal during early mouse development, it is likely that HTT plays an important role in tissue differentiation [28]. HTT is also important for neuronal health, at least in mice, since inactivation of mouse huntingtin in adult mice causes neurodegeneration [29]. Evolutionary,

CAG repeat length in the normal range in the N-terminus increases from simple organisms to the human being. Thus, *HTT* with its longer CAG repeat stretch in humans could have contributed to the development of more complex nervous systems [30]. In addition to regulating transcription, HTT may serve as a scaffolding protein, contribute to vesicular transport, and aid synapse function [24]. However, the cellular functions of normal HTT remain little understood.

The CAG repeat expansion mutation in the *HTT* gene translates to an expanded polyglutamine stretch in the HTT protein. There is good evidence from HD model systems and human HD to suggest that mutant HTT confers a toxic gain of function [4]. A pathological hallmark of HD is abnormal conformation of mutant HTT giving rise to various forms of mutant HTT including soluble, intermediate, monomeric, and oligomeric forms that in themselves may be toxic. Eventually, this could result in intranuclear and cytoplasmic inclusions composed of aggregated N-terminal HTT fragments and a variety of ubiquitinated proteins [4, 31]. Recent evidence in mouse models (both R 6/2 exon 1 *HTT* transgenes and *Hdh*Q150 knock-in mice) and human HD suggests that abnormal splicing of exon 1 *HTT* results in N-terminal exon 1 HTT protein fragments [32]. Thus, exon 1 HTT protein fragments may be a common important denominator in HD pathogenesis [32]. Normal huntingtin can be caught up in these inclusions resulting in an additional loss of function [33]. The inclusions challenge the cell's clearance systems, in particular the ubiquitin-proteasome system and autophagy pathways [26, 34, 35]. It remains an open question if inclusions themselves are toxic or a response of the cell to protect itself. However, promoting clearance of mutant HTT would be expected to be beneficial for patients. Hyperacetylation, e.g., by inhibiting deacetylation with histone deacetylase (HDAC) inhibitors, can help targeting the mutant HTT protein to autophagosomes to facilitate its degradation and thus clearance [36]. For a more in-depth review of the pathophysiology, the reader is referred to reviews dedicated to this topic [4, 26].

Much of what we know about the pathophysiology of HD comes from model systems. This certainly has contributed substantially to our understanding of HD pathophysiology. However, hypotheses derived from model systems, in particular those relevant for novel therapies, need to be tested in humans with the disease. The HD gene mutation can be identified reliably; thus the molecular changes underlying the pathophysiology of HD can be investigated in manifest HD patients but also in expansion mutation carriers many years before they develop unequivocal signs of HD. Understanding the evolution of HD biology independent of, and in conjunction with, the clinical phenotype can help identify targets for treatments that may prevent or delay the emergence of HD signs and slow disease progression. Such changes can serve as biomarkers of the activity of HD biology and, in the future, may help indicate when to initiate treatment and how to assess novel HD therapeutics in clinical trials [37].

A key finding in HD is the pronounced, and selective, loss of GABAergic medium spiny striatal neurons projecting to the substantia nigra and the globus pallidus [38]. MRI demonstrates loss of striatal volume that is evident even before the emergence of HD signs but also changes in white matter (Fig. 5.2a); with time striatal volumes diminish faster than in controls [39, 40]. However, there is also good evidence from neuroimaging studies for early cortical involvement in particular the

Gray matter volume: controls > HD

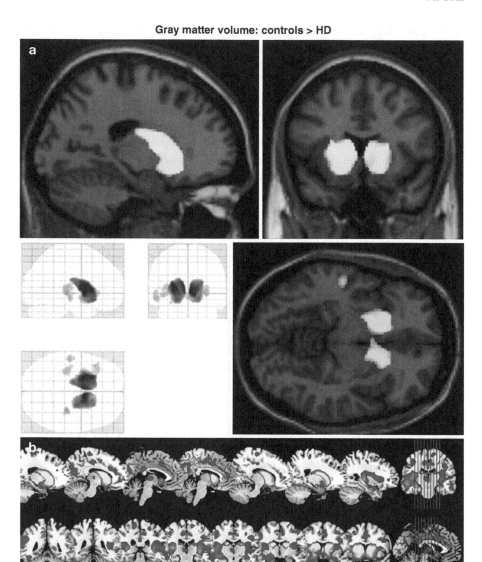

Fig. 5.2 Three Tesla structural MRI in manifest HD. (**a**) The striatum shows decreased gray matter volume (GMV) in manifest HD individuals compared to healthy controls. Results of the 2nd-level ANOVA, $p < 0.05$, FWE corrected. The figure displays the SPM5 "glass brain" output (*bottom, left*) together with maps rendered onto the anatomical template implemented in SPM5. (**b**) More widespread atrophy extending beyond the striatum in manifest HD. Results of the 2nd-level ANOVA, $p < 0.001$ (uncorrected at the voxel level), $p < 0.05$ corrected for spatial extent. For illustration purposes, the 2nd-level maps were thresholded at $t = 3.4$ (corresponding $p < 0.001$, uncorrected for height) and rendered onto the anatomical template implemented in MRIcron (http://www.mccauslandcenter.sc.edu/mricro/mricron/) (Images courtesy of Dr C Wolf)

motor cortex and the occipital lobe [41, 42] indicating that HD is not confined to a single brain region but leads to widespread pathology [38, 43] (Fig. 5.2b). The HTT mutation is present in every tissue. HD predominantly, but not exclusively, affects the brain [44, 45]. This is important because evidence for HD pathology beyond the brain means that peripheral tissues may serve as a source for biomarkers.

As with clinical trials the study of human HD benefits from multi-site efforts following a standard protocol with appropriate quality assessment and quality control. Examples for such efforts are the observational studies COHORT [6], REGISTRY [5], and PREDICT-HD [46], as well as TRACK-HD and TRACK-ON HD [39]. These studies have already contributed enormously to our understanding of human HD.

Diagnosis and Age-at-Onset

The clinical diagnosis of manifest HD can be made with certainty if unequivocal signs of HD are present in an individual with a CAG repeat expansion mutation in the *HTT* gene. A clinical diagnosis of manifest HD needs to be distinguished from a genetic test result demonstrating a CAG repeat expansion in the *HTT* gene in someone who is completely well. The result of a genetic test is not a disease; it can be a predictor of a disease in the individual's future, which is why this is referred to as predictive testing (see above). It is usually very difficult, if not impossible, to precisely determine the age-at-onset of clinical signs of HD. The concept of age-at-onset refers to when a carrier of the mutated *HTT* gene develops unequivocal signs of HD. The accurate determination of age-at-onset is critical to identify factors that modify age-at-onset and to develop and evaluate therapies that aim to delay it. If a manifest HD patient attends an HD clinic, the clinician estimates age-at-onset retrospectively based on information from the patient, relatives, and carers. Age-at-onset is most commonly defined as the age-at-onset of motor signs; however, in many patients the first sign of HD may be a non-motor sign with motor signs appearing later [47].

Predicting age-at-onset accurately in the prodromal phase – when the person shows no sign of disease – is also very important with a view to future clinical trials that evaluate the effects of therapeutics that aim to delay age-at-onset. By and large, longer CAG repeat expansions are associated with an earlier onset, so that most, but not all, patients with juvenile HD (see below) have more than 60 CAG repeats; carriers of shorter CAG repeat expansions tend to develop signs of HD later in life (Fig. 5.3). However, the variation of CAG repeat size only explains about 50–60 % of the variability of age-at-onset [10]. This means that in a group of individuals with the same CAG repeat length, the age-at-onset can differ by more than 10 years (Fig. 5.3). A number of algorithms have been devised to help estimate age-at-onset in prodromal *HTT* expansion mutation carriers (reviewed in [48]). The Langbehn formula [49] uses CAG repeat length and age because of their well-known influences on age-at-onset and calculates the time to a predefined degree of probability of manifesting signs of HD. However, CAG repeat length accounts for only about

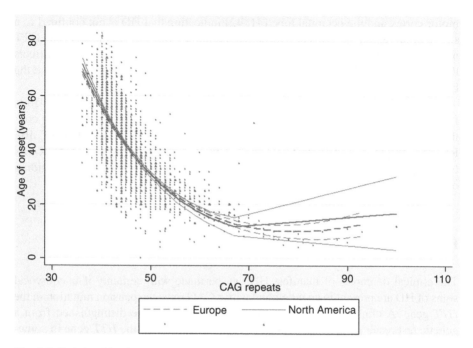

Fig. 5.3 Relationship of age-at-onset and CAG repeat length. The CAG repeat and age-at-onset relationship is similar in patients from the REGISTRY study in Europe and the North American and Australian COHORT study (Graph courtesy of Dr A Gemperli)

50–60 % of the variability, so other factors not modeled in this formula likely influence AAO. For patients very close to onset, i.e., persons at greatest risk, the Langbehn formula may be limited, and it is not helpful comparing prodromal with manifest HD. It may also not be suitable for all CAG repeat lengths and has been established mainly in North American patients. Another formula uses CAG repeat length and parental onset age to estimate AAO. This may accommodate for some other inherited factors as an advantage over the Langbehn formula. However, it was derived from a small sample of affected parent-child pairs and needs to be validated in larger numbers of patients.

In addition to the CAG repeat expansion in the *HTT* gene, other factors, e.g., genetic or environmental, may influence when signs of HD develop, in which domain the first signs occur and how they evolve over time. At present, many studies of genetic modifiers relate their effect to a general onset of HD. It is possible that there are domain-specific onset modifiers. Such domain-specific modifying effects may be overlooked unless domain-specific onsets are defined.

Clinical Manifestations

Most carriers of the *HTT* mutation develop clinical signs of HD between the ages of 30 and 50. However, there are very early, so-called juvenile, or even childhood onset, forms of HD [50], as well as late-onset variants with onset later than age 60 [51]. The clinical spectrum of HD comprises progressive motor dysfunction with a mixed movement disorder featuring chorea, dystonia, bradykinesia and rigidity, clumsiness, and gait and balance abnormalities. Patients invariably develop dementia with personality changes and progressive loss of autonomy, while behavioral problems such as depression or irritability are present in many, but not all, patients. In addition, urge incontinence and bowl problems, weight loss, as well as insensitivity to pain may develop; in the absence of an alternative explanation, these clinical features are probably also due to HD. In the following the clinical features of HD will be divided into clinical domains bearing in mind that in practice this may not always be so clear.

Motor Domain

The motor phenotype consists in most patients of chorea and dystonia. Hence HD is considered a predominantly hyperkinetic movement disorder. Chorea as the sole feature of the motor manifestations is, in the author's experience, the exception rather than the rule. In most patients, the movement disorder is mixed. Chorea is generalized and often involves the face, mouth, tongue, trunk, and arms more than the legs. Patients find it difficult to maintain a posture; motor impersistence affects the eyes, the tongue, or the limbs. Almost always chorea is accompanied by a degree of dystonia; in particular cervical dystonia is common. Parkinsonian features – bradykinesia and rigidity – are also not uncommon and can be the most prominent motor sign in some patients, e.g., those with juvenile onset but also in other patients with a later onset. The type of movements that predominate may reflect inherent differences in other domains, e.g., cognition, with predominantly hypokinetic–rigid patients performing worse on cognitive assessments and being more impaired in day-to-day function [52]. However, cognitive assessments often have a motor component and time limits so that simply being slower may mean that hypokinetic–rigid patients cannot complete tasks with the same speed as predominantly hyperkinetic patients.

Patients become increasingly clumsy because of involuntary movements and declining motor coordination. Household items break more frequently, e.g., crockery, or impaired fine motor skills may affect work performance in particular in jobs that require those skills.

A characteristic feature of HD is the oculomotor disorder. In addition to gaze impersistence – the patient finds it difficult to maintain eye contact, and the eyes seem choreatic – patients may have to induce voluntary eye movements with a blink or head movements. These head movements or eye blinking can be suppressed to a certain degree, but eventually patients may be incapable of initiating saccades without turning their head. This can resemble oculomotor apraxia.

A swallowing disorder develops in most patients with HD. The movement disorder affects muscles responsible for all phases of swallowing. The swallowing disorder often develops from first the incoordination of tongue and pharyngeal muscles so that patients and family report a tendency to swallow too much at a time. The patient may find it difficult to move food in a well-coordinated fashion with the tongue towards the pharynx. Later even swallowing saliva may become difficult with frequent aspiration followed by bouts of coughing. Eventually, the risk for aspiration may increase to a degree that necessitates the placement of a gastric feeding tube.

The assessment of the motor phenotype relies on the motor subscale of the Unified Huntington's Disease Rating Scale (UHDRS) [53], a categorical, semi-quantitative scale susceptible to error and with substantial inter-rater variability [54]. The scale was designed for manifest HD even though it was able to detect small signal changes in preclinical HD [55]. There is considerable interest to develop tools that quantitate the motor assessment objectively including in the pre-clinical phase of the disease. To this end computer-assisted measurements of a vari-ety of motor acts such as finger tapping [56], grip force, tongue force, or chorea have been developed that are sensitive to change over time [39]. Subtle motor signs may precede the diagnosis of manifest HD by many years [46, 55]. The develop-ment of objective quantifiable measures of HD motor signs is important because, in the future, treatments may aim to delay or avoid onset of HD. Thus, monitoring subtle clinical signs that are present in the preclinical phase of HD could be very valuable.

Cognitive Domain

Cognition is invariably affected over the course of HD (for a review see [57]). Some people develop cognitive impairment early on before a diagnosis of motor manifest HD is made [58]. In some it may be the first sign of manifest HD, while others seem less impaired until they are further advanced. Once evident, cognitive skills decline until every HD patient has developed dementia. This very likely reflects ongoing degeneration in the underlying brain structure, in particular fronto-striatal circuits. Executive dysfunction is typical in HD. Patients find it more and more difficult to plan and organize their daily activities and chores and are easily overwhelmed when mental flexibility is required, e.g., in dealing with multiple tasks that they have to attend to at the same time. This can be a complaint at the workplace where it may be noted that the affected individual has slowed down, achieving less and less even when working more hours. The patient may only be able to deal with one task at a time having difficulty to switch attention efficiently. The patient may note these shortcomings himself; as a consequence self-esteem may drop, frustration may rise, and people can become irritable, angry, or depressed. Mental inflexibility can develop into perseveration, and a lack of will and executive dysfunction can develop into apathy (see below).

As the disease progresses, patients become more self-centered and fail to see other people's viewpoint. The perception of self and others changes, as does the ability to monitor self-appearance and the consequences of actions, e.g., regarding the patient's social surroundings. Patients progressively neglect themselves. This includes grooming, personal hygiene, and eating but also their relationships with others, in particular their family and carers. This may be related to deficits of emo-tion recognition abilities [59], in particular concerning negative emotions, as a result of which the relationship with family and carers can suffer a loss of empathy and connectedness when the patient increasingly needs support and, at the same time, changes as a person (for a review see [60]). Not all cognitive abilities decline with the same speed. Language skills, for instance, may be preserved much longer than executive skills. However, as in other forms of dementia, HD patients gradually lose autonomy and become dependent on carers and family.

It is important to distinguish attention deficits with subsequent cognitive impairment from primary cognitive deficits [61]. Inattention and loss of energy and drive can be important features of depression and can lead to reduced cognitive performance. In any HD patient with cognitive impairment, in particular when this evolves rapidly at early disease stages, a mood disorder has to be considered and treated. Psychosis may be another explanation for a (treatable) loss of cognitive skills and can be difficult to diagnose clinically (see below).

Neuropsychiatric Manifestations

The motor signs of HD may be the most noticeable. However, together with the cognitive features, the main neuropsychiatric aspects of HD – depression, anxiety, irritability and aggression, perseveration, apathy, and psychosis –are very often more troublesome. They have a higher negative impact on quality of life for patients, families, and caregivers than the motor manifestations [57, 62]. Consistent with clinical impression, systematic data analysis, using, e.g., principal component analysis, confirms that neuropsychiatric symptoms in HD can be differentiated into those that pertain to affect, irritability/aggression, and apathy [63, 64]. With the exception of apathy, behavioral abnormalities may be common but are not invariably part of HD. This may relate to the episodic nature of many behavioral problems, e.g., depression or psychosis, which are amenable to treatment. HD and depression may be two separate disorders [65]. In this context, it was recently shown that current sub-threshold depressive symptoms in early HD were associated with microstructural changes – without concomitant brain volume loss – in brain regions known to be involved in major depressive disorder (MDD), but not those typically associated with HD pathology [66]. Apathy, in contrast, may reflect degeneration within fronto-striatal-cortical networks as HD advances [39, 67].

Affective Disorders

The most common affective disorder in HD is depression. Anxiety may also be very common, but it can be difficult to say whether anxiety goes beyond the uncertainty and worrying when expecting a terrible disease to strike. Anxiety is very often part of a depressive disorder from which it is therefore difficult to disentangle. Depression can occur at any point in time during the course of HD, with its severity ranging from sub-threshold depressive symptoms to MDD [68]. The estimated prevalence of depression in symptomatic HD varies between 30 and 45 % [46] with an estimated prevalence rate of 16.5 % compared to 5.5 % in the normal population [69]. In HD, depression is highly debilitating and a key determinant of social functioning, life satisfaction, and well-being [62]. The need for complex adjustments coupled with preserved insight into the significance of HD symptoms may be one factor explaining why depressive symptoms seem to be particularly common in early HD [70]. Adverse life circumstances may therefore play an important role in triggering and maintaining depression in individuals with HD. The diagnosis of MDD may be challenging in HD because many of the physical symptoms required to make a diagnosis of MDD according to DSM-IV-TR can also be part of HD even in the absence of the core symptoms of sadness and anhedonia. It is therefore important to use appropriate assessment tools such as the clinician-rated Problem Behaviours Assessment (PBA) [71] and the Hospital Anxiety and Depression Scale [72] self-rating scale. These instruments focus on the core symptoms of depression without any of the physical signs and symptoms that may simply be the result of a degenerative disorder.

The description of George Huntington already emphasized that suicide was a serious risk in HD [1]. Suicidal ideation is probably very common in HD with data suggesting that maybe a third of HD patients entertain suicidal ideas in their lifetime [47, 70, 73]. Low mood is a predictor for suicidal ideas; hence screening for the core symptoms of depression is important. However, many people with HD, in particular in the preclinical phase, consider suicide an option should they lose their autonomy when the disease manifests. Considering the inevitability of a relentlessly progressive loss of abilities once HD symptoms become obvious, this is understandable; suicide as a way out of this conundrum can reassure people with HD that they retain a degree of control over their fate. Once affected by the signs of manifest HD, the increasing loss of insight, energy, and motivation means that very few patients actually go on and kill themselves. In the author's experience it can be very helpful if patients feel that their physician acknowledges their predicament and does not think that such thoughts are wrong and a sign of illness.

Irritability and Aggression

Relatives and carers of HD patients often report that irritability and outbursts of verbal and physical aggression are the most difficult behavioral challenge. HD patients, particularly when insight is preserved, develop mood swings where they feel extreme anger irresistibly welling up within them leading to an explosive outburst with verbal abuse and aggression towards objects or even towards other people. This is then followed by remorse, sadness, and, sometimes, even ideas of suicide. This resembles rage attacks in the context of impaired impulse control in other circumstances. Often the hostility and aggression are directed towards those closest to the patient. Once insight is lost, it is then mainly the family and carers of patients that report such behavior. The behavior can go on for a long time, often hours or even days, and is out of proportion to the preceding, often only very minor, provocation or inconvenience. A simple reminder of a trivial task can suffice to spark an outburst. People around the patient have to change their behavior, and, when untreated, irritability and aggression can estrange the patient emotionally from those he or she most depends on. As a consequence family and carers may feel that they cannot cope with the situation so that the patient may no longer be able to live at home.

Some degree of irritability is probably quite common in HD [57, 74]. It can be assessed using the clinician-rated PBA, or PBAs, or the Snaith irritability self-rating scale [71, 75]. For management decisions it is important to carefully assess whether other signs and symptoms such as low mood, sleep problems, motor signs, and the patient's life circumstances can contribute towards irritability. Depression, for instance, can be associated with irritability, and the everyday frustrations of being clumsy or not having anything useful to do can lower self-esteem, impact on mood, and thus contribute to irritability.

Apathy

Apathy denotes a lack of interest, feeling, drive, emotion, or concern. Some patients experience this as very unpleasant and suffer from apathy, e.g., in the context of depression. Others may have lost insight and are not concerned, quite in contrast to their relatives. If pronounced this cluster of symptoms resembles abulia, and patients lack the will to do anything. In HD, apathy develops insidiously so that it may take some time for the symptoms to become noticeable. When it does, however, apathy can be a very frustrating sign of HD for family and carers [57]. The patient has lost his interest in hobbies, spends more time doing very little, and needs a push from those around him. If not reminded by others, the patient may neglect personal hygiene, grooming, and even eating. He may spend a lot of time watching TV and may not even be bothered to change the channel. In contrast to family and carers, the patient is not unduly concerned by his lack of will.

The onset of apathy may predate the onset of unequivocal motor signs of HD and probably reflects degeneration in fronto-striatal-cortical networks [39]. Apathy is thus part of the personality changes that develop with dementia in HD. However, it may also be part of several different underlying psychopathologies including depression and, as a negative syndrome, psychosis [76, 77]. Current concepts of what constitutes apathy, and how to diagnose it, have recently been reappraised [76, 77]. Since apathy is currently not well defined, and its origins may be quite different, it is perhaps not surprising that the association of apathy with the biological load of HD (disease burden, [78]) is not as strong as that of motor or cognitive signs [47].

Perseveration and Obsessive–Compulsive Disorder

Obsessive–compulsive disorder (OCD) is an anxiety disorder. Patients feel they have to do certain things, dwell on a thought, recall an experience, or ruminate on something abstract (ICD-10; http://www.who.int/classifications/icd/en/). Characteristically, at the expense of mounting levels of anxiety, patients need to resist such behavior or thinking, which is qualified as alien to his or her personality. Some of the obsessive behaviors aim to relieve anxiety, worry, fear, or uneasiness in a ritualistic way, e.g., washing in response to fear of contamination. OCD in patients with HD may not be more common than in the general population. However, repetitive behaviors that do not follow on from anxiety-provoking thoughts and are characteristically not perceived as alien or abnormal are very common in HD [57]. The distinction from OCD is important because the underlying pathophysiological concepts and management differ between OCD and perseveration. Perseverative behaviors are most commonly reported by family and carers. The patient, very often when a degree of dementia has already developed, gets stuck on certain ideas or behaviors and is not easily redirected. He prefers certain routines in his day-to-day life and may react angrily if forced to vary from this routine. Simple things such as an appointment with the doctor can provoke discomfort and sleepless nights, and together with a loss of the sense of time, the patient may be restless and urging for the departure for that appointment hours before it would actually be necessary. He may repeatedly ask the same questions or get stuck on a certain topic about which he will go on and on. Without prompting he may return to this topic later in the day, or sometimes even the next day, and will again dwell on this for a long time. While the patient may be unfazed, such perseverative behaviors can cause major distress to family and carers.

Psychosis

Psychosis in HD strikingly resembles schizophrenia with delusions, auditory hallucinations, disordered thinking, social withdrawal, and emotional blunting. The prevalence figures for psychosis in HD range from 3 to 11 % (see [57, 69]). However, the prevalence of psychosis in HD is much lower than for other neuropsychiatric symptoms (see above). Similar to major depressive episodes, the diagnosis of schizophrenia in HD is difficult and, according to the various diagnostic guidelines such as ICD or DSM, not recommended because of the organic basis of HD. Clinically, it can be challenging to distinguish apathy and emotional changes in the context of the degenerative dementia in HD from negative symptoms as a sign of a psychotic episode in particular in the absence of clear evidence of delusions or/and hallucinations. Psychosis can occur at any time in HD including the motor premanifest phase. Hence, prominent negative symptoms and disordered thinking with a rather abrupt onset in an HD patient in whom cognition seemed, until then, fairly intact raise the possibility of a psychosis, in particular if there is no evidence to suggest the presence of a mood disorder. Considering the impact on quality of life, a suspicion of psychosis merits empirical treatment, which can sometimes clarify that distinction.

Juvenile HD

Juvenile HD (JHD) is often defined as HD with an arbitrary age-at-onset before age 20. More importantly, JHD affects individuals who are still developing intellectually, emotionally, and as socially competent independent persons. This may help understand some of the clinical phenomena but also the enormous implications of JHD for the patient and the affected family [50].

Fortunately, JHD is rare. Depending on the epidemiological study, between 1 and 10 % of HD patients have JHD, while childhood HD (onset before age 10) may be even rarer [79]. JHD can manifest before unequivocal signs of HD become apparent in a parent; most commonly, JHD is inherited from an affected father (anticipation, see below).

Clinically, as in adult-onset HD, patients have a mixed movement disorder, cognitive impairment, and behavioral abnormalities. However, there are important differences between the phenotype of adult-onset HD and JHD. Instead of chorea, which, if present, appears fairly late, a hypokinetic–rigid syndrome, dystonia, ataxia, and tremor predominate. A considerable number of patients with JHD (about 30 %) have epileptic seizures. The cognitive changes and behavioral abnormalities can be particularly troubling. Depending on the age of the child, this becomes apparent as developmental delay or, in older children, as a loss of cognitive abilities and dementia. Children may show a change of character and personality with aggressive, hostile, and oppositional and antisocial behavior; as in adults with HD, depression and apathy are common. It is sometimes difficult to differentiate signs and symptoms that are directly related to the biology of HD from those that may arise as a consequence of the psychologically and emotionally challenging responses from the environment.

The management and treatment of JHD is even more complex than in adult HD. It always requires close interdisciplinary collaboration of patients and their family and carers, medical doctors, psychologists, teachers, and others around the patient and his family. The interested reader is referred to a very good book on JHD for further reading [50].

Treatment and Management of HD

Standards of Care and Multidisciplinary Clinic

HD management is not limited to the HD patient as the situation affects the whole family. It includes relatives being at risk of having inherited the HD mutation, those knowing they carry the HD gene, carers, and symptomatic family members. This adds to the complexity of HD. Therefore, HD care requires a multidisciplinary approach involving a wide range of services that can support the symptomatic individual in each stage of HD as well as addressing the needs of those around him [80]. A multidisciplinary approach to the family with HD comprises a variety of specialized services, such as neurology, psychiatry, neuropsychology, clinical genetics, physiotherapy, speech and language therapy, dietician, social services, and dentistry. An HD management clinic can serve as the hub where these specialized services come together to benefit the HD family. The role of the HD clinic is to provide information and to establish a management plan in collaboration with partner agencies outside the clinic. These partner agencies include, e.g., acute services and inpatient care, general physicians and primary care services, psychiatric care, psychiatric nursing services, social worker and welfare rights, financial advisors, disability employment advisors, peer support groups, housing support services, day-care services, personal care, occupational therapy services, and drivers' licensing authorities. For examples of multidisciplinary HD clinics and guidelines, the reader is referred to the websites of the Huntington's Disease Society of America (www. hdsa.org) or the European Huntington's Disease Network's Standards of Care working group (EHDN; www.euro-hd.net).

Management of Clinical Manifestations

The majority of patients attending an HD clinic will have manifest HD. However, it is important to recognize that living with the knowledge of carrying a mutated *HTT* gene can be very difficult. Even though there are no clinical signs of manifest HD, people with premanifest HD may be in need of support or pharmacological or non-pharmacological treatment. In the absence of causal, disease-modifying treatment options, the approach to treatment will need to be guided by first establishing a hierarchy of problems. This requires taking a careful history from the patient and, importantly, family and caregivers. It is important to identify who has which problem. In particular, behavioral problems, such as apathy or irritability, are sometimes much more troublesome for the family and carers than for the patient who may even deny having any problems at all. Listing the problems in their order of relevance determines in which order the problems need to be addressed. It is very important to explain carefully to the patient and caregivers what the pharmacological and non-pharmacological management options are. This should lead to the definition of the treatment goals and how to measure treatment effects. As a general principle for therapy, one should try and address as many problems from the list with a single intervention. In the course of the evaluation of treatment effects, one has to be mindful of drug interactions when using more than one drug and of trying to differentiate signs of HD from side effects of medication. It is sometimes useful to consider reducing the amount of pharmacotherapy rather than adding yet another drug or increasing the doses since side effects of medication can be mistaken as signs of HD. Finally, it is always important to consider causes other than HD for a particular problem. There is no evidence to suggest that HD patients are protected from other ailments that may thus affect them just as they affect the rest of the population.

Motor Signs and Symptoms

Once it has been established that motor signs and symptoms impair the patient's day-to-day activities, there are pharmacological and non-pharmacological treatment options (Table 5.1). When considering the treatment options, it is important to be clear about the type of motor symptom – hyperkinetic, hypokinetic–rigid, gait

Table 5.1 Treatment options in HD

Clinical problem	Treatment options	
	Pharmacological	Non-pharmacological
Chorea	D-RA (tiapride, olanzapine); TBZ	If severe and intractable Huntington's chair
	↓SSRI; mirtazapine if restless	
Dystonia	↓D-RA, ↓TBZ	Physiotherapy
Hypokinesia–rigidity	↓D-RA, ↓TBZ	Physiotherapy
Gait and balance	↓D-RA, ↓TBZ (if dystonic or rigid); D-RA, TBZ if choreic	Physiotherapy
Dysphagia	↓D-RA, ↓TBZ (if dystonic or rigid)	Speech and language therapy, gastric feeding tube, nutritional advice
Weight loss	D-RA (if also choreic or irritable)	Speech and language therapy, hyper-caloric food supplements
Depression/anxiety	SSRI, mirtazapine, agomelatine, D-RA (enhancing SSRI)	Psychotherapy (if cognitively not too impaired)
Irritability	SSRI, mirtazapine, agomelatine, D-RA	
Aggression, verbal	SSRI, mirtazapine, agomelatine, D-RA	
Aggression, physical	D-RA (olanzapine, quetiapine)	
Perseveration	Try SSRI	Behavior modification
Apathy	SSRI, aripiprazole, bupropion ↓D-RA, ↓TBZ	Establish routines, day care
Psychosis	D-RA	
Sleep–wake cycle	Mirtazapine, agomelatine	Sleep hygiene

Abbreviations: *D-RA* dopamine receptor antagonist, *TBZ* tetrabenazine, *SSRI* selective serotonin reuptake inhibitor, ↓ reduce

and balance problems, and swallowing – that is to be treated. Symptomatic pharmacological treatment of hyperkinetic motor manifestations most commonly targets the dopaminergic system. Available drugs include dopamine receptor antagonists ("antipsychotics") that target postsynaptic dopamine receptors and tetrabenazine, a reversible inhibitor of the vesicular monoamine transporter 2 (VMAT-2) that concentrates monoamines such as dopamine within presynaptic vesicles. VMAT-2 inhibition leads to the depletion of presynaptic dopamine. The evidence indicating efficacy of these agents is largely empirical even though small clinical trials suggest antipsychotics are effective, and a placebo-controlled trial demonstrated that tetrabenazine was better than placebo in treating chorea [81, 82]. However, a systematic Cochrane Review of the available data was unable to recommend any drug solely based on the available evidence [83]. It is further important to note that antipsychotics and tetrabenazine can have substantial side effects; adverse affects on cognition, mood, or alertness can outweigh the beneficial effects on chorea and were even recorded in the tetrabenazine trial [82]. It may therefore be premature to derive firm evidence-based guidelines for clinical practice.

In the absence of evidence-based guidelines, recent efforts benefited from large clinical networks in North America and Australia (Huntington Study Group, www. hsg.org) and Europe (European Huntington's Network, EHDN; www.euro-hd.net) in capturing expert's experience in the management of HD motor symptoms [84]. This initiative has published an algorithm for the treatment of chorea [84]. This algorithm takes into account that abnormal movements are often accompanied by neuropsychiatric manifestations (see next section). Hence, the choice of treatment depends also on non-motor problems (Table 5.1). In addition, the movement disorder in HD is mixed. While chorea may respond to some degree to treatment, dystonia and hypokinetic–rigid symptoms are more difficult to treat with medication. While it is always worth a trial, hypokinetic–rigid symptoms often do not respond satisfactorily to dopaminergic therapy, at least in the author's experience. Hypokinetic–rigid symptoms and dystonia can be a side effect of medication, in particular from antipsychotics and tetrabenazine. Dystonia and hypokinetic–rigid symptoms can also contribute to gait and balance disorders, as well as speech and swallowing difficulties. Sometimes, reducing the dose of these medications can have a beneficial effect on these motor symptoms.

Non-pharmacological approaches such as physiotherapy, occupational therapy, and speech and language therapy are very important in the treatment of these motor manifestations. A guideline for physiotherapy published by the EHDN physiotherapy working group can be found on www.euro-hd.net/html/network/groups/physio.

Dysphagia puts patients at risk for aspiration pneumonia. Treatment and prevention of dysphagia remain empirical in the absence of any evidence to support a particular therapeutic strategy [85]. Frequent coughing at meal times suggests the presence of swallowing difficulties. However, it remains unclear if, e.g., videofluoroscopy assessment is more sensitive to predict aspiration than common clinical sense. The option of placing a gastric feeding tube needs to be discussed with the patient and carers/family when, in the judgment of the physician, the risk of aspiration exists. Patients can benefit greatly from counseling by a dietician whenever dysphagia and/or nutrition (including weight loss) become an issue.

Neuropsychiatric and Cognitive Manifestations

For depression, the treatment and management recommendations are essentially the same as for depression in other contexts [86]. Depending on the severity of the depressive disorder, this can include medication and psychotherapy (Table 5.1). If anxiety and loss of energy predominate, the drug of first choice may be a serotonin reuptake inhibitor (SSRI) such as citalopram or a mixed SSRI/noradrenalin uptake inhibitor (NARI) such as venlafaxine. If insomnia, anxiety, and restlessness are prominent, a sleep-inducing drug such as mirtazapine or agomelatine may be a good first choice. Some patients benefit from a combination of an SSRI, or combined SSRI/NARI, and a sleep-inducing antidepressant. An important distinction is that between insomnia as a biological feature of depression and the fairly common disturbance of the sleep–wake cycle in HD. A sleep-inducing antidepressant can be useful in both situations to improve the quality of sleep. This can have beneficial effects on concentration and thus cognition. It remains to be shown whether improved sleep quality may also contribute to the clearance of unwanted proteins and thus have a biological effect [87].

Apathy can be a particularly troublesome problem in HD since there is no good treatment. Apathy can always, at least in part, occur in the context of depression even if there are no obvious signs of depressed mood. Thus, a trial of an antidepressant, e.g., an SSRI or combined SSRI/NARI like venlafaxine, is warranted (see Table 5.1). If this does not improve the situation, aripiprazole should be tried. In the author's experience, aripiprazole sometimes improves drive and concentration with beneficial effects on cognition even though this needs to be investigated more systematically. The same is true for bupropion and other pharmacological treatments that are used in other neurodegenerative disorders [88]. It is important to bear in mind that sedation and apathy can be side effects of medication in particular antipsychotics and tetrabenazine. Reducing the dose of these drugs can sometimes have a big effect on apathy. Non-pharmacological interventions are very important even though there is no evidence to advocate any one in particular. These can include scheduled activities, such as physiotherapy and diverse different tasks so that the patient, with the help of a carer who provides support, accepts to adopt a routine.

Irritability and aggression sooner or later cause distress for the patient and his family. Similar to motor treatments and apathy treatments, there is no evidence base that could serve as a treatment guideline. For the same reasons as for chorea, expert opinion has been synthesized into a treatment algorithm for irritability/aggression [89]. Depending on the severity of outbursts, the first choice may be an SSRI if outbursts are verbal but not physical. If depression and insomnia also pose problems, mirtazapine or agomelatine may be good alternatives. Severe aggression with or without physical violent behavior and impulsivity warrant an antipsychotic as first choice (see Table 5.1). The presence of other symptoms also plays a role in the choice of treatment. Concurrent depression, anxiety, or OCD may also respond to an antidepressant, whereas concurrent psychosis requires antipsychotics. In some cases the severity of aggression may require an admission to a psychiatric institution.

Taken together, treatment should be customized to fit the set of problems the patient has. Multiple input including pharmacological treatments and environmental modifications and also psychosocial support and education for caregivers can help them understand and cope with the situation to the benefit of the HD family. Psychological support for patients and relatives is very important, and psychotherapy, in particular in the preclinical phase of HD and in early manifest HD, can be very helpful.

Lifestyle Measures

Environmental enrichment was able to delay the onset, and slow progression, of phenotypical manifestations in HD models [90, 91]. In humans, small effects on age-at-onset were observed in a retrospective analysis of participants' daily activities. The less active the participants were, the earlier signs and symptoms of HD emerged [92]. Similar to what has been observed in other neurodegenerative disorders such as Parkinson's disease or Alzheimer's disease this supports the notion that mental and physical activity has beneficial effects on health [93]. In contrast, the supplementation of essential fatty acids has only had an effect on motor performance in an HD animal model without influencing the degenerative process [94]. In human HD patients, two double-blind placebo-controlled trials also showed no effect of ethyl EPA, an omega-3 fatty acid, over 6 or 12 months [95, 96]. There is data to support the beneficial effects of a Mediterranean diet on cardiovascular disease [97]; however, in the absence of such evidence in HD, no clear recommendations can be given for specific diets in HD.

Outlook for Causal Treatment

HD is a monogenic disease in which the mutant allele causes the disease. Thus, lowering levels of mutant HTT should have a beneficial effect on HD and is thus a very attractive therapeutic option. In conditional model systems of HD, turning off mutant *HTT* can reverse the phenotype and neuropathology of HD [98, 99]. However, given that in a knockout model abolishing HTT is lethal in the uterus in mice [28], great care has to be taken to ensure that the normal HTT is not interfered with. Targeting the protein is difficult because of its many and largely unknown interactions with other proteins. For this reason inhibiting the expression of mutant *HTT* at the gene or mRNA level seems particularly promising. To this end, different strategies are being pursued. The first used an approach that was not allele specific. In mouse models, and nonhuman primates, antisense oligonucleotides injected into the brain reduced expression of the mutant and wild-type HTT by 50 %, which was well tolerated [100, 101].

Other approaches are the use of small interfering RNAs or micro RNAs, molecules that contribute to the regulation of gene expression at the RNA level [102]. If targeting the mutant and the normal HTT, there is always the risk that not enough mutant HTT is suppressed, while too little of the normal HTT is left. Even though in nonhuman primates suppression of normal HTT by about 30–50 % seemed safe [103], it is difficult to say how this would be in humans. There is always the possibility that the desired effects may not suffice, while the function of the normal HTT is compromised and causes harm. For this reason, allele-specific treatments would be preferable. Already there is evidence that using, e.g., CAG-directed zinc finger protein (ZFP) repressors [104], or RNAi [105], could do exactly this. With such approaches care has to be taken to ensure that other genes with CAG repeats are not targeted. Apart from the choice of novel therapeutics, two main questions remain before clinical trials can be considered. The first concerns the type of delivery. Delivery of ASO therapeutics given intrathecally would mean the ASOs reach the cortex but not deeper brain areas. The aim would thus be to lower HTT expression in cortical areas. In transgenic animals, such treatment was able to exert a beneficial effect [100]. siRNAs could be administered via a pump implanted into the putamen [106]. ZFPs and miRNAs could be delivered using viral vectors, e.g., adeno-associated viral vectors (AAV) [107]. AAV transduction has the advantage, and maybe the risk, that it only needs to be done once per patient. As AAV viruses integrate into the host genome, presumably a lifetime expression of the HTT suppressing agent is envisioned. Such an approach has been used in ALS, in Parkinson's disease, and in congenital blindness [108, 109]. In Parkinson's disease, neurturin AAV delivery was well tolerated, and expression of transgenes was stable several years after the initial transduction [110]. However, depending on the AAV used, an immune response to the transduction of neurons and astrocytes remains a potential risk [111].

The second question concerns the appropriate readouts to use so that the effectiveness of such treatments could be assessed. In models, one can use brain tissue, but for obvious reasons this is not a possibility in humans. Therefore the development and validation of HTT-dependent biological readouts focuses on methods to measure HTT levels, both normal and mutant, or HTT aggregates [112], while at the same time evaluating other potential biomarkers [37].

Conclusions

Twenty years after the discovery of the causal CAG repeat expansion mutation in the *HTT* gene, Huntington's disease remains an incurable devastating disorder. However, great progress has been made in understanding the pathophysiology of HD. This has led to the development of the first gene therapy approaches with promising results in HD model systems. This raises hopes that HD, a monogenetic fully penetrant autosomal dominant disorder, may be a model for novel therapeutics in neurodegenerative disorders. In addition, thanks to the efforts of the HD community – families, clinicians, health professionals, and researchers – standards of care are improving patients' and families' quality of life. HD networks (HSG, EHDN, RLAH), and their observational studies, have further laid the groundwork for conducting clinical trials of high quality on a global stage. This includes collaboration with clinical trial sponsors in designing and conducting clinical trials, preparing and training investigators, and developing the right assessment tools [113]. The time seems right for the clinical trials of the future that hopefully will change our treatment options to relieve the plight of all those affected by HD.

References

1. Huntington G. On chorea. Med Surg Rep. 1872;26:320–1.
2. Gusella JF, Wexler NS, Conneally PM, Naylor SL, Anderson MA, Tanzi RE, et al. A polymorphic DNA marker genetically linked to Huntington's disease. Nature. 1983;306:234–8.
3. The Huntington's Disease Collaborative Research Group. A novel gene containing a trinucleotide repeat that is expanded and unstable on Huntington's disease chromosomes. Cell. 1993;72:971–83.

4. Ross CA, Tabrizi SJ. Huntington's disease: from molecular pathogenesis to clinical treatment. Lancet Neurol. 2011;10:83–98.
5. Orth M, Handley OJ, Schwenke C, Dunnett S, Wild EJ, Tabrizi SJ, et al. Observing Huntington's disease: the European Huntington's Disease Network's REGISTRY. J Neurol Neurosurg Psychiatry. 2011;82:1409–12.
6. Dorsey ER, Beck CA, Darwin K, Nichols P, Brocht AF, Biglan KM, et al. Natural history of Huntington disease. JAMA Neurol. 2013;70(12):1520–30.
7. Pringsheim T, Wiltshire K, Day L, Dykeman J, Steeves T, Jette N. The incidence and prevalence of Huntington's disease: a systematic review and meta-analysis. Mov Disord. 2012;27:1083–91.
8. Hayden MR, MacGregor JM, Beighton PH. The prevalence of Huntington's chorea in South Africa. S Afr Med J. 1980;58:193–6.
9. Wright HH, Still CN, Abramson RK. Huntington's disease in black kindreds in South Carolina. Arch Neurol. 1981;38:412–4.
10. Wexler NS, Lorimer J, Porter J, Gomez F, Moskowitz C, Shackell E, et al. Venezuelan kindreds reveal that genetic and environmental factors modulate Huntington's disease age of onset. Proc Natl Acad Sci U S A. 2004;101:3498–503.
11. Harper P. Huntington's disease: genetic and molecular studies. In: Bates G, Harper P, Jones L, editors. Huntington's disease: genetic and molecular studies. 3rd ed. Oxford: Oxford University Press; 2002. p. 113–58.
12. Gonitel R, Squitieri F. Molecular mechanisms in juvenile Huntington's disease. In: Quarrell OWJ, Brewer H, Squitieri F, Barker RA, Nance MA, Landwehrmeyer GB, editors. Juvenile Huntington's disease. Oxford: Oxford University Press; 2009. p. 79–100.
13. Warner JP, Barron LH, Brock DJ. A new polymerase chain reaction (PCR) assay for the trinucleotide repeat that is unstable and expanded on Huntington's disease chromosomes. Mol Cell Probes. 1993;7:235–9.
14. Riess O, Noerremoelle A, Soerensen SA, Epplen JT. Improved PCR conditions for the stretch of (CAG)n repeats causing Huntington's disease. Hum Mol Genet. 1993;2:637.
15. Losekoot M, van Belzen MJ, Seneca S, Bauer P, Stenhouse SA, Barton DE. EMQN/CMGS best practice guidelines for the molecular genetic testing of Huntington disease. Eur J Hum Genet. 2013;21:480–6.
16. Quarrell OW, Handley O, O'Donovan K, Dumoulin C, Ramos-Arroyo M, Biunno I, et al. Discrepancies in reporting the CAG repeat lengths for Huntington's disease. Eur J Hum Genet. 2012;20:20–6.
17. MacLeod R, Tibben A, Frontali M, Evers-Kiebooms G, Jones A, Martinez-Descales A, et al. Recommendations for the predictive genetic test in Huntington's disease. Clin Genet. 2013;83:221–31.
18. Tibben A. Genetic counselling and presymptomatic testing. In: Bates G, Harper P, Jones L, editors. Huntington's disease. 3rd ed. Oxford: Oxford University Press; 2002. p. 198–250.
19. Codori AM, Slavney PR, Rosenblatt A, Brandt J. Prevalence of major depression one year after predictive testing for Huntington's disease. Genet Test. 2004;8:114–9.
20. de Die-Smulders CE, de Wert GM, Liebaers I, Tibben A, Evers-Kiebooms G. Reproductive options for prospective parents in families with Huntington's disease: clinical, psychological and ethical reflections. Hum Reprod Update. 2013;19:304–15.
21. Van Rij MC, De Rademaeker M, Moutou C, Dreesen JC, De Rycke M, Liebaers I, et al. Preimplantation genetic diagnosis (PGD) for Huntington's disease: the experience of three European centres. Eur J Hum Genet. 2012;20:368–75.
22. Andrade MA, Bork P. HEAT repeats in the Huntington's disease protein. Nat Genet. 1995;11:115–6.
23. Li W, Serpell LC, Carter WJ, Rubinsztein DC, Huntington JA. Expression and characterization of full-length human huntingtin, an elongated HEAT repeat protein. J Biol Chem. 2006;281:15916–22.

24. Li SH, Li XJ. Huntingtin-protein interactions and the pathogenesis of Huntington's disease. Trends Genet. 2004;20:146–54.
25. Shirasaki DI, Greiner ER, Al-Ramahi I, Gray M, Boontheung P, Geschwind DH, et al. Network organization of the huntingtin proteomic interactome in mammalian brain. Neuron. 2012;75:41–57.
26. La Spada AR, Taylor JP. Repeat expansion disease: progress and puzzles in disease pathogenesis. Nat Rev Genet. 2010;11:247–58.
27. Cattaneo E, Zuccato C, Tartari M. Normal huntingtin function: an alternative approach to Huntington's disease. Nat Rev Neurosci. 2005;6:919–30.
28. Zeitlin S, Liu JP, Chapman DL, Papaioannou VE, Efstratiadis A. Increased apoptosis and early embryonic lethality in mice nullizygous for the Huntington's disease gene homologue. Nat Genet. 1995;11:155–63.
29. Dragatsis I, Levine MS, Zeitlin S. Inactivation of Hdh in the brain and testis results in progressive neurodegeneration and sterility in mice. Nat Genet. 2000;26:300–6.
30. Lo Sardo V, Zuccato C, Gaudenzi G, Vitali B, Ramos C, Tartari M, et al. An evolutionary recent neuroepithelial cell adhesion function of huntingtin implicates ADAM10-Ncadherin. Nat Neurosci. 2012;15:713–21.
31. Olshina MA, Angley LM, Ramdzan YM, Tang J, Bailey MF, Hill AF, et al. Tracking mutant huntingtin aggregation kinetics in cells reveals three major populations that include an invariant oligomer pool. J Biol Chem. 2010;285:21807–16.
32. Sathasivam K, Neueder A, Gipson TA, Landles C, Benjamin AC, Bondulich MK, et al. Aberrant splicing of HTT generates the pathogenic exon 1 protein in Huntington disease. Proc Natl Acad Sci U S A. 2013;110:2366–70.
33. Busch A, Engemann S, Lurz R, Okazawa H, Lehrach H, Wanker EE. Mutant huntingtin promotes the fibrillogenesis of wild-type huntingtin: a potential mechanism for loss of huntingtin function in Huntington's disease. J Biol Chem. 2003;278:41452–61.
34. Rubinsztein DC. The roles of intracellular protein-degradation pathways in neurodegeneration. Nature. 2006;443:780–6.
35. Krainc D. Clearance of mutant proteins as a therapeutic target in neurodegenerative diseases. Arch Neurol. 2010;67:388–92.
36. Jeong H, Then F, Melia Jr TJ, Mazzulli JR, Cui L, Savas JN, et al. Acetylation targets mutant huntingtin to autophagosomes for degradation. Cell. 2009;137:60–72.
37. Weir DW, Sturrock A, Leavitt BR. Development of biomarkers for Huntington's disease. Lancet Neurol. 2011;10:573–90.
38. Vonsattel JP, DiFiglia M. Huntington disease. J Neuropathol Exp Neurol. 1998;57:369–84.
39. Tabrizi SJ, Scahill RI, Owen G, Durr A, Leavitt BR, Roos RA, et al. Predictors of phenotypic progression and disease onset in premanifest and early-stage Huntington's disease in the TRACK-HD study: analysis of 36-month observational data. Lancet Neurol. 2013;12:637–49.
40. Paulsen JS, Magnotta VA, Mikos AE, Paulson HL, Penziner E, Andreasen NC, et al. Brain structure in preclinical Huntington's disease. Biol Psychiatry. 2006;59:57–63.
41. Rosas HD, Salat DH, Lee SY, Zaleta AK, Pappu V, Fischl B, et al. Cerebral cortex and the clinical expression of Huntington's disease: complexity and heterogeneity. Brain. 2008; 131:1057–68.
42. Tabrizi SJ, Langbehn DR, Leavitt BR, Roos RA, Durr A, Craufurd D, et al. Biological and clinical manifestations of Huntington's disease in the longitudinal TRACK-HD study: cross-sectional analysis of baseline data. Lancet Neurol. 2009;8:791–801.
43. Thu DC, Oorschot DE, Tippett LJ, Nana AL, Hogg VM, Synek BJ, et al. Cell loss in the motor and cingulate cortex correlates with symptomatology in Huntington's disease. Brain. 2010;133:1094–110.
44. Sassone J, Colciago C, Cislaghi G, Silani V, Ciammola A. Huntington's disease: the current state of research with peripheral tissues. Exp Neurol. 2009;219:385–97.
45. van der Burg JM, Bjorkqvist M, Brundin P. Beyond the brain: widespread pathology in Huntington's disease. Lancet Neurol. 2009;8:765–74.

46. Paulsen JS, Langbehn DR, Stout JC, Aylward E, Ross CA, Nance M, et al. Detection of Huntington's disease decades before diagnosis: the Predict-HD study. J Neurol Neurosurg Psychiatry. 2008;79:874–80.
47. Orth M, Handley OJ, Schwenke C, Dunnett SB, Craufurd D, Ho AK, et al. Observing Huntington's disease: the European Huntington's Disease Network's REGISTRY. PLoS Curr. 2010;2:RRN1184.
48. Langbehn DR, Hayden MR, Paulsen JS. CAG-repeat length and the age of onset in Huntington disease (HD): a review and validation study of statistical approaches. Am J Med Genet B Neuropsychiatr Genet. 2010;153B:397–408.
49. Langbehn DR, Brinkman RR, Falush D, Paulsen JS, Hayden MR. A new model for prediction of the age of onset and penetrance for Huntington's disease based on CAG length. Clin Genet. 2004;65:267–77.
50. Quarrell OWJ, Brewer HM, Squitieri F, Barker RA, Nance MA, Landwehrmeyer GB. Juvenile Huntington's disease. Oxford: Oxford University Press; 2009.
51. Kremer B. Clinical neurology of Huntington's disease. In: Bates G, Harper P, Jones L, editors. Huntington's disease. 3rd ed. Oxford: Oxford University Press; 2002. p. 28–61.
52. Hart EP, Marinus J, Burgunder JM, Bentivoglio AR, Craufurd D, Reilmann R, et al. Better global and cognitive functioning in choreatic versus hypokinetic-rigid Huntington's disease. Mov Disord. 2013;28:1142–5.
53. Huntington Study Group. Unified Huntington's disease rating scale: reliability and consistency. Mov Disord. 1996;11:136–42.
54. Hogarth P, Kayson E, Kieburtz K, Marder K, Oakes D, Rosas D, et al. Interrater agreement in the assessment of motor manifestations of Huntington's disease. Mov Disord. 2005;20:293–7.
55. Biglan KM, Ross CA, Langbehn DR, Aylward EH, Stout JC, Queller S, et al. Motor abnormalities in premanifest persons with Huntington's disease: the PREDICT-HD study. Mov Disord. 2009;24:1763–72.
56. Bechtel N, Scahill RI, Rosas HD, Acharya T, van den Bogaard SJ, Jauffret C, et al. Tapping linked to function and structure in premanifest and symptomatic Huntington disease. Neurology. 2010;75:2150–60.
57. Craufurd D, Snowden J. Neuropsychological and neuropsychiatric aspects of Huntington's disease. In: Bates G, Harper P, Jones L, editors. Huntington's disease. 3rd ed. Oxford: Oxford University Press; 2002. p. 62–94.
58. Paulsen JS, Smith MM, Long JD. Cognitive decline in prodromal Huntington disease: implications for clinical trials. J Neurol Neurosurg Psychiatry. 2013;84:1233–9.
59. Sprengelmeyer R, Young AW, Calder AJ, Karnat A, Lange H, Homberg V, et al. Loss of disgust. Perception of faces and emotions in Huntington's disease. Brain. 1996;119(Pt 5): 1647–65.
60. Henley SM, Novak MJ, Frost C, King J, Tabrizi SJ, Warren JD. Emotion recognition in Huntington's disease: a systematic review. Neurosci Biobehav Rev. 2012;36:237–53.
61. Wolf RC, Gron G, Sambataro F, Vasic N, Wolf ND, Thomann PA, et al. Brain activation and functional connectivity in premanifest Huntington's disease during states of intrinsic and phasic alertness. Hum Brain Mapp. 2012;33:2161–73.
62. Ho AK, Gilbert AS, Mason SL, Goodman AO, Barker RA. Health-related quality of life in Huntington's disease: Which factors matter most? Mov Disord. 2009;24:574–8.
63. Kingma EM, van Duijn E, Timman R, van der Mast RC, Roos RA. Behavioural problems in Huntington's disease using the Problem Behaviours Assessment. Gen Hosp Psychiatry. 2008;30:155–61.
64. Rickards H, De Souza J, van Walsem M, van Duijn E, Simpson SA, Squitieri F, et al. Factor analysis of behavioural symptoms in Huntington's disease. J Neurol Neurosurg Psychiatry. 2011;82:411–2.
65. Folstein S, Abbott MH, Chase GA, Jensen BA, Folstein MF. The association of affective disorder with Huntington's disease in a case series and in families. Psychol Med. 1983; 13:537–42.

66. Sprengelmeyer R, Orth M, Müller HP, Wolf RC, Grön G, Depping MS, et al. The neuroanatomy of subthreshold depressive symptoms in Huntington's disease: a combined diffusion tensor imaging (DTI) and voxel-based morphometry (VBM) study. Psychol Med. 2013;7:1–12.
67. Thompson JC, Snowden JS, Craufurd D, Neary D. Behavior in Huntington's disease: dissociating cognition-based and mood-based changes. J Neuropsychiatry Clin Neurosci. 2002;14:37–43.
68. Epping EA, Paulsen JS. Depression in the early stages of Huntington disease. Neurodegener Dis Manag. 2011;1:407–14.
69. van Duijn E, Kingma EM, Timman R, Zitman FG, Tibben A, Roos RA, et al. Cross-sectional study on prevalences of psychiatric disorders in mutation carriers of Huntington's disease compared with mutation-negative first-degree relatives. J Clin Psychiatry. 2008;69:1804–10.
70. Paulsen JS, Hoth KF, Nehl C, Stierman L. Critical periods of suicide risk in Huntington's disease. Am J Psychiatry. 2005;162:725–31.
71. Craufurd D, Thompson JC, Snowden JS. Behavioral changes in Huntington disease. Neuropsychiatry Neuropsychol Behav Neurol. 2001;14:219–26.
72. Zigmond AS, Snaith RP. The hospital anxiety and depression scale. Acta Psychiatr Scand. 1983;67:361–70.
73. Hubers AA, van Duijn E, Roos RA, Craufurd D, Rickards H, Bernhard Landwehrmeyer G, et al. Suicidal ideation in a European Huntington's disease population. J Affect Disord. 2013;151:248–58.
74. Reedeker N, Bouwens JA, Giltay EJ, Le Mair SE, Roos RA, van der Mast RC, et al. Irritability in Huntington's disease. Psychiatry Res. 2012;200:813–8.
75. Snaith RP, Constantopoulos AA, Jardine MY, McGuffin P. A clinical scale for the self-assessment of irritability. Br J Psychiatry. 1978;132:164–71.
76. Starkstein SE, Leentjens AF. The nosological position of apathy in clinical practice. J Neurol Neurosurg Psychiatry. 2008;79:1088–92.
77. Robert P, Onyike CU, Leentjens AF, Dujardin K, Aalten P, Starkstein S, et al. Proposed diagnostic criteria for apathy in Alzheimer's disease and other neuropsychiatric disorders. Eur Psychiatry. 2009;24:98–104.
78. Penney Jr JB, Vonsattel JP, MacDonald ME, Gusella JF, Myers RH. CAG repeat number governs the development rate of pathology in Huntington's disease. Ann Neurol. 1997;41:689–92.
79. Quarrell O, O'Donovan KL, Bandmann O, Strong M. The prevalence of Juvenile Huntington's disease: a review of the literature and meta-analysis. PLoS Curr. 2012;4:e4f8606b742ef3.
80. Nance MA, Westphal B. Comprehensive care in Huntington's disease. In: Bates G, Harper P, Jones L, editors. Huntington's disease. 3rd ed. Oxford: Oxford University Press; 2002. p. 475–500.
81. Venuto CS, McGarry A, Ma Q, Kieburtz K. Pharmacologic approaches to the treatment of Huntington's disease. Mov Disord. 2012;27:31–41.
82. Huntington Study Group. Tetrabenazine as antichorea therapy in Huntington disease: a randomized controlled trial. Neurology. 2006;66:366–72.
83. Mestre T, Ferreira J, Coelho MM, Rosa M, Sampaio C. Therapeutic interventions for symptomatic treatment in Huntington's disease. Cochrane Database Syst Rev. 2009;(3):CD006456.
84. Burgunder JM, Guttman M, Perlman S, Goodman N, van Kammen DP, Goodman L. An international survey-based algorithm for the pharmacologic treatment of chorea in Huntington's disease. PLoS Curr. 2011;3, RRN1260.
85. Heemskerk AW, Roos RA. Dysphagia in Huntington's disease: a review. Dysphagia. 2011;26:62–6.
86. Geddes JR, Miklowitz DJ. Treatment of bipolar disorder. Lancet. 2013;381:1672–82.
87. Xie L, Kang H, Xu Q, Chen MJ, Liao Y, Thiyagarajan M, et al. Sleep drives metabolite clearance from the adult brain. Science. 2013;342:373–7.
88. Krishnamoorthy A, Craufurd D. Treatment of apathy in Huntington's disease and other movement disorders. Curr Treat Options Neurol. 2011;13:508–19.

89. Groves M, van Duijn E, Anderson K, Craufurd D, Edmondson MC, Goodman N, et al. An international survey-based algorithm for the pharmacologic treatment of irritability in Huntington's disease. PLoS Curr. 2011;3, RRN1259.
90. van Dellen A, Blakemore C, Deacon R, York D, Hannan AJ. Delaying the onset of Huntington's in mice. Nature. 2000;404:721–2.
91. Hockly E, Cordery PM, Woodman B, Mahal A, van Dellen A, Blakemore C, et al. Environmental enrichment slows disease progression in R6/2 Huntington's disease mice. Ann Neurol. 2002;51:235–42.
92. Trembath MK, Horton ZA, Tippett L, Hogg V, Collins VR, Churchyard A, et al. A retrospective study of the impact of lifestyle on age at onset of Huntington disease. Mov Disord. 2010;25:1444–50.
93. Spires TL, Hannan AJ. Nature, nurture and neurology: gene-environment interactions in neurodegenerative disease. FEBS Anniversary Prize Lecture delivered on 27 June 2004 at the 29th FEBS Congress in Warsaw. FEBS J. 2005;272:2347–61.
94. Van Raamsdonk JM, Pearson J, Rogers DA, Lu G, Barakauskas VE, Barr AM, et al. Ethyl-EPA treatment improves motor dysfunction, but not neurodegeneration in the YAC128 mouse model of Huntington disease. Exp Neurol. 2005;196:266–72.
95. Huntington Study Group TEND-HD Investigators. Randomized controlled trial of ethyl-eicosapentaenoic acid in Huntington disease: the TREND-HD study. Arch Neurol. 2008; 65:1582–9.
96. Puri BK, Leavitt BR, Hayden MR, Ross CA, Rosenblatt A, Greenamyre JT, et al. Ethyl-EPA in Huntington disease: a double-blind, randomized, placebo-controlled trial. Neurology. 2005;65:286–92.
97. Estruch R, Ros E, Salas-Salvado J, Covas MI, Corella D, Aros F, et al. Primary prevention of cardiovascular disease with a Mediterranean diet. N Engl J Med. 2013;368:1279–90.
98. Yamamoto A, Lucas JJ, Hen R. Reversal of neuropathology and motor dysfunction in a conditional model of Huntington's disease. Cell. 2000;101:57–66.
99. Martin-Aparicio E, Yamamoto A, Hernandez F, Hen R, Avila J, Lucas JJ. Proteasomal-dependent aggregate reversal and absence of cell death in a conditional mouse model of Huntington's disease. J Neurosci. 2001;21:8772–81.
100. Kordasiewicz HB, Stanek LM, Wancewicz EV, Mazur C, McAlonis MM, Pytel KA, et al. Sustained therapeutic reversal of Huntington's disease by transient repression of huntingtin synthesis. Neuron. 2012;74:1031–44.
101. Boudreau RL, McBride JL, Martins I, Shen S, Xing Y, Carter BJ, et al. Nonallele-specific silencing of mutant and wild-type huntingtin demonstrates therapeutic efficacy in Huntington's disease mice. Mol Ther. 2009;17:1053–63.
102. Davidson BL, Monteys AM. Singles engage the RNA interference pathway. Cell. 2012; 150:873–5.
103. Grondin R, Kaytor MD, Ai Y, Nelson PT, Thakker DR, Heisel J, et al. Six-month partial suppression of Huntingtin is well tolerated in the adult rhesus striatum. Brain. 2012;135:1197–209.
104. Garriga-Canut M, Agustin-Pavon C, Herrmann F, Sanchez A, Dierssen M, Fillat C, et al. Synthetic zinc finger repressors reduce mutant huntingtin expression in the brain of R6/2 mice. Proc Natl Acad Sci U S A. 2012;109:E3136–45.
105. Yu D, Pendergraff H, Liu J, Kordasiewicz HB, Cleveland DW, Swayze EE, et al. Single-stranded RNAs use RNAi to potently and allele-selectively inhibit mutant huntingtin expression. Cell. 2012;150:895–908.
106. Boudreau RL, Rodriguez-Lebron E, Davidson BL. RNAi medicine for the brain: progresses and challenges. Hum Mol Genet. 2011;20:R21–7.
107. Grieger JC, Samulski RJ. Adeno-associated virus vectorology, manufacturing, and clinical applications. Methods Enzymol. 2012;507:229–54.
108. Mittermeyer G, Christine CW, Rosenbluth KH, Baker SL, Starr P, Larson P, et al. Long-term evaluation of a phase 1 study of AADC gene therapy for Parkinson's disease. Hum Gene Ther. 2012;23:377–81.

109. Bennett J, Ashtari M, Wellman J, Marshall KA, Cyckowski LL, Chung DC, et al. AAV2 gene therapy readministration in three adults with congenital blindness. Sci Transl Med. 2012;4:120ra15.
110. Bartus RT, Baumann TL, Brown L, Kruegel BR, Ostrove JM, Herzog CD. Advancing neurotrophic factors as treatments for age-related neurodegenerative diseases: developing and demonstrating "clinical proof-of-concept" for AAV-neurturin (CERE-120) in Parkinson's disease. Neurobiol Aging. 2013;34:35–61.
111. Ciesielska A, Hadaczek P, Mittermeyer G, Zhou S, Wright JF, Bankiewicz KS, et al. Cerebral infusion of AAV9 vector-encoding non-self proteins can elicit cell-mediated immune responses. Mol Ther. 2013;21:158–66.
112. Baldo B, Paganetti P, Grueninger S, Marcellin D, Kaltenbach LS, Lo DC, et al. TR-FRET-based duplex immunoassay reveals an inverse correlation of soluble and aggregated mutant huntingtin in Huntington's disease. Chem Biol. 2012;19:264–75.
113. HORIZON Investigators of the Huntington Study Group and European Huntington's Disease Network. A randomized, double-blind, placebo-controlled study of latrepirdine in patients with mild to moderate Huntington disease. JAMA Neurol. 2013;70:25–33.

Chapter 6
Huntington's Disease: Molecular Pathogenesis and New Therapeutic Perspectives

Claudia Perandones and Ignacio Muñoz-Sanjuan

Abstract Huntington's disease (HD) is an autosomal dominant, progressive neuro-degenerative disorder with a clinical spectrum that includes chorea, incoordination, cognitive decline, and behavioral difficulties. The underlying genetic defect responsible for the disease is the expansion of a CAG repeat in the huntingtin (*HTT*) gene. This repeat is unstable and its length is inversely correlated with the age at onset of the disease. Despite its widespread distribution, mutant HTT causes neurodegeneration, which occurs preferentially in the striatum and deeper layers of the cortex. Mechanisms implicated in HD include those relevant to DNA repair, transcriptional and translational modulation of expanded trinucleotide repeats (including somatic expansion), mitochondria and energy homeostasis, vesicular trafficking dynamics, oligomerization of mHTT (chaperone biology), autophagy, epigenetic mechanisms, and synaptic signaling. Notably, not all the effects of mutant HTT are cell autonomous. The present review focuses on the molecular pathogenesis of HD and the current state of therapeutic development for the treatment of HD. We review the preclinical and clinical development molecular therapies targeting HTT expression and the modulation of biological mechanisms thought to contribute to disease pathogenesis via novel therapeutic agents.

Keywords Huntington's disease • Huntingtin • CAG repeats • Kynurenine • PDE • HDACs • ASO • siRNA • Zinc finger repressors • Dopamine • Glutamate • GABA • Adenosine

C. Perandones
National Center for Medical Genetics, National Agency of Laboratories
and Health Institutes of Argentina (ANLIS) Dr. Carlos G. Malbrán,
Av. Vélez Sarsfield 563, Buenos Aires 1281, Argentina
e-mail: claudia.perandones@gmail.com

I. Muñoz-Sanjuan (✉)
CHDI Management/CHDI Foundation Inc.,
6080 Center Drive, Suite 100, Los Angeles, CA 90045, USA
e-mail: ignacio.munoz@chdifoundation.org; www.chdifoundation.org

F.E. Micheli, P.A. LeWitt (eds.), *Chorea*,
DOI 10.1007/978-1-4471-6455-5_6, © Springer-Verlag London 2014

Introduction to Huntington's Disease

Huntington's disease (HD) is a devastating, autosomal dominant, neurodegenerative disorder, which is caused by a CAG triplet repeat expansion in exon 1 of the huntingtin (*HTT*) gene, encoding an abnormally expanded polyglutamine (polyQ) tract [1, 2]. HD is the most prevalent of all triplet repeat diseases. The CAG repeat length (>35 CAGs) is inversely correlated with age at diagnosis of clinical symptoms (typically motor symptoms), accounting for 60–70 % of this variation (see Figures in Dr. Orth's Chap. 5) [3]. Although the disease can manifest at any age, clinically relevant symptoms begin in mid-life (35–50 years) and progression is slow, taking ~20 years to death. In juvenile cases of HD (<20 years of age, with CAG repeats typically >60), the disease is accelerated and symptoms include more severe bradykinesia, rigidity, seizures, and severe dementia with little to no chorea [4].

Clinically, adult HD is characterized by motor, cognitive, and psychiatric disturbances. These include impairments in executive function, planning, and working memory, impulsivity, loss of attention, motivation and self-care, and deficits in movement control (chorea, dyskinesias). In later stages the patients develop rigidity, bradykinesia, and dementia [5, 6].

There is also loss of body and muscle weight even when patients are on a high-calorie intake [7]. Difficulty in swallowing leading to choking, pneumonia, and complications with lack of mobility generally results in death. By the time a patient is clinically diagnosed with onset of disease, there is an estimated 60–80 % loss of the striatum [8]. The indirect pathway medium spiny neurons of the dorsal striatum appear to be the most vulnerable to cell death in HD, even though HTT is ubiquitously expressed. The discovery of the disease-causing mutation in *HTT* and the development of genetically engineered rodent models have enabled investigation of the potential pathogenic mechanisms, through pharmacological/molecular strategies as well as genetic manipulations [9, 10].

Most of our current understanding of HD pathogenesis comes from these animal models. HTT is a 3144 amino acid protein with a polyglutamine stretch and a poly-proline domain located in the N-terminus, nuclear import and export signals, numerous regions predicted to form highly structured HEAT repeats, and additional aggregation motifs outside of the polyglutamine tract itself [11, 12]. HTT can be found in the cytosol, in cell membranes (endoplasmic reticulum and Golgi apparatus), and in the nucleus. A pathological hallmark of HD is the presence of intranuclear and neuropil inclusions composed of aggregated N-terminal HTT fragments and several ubiquitinated proteins [13–15].

HTT Distribution, Expression Levels, and Somatic Instability

After the *HTT* gene was discovered, a number of studies focused on determining whether heterogeneities in *HTT* mRNA and/or protein expression could explain the increased vulnerability of MSNs and cortical neurons in HD. However, as extensive mRNA and protein expression analyses indicate, *HTT* is expressed in nearly all tissues and no significant differences have been reported in *HTT* mRNA expression in patients versus unaffected subjects, even though there is evidence for altered translation rates and differing half-lives for HTT alleles with large CAG expansions. In fact, striatal interneurons, which do not undergo cell death in HD, have been shown to express higher levels of *HTT* than MSNs [16]. This would indicate that increased vulnerability of MSNs does not result from different levels in mutant *HTT* mRNA expression. In recent studies in patients and rodent HD models, the mutant CAG repeat tract of *HTT* was seen to undergo both inter- and intragenerational variability in expansion size [17]. Studies on end-stage human HD autopsy material have shown that hyperexpansions of over 200 repeats occur in approximately 10 % of sampled striatal cells [18]. Moreover, observation of the striatum and the cerebral cortex of HD mice has revealed age-dependent instability of *HTT* [19–22].

Based on these findings, increased instability of the expansion was suggested as a factor in the marked vulnerability of striatal neurons in HD [17]. However, other polyQ-expansion diseases with little or no striatal pathology have also shown greater instability of CAG repeats in the striatum [23]. Furthermore, when a bacterial artificial chromosome-based HD mouse model expressing polyQ-Htt with a stable CAA-CAG tract was studied, it failed to indicate a role of somatic repeat instability in HD pathogenesis [24].

As a whole, study findings are not sufficiently conclusive to demonstrate that the differential vulnerability of striatal neurons affected in HD patients is influenced by heterogeneities in *HTT* expression levels or somatic instability.

Mechanisms Associated with HTT Pathogenesis

The mechanisms associated with HD pathogenesis share similar biology with other neurodegenerative diseases, and in general we lack a deep understanding of the specificity of the molecular networks triggered by expression of mHTT. Mechanisms implicated in HD include those relevant to DNA repair, transcriptional and translational modulation of expanded trinucleotide repeats (including somatic expansion), mitochondria and energy homeostasis, vesicular trafficking dynamics, oligomerization of mHTT (chaperone biology), autophagy, epigenetic mechanisms, and synaptic signaling. Although these mechanisms are of significant interest, most studies have not provided sufficient understanding of the dysfunction to enable a drug discovery program. Existing therapeutic programs target either HTT expression, its aggregation, or mechanisms thought to play a major role in progression and which are amenable for traditional drug development: drugs aimed at restoring the circuitry changes observed in HD, and thought to underlie its symptomatology, or drugs aimed at restoring energetic deficits, autophagy induction in brain cells, or deficits in BDNF/TrkB signaling. Below we review existing approaches to the treatment of HD based on our understanding of the molecular and synaptic pathogenic mechanisms. Table 6.1 shows ongoing and recent clinical trials in HD.

Table 6.1 Clinical trials in HD

Name or sponsor	Intervention	Starting date	Size	Current status
TREND-HD Amarin Neuroscience Ltd	Ethyl-EPA (Miraxion™)	2005	300	Completed
2CARE MGH	Coenzyme Q10	2008	608	Completed
MIG-HD	Fetal transplantation	2008	60	Completed
CREST-E	Creatine	2008	650	Recruiting
NEUROHD	Tetrabenazine Olanzapine Tiapride	2008	180	Completed
Teva Pharmaceutical Industries	ACR16	2008	437	Completed
Heinrich-Heine University (Germany)	DBS of the globus pallidus	2009	6	Completed
Raptor	Cysteamine	2010	98	Completed
Charite University (Germany)	Bupropion	2010	90	Recruiting
REACH2	PBT2	2011	100	Completed
University of British Columbia	Memantine	2011	25	Recruiting
Charite University (Germany) ETON study	Epigallocatechin gallate	2011	54	Recruiting
GW Pharmaceuticals Ltd.	Delta-9-tetrahydrocannabinol (THC) and cannabidiol (CBD)	2011	25	Completed
Auspex	SD-809 ER	2013	100	Recruiting
Pfizer	PF-02545920	2013	65	Recruiting

Trials are shown in order of starting date. We have included completed studies that have not yet reported findings

Due to length constraints we refer readers to several recent reviews on specific therapeutic programs still in preclinical stages.

Role of HTT in Vesicular Trafficking and Axonal Transport Mechanisms

Accumulating data on roles for HTT in endocytosis, endosomal motility, and axonal transport has led to an emerging model for huntingtin as an integrator of transport along the cellular cytoskeleton. HTT is able to coordinate the binding of multiple types of motor proteins to vesicular cargo, most likely by acting as a scaffold that can differentially bind to many proteins. HTT may play a key role in modulating vesicle transport between the actin and microtubule cytoskeletons through its inter-action with various partners (Fig. 6.1). Huntingtin/HAP40 may regulate vesicle association with actin [25]. Through associations with a myosin VI linker protein, such as optineurin [26], HTT may play a role in actin-based transport of endocytic vesicles. After undergoing short-range transport along the actin cytoskeleton, endo-somes are delivered to the microtubules for long-distance transport. HTT may medi-ate this cytoskeletal switching through the dissociation of HAP40, perhaps followed by a concomitant increase in HAP1 binding, which would favor microtubule asso-ciation. HTT also facilitates dynein-mediated vesicular transport along microtu-bules through direct binding to dynein and indirect binding to dynactin via HAP1, possibly leading to the formation of a quaternary complex that enhances interac-tions of various effectors [27]. Continuing with the example of endosomal cargo, once an internalized receptor that is fated to be recycled back to the plasma membrane has been sorted to a recycling compartment, it requires a change in microtubule motors to affect a switch in the direction of vesicle motility. HTT could serve as this switch through differential binding to downstream effectors. HTT-associated vesicles might also be able to switch direction if increased levels of HAP1 cause an enhanced recruitment of kinesin through the HAP1–kinesin interac-tion [28]. Alternatively, phosphorylation of HTT may cause a recruitment of kinesin to vesicles and a switch in the preferred direction of vesicular motility [29]. Depletion of endogenous HTT or expression of mHTT has been shown to affect numerous aspects of intracellular trafficking and transport. By integrating the inter-play of various binding partners as well as regulatory pathways such as phosphory-lation, HTT can coordinate vesicle binding to the appropriate molecular motor and enhance motility.

The "Dying-Back" Pattern of Degeneration in HD-Affected Neurons

The lack of protein synthesis machinery in axons and the enormous distances separating the cell body from the axonal and synaptic domains impose a unique set of challenges to neuronal cells. An example of such restraints is the process of transport and delivery of proteins and lipid components along axons, globally referred to as axonal transport (AT). Mature neurons run this AT process of membrane-bounded organelles (MBOs, including mitochondria, synaptic vesi-cles, and plasma membrane components) from their sites of synthesis in the neuronal cell body to their final destination in axons mainly by means of the microtubule-based motor protein conventional kinesin. On the other hand, the multi-subunit motor cytoplasmic dynein works by translocating signaling endo-somes, multivesicular bodies, and lysosomes from axons back to the neuronal cell body [30].

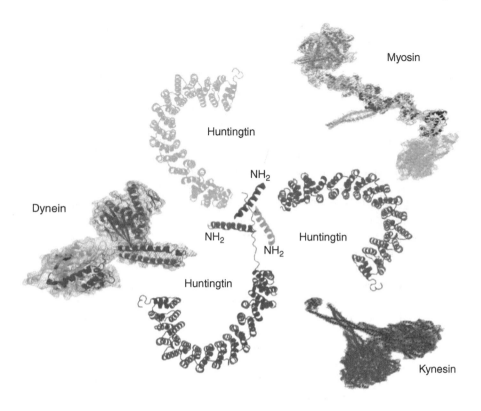

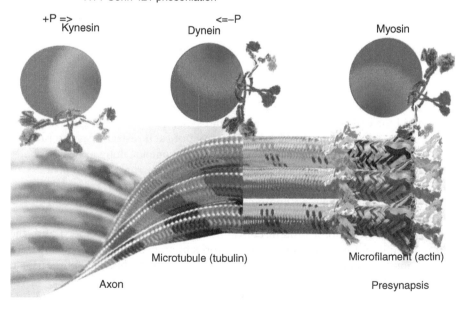

Neuronal function and axonal maintenance crucially depend on AT, since loss of function mutations in these molecular motors have been linked to various neurological diseases [31]. In this regard, pathological studies have significantly shown these diseases to have "dying-back" pattern of neuronal degeneration, which suggests that when AT is altered it might also bring about a critical pathogenic event in HD [32]. Evidence of AT deficits in HD has surfaced in multiple independent studies as well as in other polyQ-expansion diseases [33]. Mutant-HTT has first been found to inhibit AT, whether or not the neuronal cell body is altered, according to studies conducted in isolated squid axoplasm [34]. Subsequently, reductions in AT and accumulation of axonal vesicle cargos in association with polyQ-HTT expression have been experimentally observed in cultured cells [35] and *Drosophila* neurons [36]. These findings, in addition to the absolute reliance of neurons on AT and the "dying-back" pattern of neuronal degeneration seen in HD, contribute to the suggestion that AT deficits might play a role as a major pathogenic event regarding the increased vulnerability of projection neurons in this disease [37]. Various HD experimental models have shown that AT is reduced, which leads to the question of how polyQ-Htt inhibits AT. An attempt to account for this fact is that polyQ-HTT inhibits AT by promoting alterations in the activity of kinases involved in the phosphorylation of molecular motor proteins. This is supported by the fact that some well-established HD features are aberrant patterns of proteins phosphorylation including neurofilaments [38] and synapsin, as well as increased activation of kinases [39]. Polyglutamine-expanded huntingtin, when expressed, activates the SEK1-JNK pathway and induces apoptosis in a hippocampal neuronal cell line. Along these lines, polyQ-HTT has been further observed to inhibit AT through a mechanism involving activation of the JNK pathway and phosphorylation of the molecular motor proteins, in studies conducted in isolated squid axoplasm and a knock-in mouse model. Three JNK isoforms exist in mammals (JNK1, JNK2, and

---

Fig. 6.1 The upper section of the figure shows a scheme made with PyMol software from the crystallography of kinesin (Kinesin (Dimeric) From Rattus Norvegicus [Motor Protein] MMDB ID: 8768, PDB ID:3KIN), dynein (A Functional Full-Length Dynein Motor Domain from Dictyostelium discoideum. MMDB ID: 103417, PDB ID: 3VKH), and myosin VI (The Heavy Meromyosin Subfragment of Chick Smooth Muscle Myosin with Regulatory Light Chain from Gallus gallus. MMDB ID: 95020 PDB ID: 3J04.) structure. The Huntingtin molecule is represented by the PP2A regulatory subunit (Regulatory Domain Of Human Pp2a, Pr65alpha [Scaffold Protein] MMDB ID: 10153 PDB ID: 1B3U). The figure in the center shows the structure of the trimer formed by the huntingtin end in a crystal obtained in the context of another bacterial protein (Huntingtin Amino-Terminal Region with 17 Gln Residues – Crystal C99 [Signaling Protein] *Escherichia coli* K-12 MMDB ID: 77069 PDB ID: 3IOV). On the other hand, this amino-terminal end is known to also be able to form dimers [222] which are of greater affinity if one of the peptides contains expansions [223].

The lower part of the scheme illustrates the need of huntingtin so that motor proteins can be assembled, and the vesicle is transported to its subcellular destination. Microtubules are represented in the left end, and the actin filaments present in the terminals seem to have a smaller caliber in the synapses. Both microtubules and actin are on both sides of the synapses, and therefore the scheme is valid for axons and for dendrites. The phosphorylation of huntingtin on locus 421 reflects the direction of vesicle transport, and it is represented with *arrows*

JNK3) [40]. The effect by which polyQ-HTT inhibits AT is mediated by neuron-specific JNK3. Consequently, as JNK3 becomes more active in HD, AT would be more reduced in various MBO cargoes, which results in deficits in synaptic and axonal function and maintenance [41]. The more intensely active JNK pathway in HD also falls in line with the changes in gene transcription and activation of apoptotic pathways widely reported in association with polyQ-HTT expression [42]. However, how mHTT leads to the activation of JNK3 is unclear at present. Findings regarding a molecular basis underneath the increased vulnerability of MSNs and cortical projection neurons to polyQ-Htt-induced alterations in AT are currently unavailable, but they might reflect unique AT specializations of these neuronal cell types [35]. Mechanisms which can prevent axonal loss via phosphorylation of activated MLK or JNK kinases are under active investigation as potential therapeutics at CHDI.

Therapeutics Aimed at Decreasing HTT Expression

Decreasing expression of mHTT should slow disease progression, or reverse dysfunction, as evidenced by data obtained in mouse models of HD [14, 15, 43]. In hypomorphic mice, in mice with inducible expression of mutant HTT, or expressing inducible shRNAs targeting HTT, greater than 70 % knockdown of both alleles is toxic in embryonic and adult contexts. However, additional studies in adult mice and in nonhuman primates (NHPs) locally administered with HTT-lowering therapeutics have demonstrated that the reduction of both alleles to 50 % levels appears to be well tolerated [43–46]. Given that the function of HTT is unclear and that it lacks domains amenable to modulation by small-molecule therapeutics, the main approach to slow disease progression has centered on decreasing HTT expression, and four clinical approaches are in late stages of preclinical development. Lowering of HTT mRNA, mostly in a non-allele selective manner, is being pursued via a variety of strategies, including antisense oligonucleotides (ASOs; ISIS Pharmaceuticals and Roche), small interfering RNAs (siRNAs; Alnylam and Medtronic), or micro RNAs (miRNAs; Genzyme and Sanofi-Aventis) [45, 47, 48]. In addition, a DNA-targeting strategy is being developed by Sangamo and Shire, via the development of CAG-directed zinc-finger repressor (ZFP) proteins. At CHDI, our efforts currently are directed towards the development of HTT-dependent biomarkers, needed to assess early alterations after HTT suppression which can be applied in a clinical setting.

All current RNA-directed therapeutics target regions of the mRNA that are downstream of the pathogenic exon 1 sequences. A recent study shows that a mis-spliced form of HTT message is transcribed in animals and in human tissues, and is predicted to generate a truncated amino-terminal fragment of HTT, with presumed enhanced toxicity [49]. This alternatively spliced HTT mRNA would not be modulated by existing RNA-directed therapeutics, which could therefore limit their effectiveness if indeed this mRNA is important in disease pathogenesis. Similarly, second-generation RNA therapies targeting the expanded CAG tract would also target this readthrough spliced isoform of HTT. In addition to this readthrough-generated variant of HTT, there is evidence that some of the toxicity of expanded trinucleotide repeats emanates from expression of the mRNAs in Drosophila [50, 51], a molecular event which would presumably not be modulated by some of the RNA-directed therapeutics. To address these challenges of detecting mechanistic or biological changes when total or mHTT is knocked down in clinical studies, at CHDI we are exploring endpoints that will assess lowered HTT levels, or an evaluation of biological endpoints proximal to HTT itself.

For instance, one of the most conserved alterations in HD patients and HD animal models is the downregulation of several neurotransmitter receptors and enzymes critical for basal ganglia function, such as the dopamine signaling components (D1, D2 receptors, and the DAT transporter) [52, 53], the cannabinoid receptor CB1[54], the enzyme PDE10 [55–57], and the metabotropic glutamate receptor (mGluR5) [58], all of which can be visualized with existing, clinically used PET or SPECT tracers. Recent evidence suggests that lowering mHTT leads to an increase in the expression of these genes, which can presumably be monitored by imaging agents during clinical trials. Additional technologies under consideration include measurements using FDG-PET, qEEG, or fMRI measurements [59–62].

Therapeutics Targeting Neurotransmitter Pathways in HD

Alterations in HD synaptic function may be responsible for many of the early symptoms of HD. Normal HTT interacts with various cytoskeletal and synaptic vesicle proteins essential for exocytosis and endocytosis [28, 63], and altered interactions of mHTT likely contributes to abnormal synaptic transmission in HD. Synaptic alterations in corticostriatal transmission have been well documented and are one of the earliest changes detected in rodent HD models [64–66]. Although initial reports focused on the disturbances and cell death of striatal medium spiny neurons (MSNs), it is now apparent that significant alterations to striatal interneurons exist that are likely to play a very significant role in synaptic pathogenesis [60, 67–70]. Anti-dopaminergic drugs such as tetrabenazine and antipsychotics are still the most frequent therapy in the management of chorea, dyskinesias, and some psychiatric symptoms in HD [71–73]. For additional information in regards to current medications for the treatment of HD symptoms, refer to chapter by Dr. Orth. While improving chorea, tetrabenazine leads to numerous side effects, including insomnia, somnolence and depressed mood, sedation, anxiety, parkinsonism, fatigue, akathisia, drowsiness, and potential cognitive impairment during the chronic treatment phase [71]. ACR16 targets the dopamine D2 receptor and is currently in clinical development for treatment of HD [74]. In vitro the drug behaves as a low affinity, low efficacy partial agonist with fast dissociation properties [75, 76]. In vivo the drug is described as a "dopamine stabilizer" since in rodents it normalizes hyperdopaminergic states and reduces hyperlocomotion induced by amphetamine but preserves spontaneous locomotor activity and has stimulatory action in the habituated states where the dopaminergic tone is considered to be low [77, 78]. ACR16's efficacy and safety was evaluated in a small phase II [74] and a phase III randomized, double-blind, placebo-controlled MERMAiHD study [79, 80]. ACR16 was well tolerated up to 90 mg/day with a safety profile similar to placebo and, based on the phase II study indicating that ACR16 improved voluntary motor function, efficacy was evaluated in phase III using the modified motor score as the primary endpoint. This endpoint was not met. Currently, ACR16 efficacy is being evaluated by Teva Pharmaceuticals in an open label study (OPEN-HART, ClinicalTrials.gov Identifier NCT01306929) where motor function is evaluated using the UHDRS Total Motor Score as the primary endpoint.

Glutamate is the major neurotransmitter in the corticostriatal, thalamocortical, and subthalamic nucleus efferent systems [53]. A leading hypothesis in HD is that increased levels of glutamatergic neurotransmission in the striatum are neurotoxic for MSN neurons. This theory is termed the excitotoxicity hypothesis of HD and has been a major driver in the development of HD therapeutic strategies in the past [53, 81, 82]. A major hurdle in developing glutamatergic-targeted HD therapeutics is a lack of understanding of glutamatergic neurotransmission in HD. Evidence for a glutamatergic signaling impairment in HD stems from the downregulation of glial glutamate uptake [83, 84] and glutamate transporter GLUT1 (EAAT2) expression [84, 85]. These findings suggest dysregulation of extracellular glutamate levels in HD and point to GLUT1 as a potential therapeutic target. Indeed, in the R6/2 mouse model, chronic treatment with the antibiotic ceftriaxone upregulates GLUT1 expression and improves glutamate uptake [84, 86]. Antiglutamatergic therapies, such as the antagonism of NMDA receptors, are being explored clinically using amantadine, remacemide, and memantine (A randomized, placebo-controlled trial of coenzyme Q10 and remacemide in Huntington's disease 2001). The antiglutamate release agent riluzole failed to modulate HD symptoms in a clinical study [87]. Other approaches that might be explored therapeutically include the modulation of NR2B, mGluR2/3, and mGluR5 signaling [58, 88–92]. Novartis sponsored a phase II trial in HD with their mGluR5 antagonist AFQ056, although this study was terminated (ClinicalTrials.gov identifier NCT01019473).

The importance of endocannabinoid signaling in the basal ganglia circuitry is well known [93, 94]. Imaging and histological studies have highlighted a decrease of CB1 expression in the lateral and medial striatum of HD patients (although not in the cortex), and in HD rodent models there is a loss of CB1 expression in MSNs and a subset of interneurons of the striatum, suggesting that altered endocannabinoid signaling might be relevant to the psychiatric, cognitive, and motor symptoms of HD [68, 95–102].

Although the effects of cannabinoid administration in HD patients have not been rigorously explored clinically, a recent double-blind randomized, crossover study with 22 patients evaluated the effects of nabilone, a synthetic analog of THC, in HD symptoms. The study, not sufficiently powered to assess efficacy, demonstrated that nabilone has an acceptable safety profile and significantly improved psychiatric endpoints, as well as a marginal improvement in the chorea subscore of the UHDRS [103]. A second study using two natural cannabinoids is underway (Table 6.1), although the results from this study have not yet been reported. Given the alterations in this signaling mechanism and its importance in basal ganglia function, the role of this neuromodulator in HD should be further investigated.

The adenosine receptor 2a (A2a) has also been shown to be downregulated prior to HD symptom onset. Despite the loss of expression of A2a, its signaling appears to be preserved in HD models [99, 104–108]. A2a receptors are enriched in indirect-pathway striatopallidal MSNs, where they modulate dopamine D2 receptor signaling via the activation of cAMP. A2a signaling mechanisms are complex, tied to the pre- versus postsynaptic localization of the receptor, where it forms heteromeric complexes with other receptors, such as D2 and mGluR5 in the postsynaptic site, or adenosine A1 receptors in the presynaptic compartment [106, 108, 109]. In HD mouse models, recent reports suggest that both the inhibition and activation of A2a receptors are beneficial [104, 108, 110–112]. Several A2a antagonists have been developed for the treatment of Parkinson's disease. Clinical development of these molecules was stopped due to lack of efficacy or adverse side effects including dyskinesias, which is a significant concern for HD patients [113, 114].

Modulation of Phosphodiesterases (PDES)

Phosphodiesterases (PDEs) catalyze the degradation of cyclic nucleotides, and their inhibition leads to sustained signaling mediated by increased levels of cAMP, cGMP, or both. Intracellular cyclic nucleotide signaling plays a fundamental role in synaptic transmission, plasticity, and gene regulation [115–117]. Specifically, signaling pathways downstream of cAMP elevation are deregulated in HD animal models and in HD patient postmortem samples [57, 118–126]. Administration of rolipram (a PDE4 selective inhibitor) and TP-10 (a selective PDE10 inhibitor) to HD mouse models was reported to delay disease progression [119, 121, 127, 128]. PDE4 is cAMP specific, whereas PDE10 modulates signaling by both cAMP and cGMP [117, 129–131]. Interestingly, PDE10 is one of the earliest and most significantly differentially downregulated transcripts in most HD animal models and in HD patient postmortem samples [121, 132–135]. CHDI and collaborators have monitored expression of PDE10 in HD patients through the use of a specific PDE10 PET tracer, confirming the findings identified in HD rodent models (unpublished observations). As PDE10 is expressed in MSNs of the striatum, it is possible that loss of PDE10 expression is a compensatory mechanism downstream of synaptic alterations in the basal ganglia [133, 135].

Studies conducted in rodent, NHPs, and humans with rolipram and other nonselective PDE4 inhibitors have highlighted the pro-cognitive effects in a variety of tasks involving both the hippocampal and the corticostriatal systems, and this class of molecules should be investigated in HD [116, 136–138]. Our work with selective PDE inhibitors against PDE2, PDE9, and PDE10 obtained from Pfizer showed that inhibition of these enzymes, presumably via a modulation of cGMP levels intracellularly, can rescue the synaptic deficits characterized in HD mouse models [53, 64] either in acute in vitro studies using corticostriatal and hippocampal slices or after subchronic dosing studies, highlighting their potential for the treatment of HD (data unpublished). In addition, PDE10 inhibition rescues the alterations observed in subthalamic neuron (STN) firing in vivo, suggesting that the modulation of PDE10 activity can profoundly affect the circuitry in ways that are consistent with a potential beneficial clinical effect. We, in collaboration with Pfizer, are continuing to explore the preclinical and clinical efficacy of PDE9 and PDE10 inhibitors. The first phase II safety and tolerability study using MP-10 will take place in the near future (Clinicaltrials.gov #NCT01806896). Other companies, such as Omeros, are also pursuing PDE10 inhibitors for HD.

Modulation of BDNF/TRKB Signaling

The BDNF neurotrophin pathway has been extensively associated with HD pathology since it was shown that TrkB (the tyrosine kinase receptor for BDNF) is required for proper maturation and homeostasis of the striatum in rodents and that BDNF levels are decreased in HD rodent models. The effect of BDNF signaling on synaptic plasticity and axonal transport in MSNs reflects a critical role for this pathway in HD pathophysiology [88, 139–159]. From a therapeutic perspective, activation of TrkB selectively in the basal ganglia might be disease modifying, based on studies in HD rodent models. The key question is how best to modulate TrkB signaling while minimizing potential adverse effects associated with BDNF signaling through either the p75 receptor or in regions outside of the basal ganglia associated with weight loss (hypothalamus), seizure activity (hippocampus), and/or pain (amygdala) [160–163]. Although there have been several reports of small-molecule "mimetics" of BDNF and cyclic peptides presumed to activate TrkB receptors, we have not been able to reproduce any of these findings utilizing a comprehensive set of cellular assays, including TrkB dimerization, phosphorylation, and signaling via ERK and pI3K pathways, calling into question whether the reported effects truly reflect a direct modulation of the TrkB signaling pathway [150, 155, 164–167]. At CHDI we are exploring the efficacy of TrkB monoclonal antibodies which confer selectivity over other trk receptors and p75.

Small Molecules Targeting Cellular Energetic Mechanisms

Energetic alterations are commonly described in every neurodegenerative condition, including HD. In HD animal models and in premanifest and manifest humans, there appear to be signs of metabolic and energetic disturbances, both in muscle and brain [64, 168–174]. In general, the findings obtained in HD patients and HD rodent models are not always consistent, although this is likely to be confounded by tissue measurements as opposed to the measurement of specific cell populations during disease progression. Glycolytic and mitochondrial energetic mechanisms in vulnerable neuronal and glial populations have not been rigorously investigated. In humans, the most notable findings are alterations in creatine/phosphocreatine levels quantified by magnetic resonance spectroscopy (MRS), although these studies involved few patients and did not follow them longitudinally [171, 174].

Evidence of abnormal energy utilization has come from monitoring energetic endpoints and lactate production in muscle during an exercise test in 25 HD patients. This small study showed elevated lactate levels and lower anaerobic threshold in HD subjects, for manifest and premanifest phases of disease [175]. Other studies have revealed variable effects in brain lactate levels and altered glutamate/glutamine ratios in MRS studies [171–174, 176]. Creatine kinase (CK) deficiency has been linked to HD, and supplementation of brain isoforms of CK shows positive effects in rodent models of HD [177–179].

Several interventional clinical trials are underway to probe the hypothesis that elevated reactive oxygen species, or deficits in energy homeostasis, contribute to HD progression. The clinical studies include the evaluation of cysteamine (CYTE-I-HD study), creatine (CREST-E study), ethyl-EPA (TREND-HD study), and coenzyme Q10 (2CARE study; see Table 6.1). Most of these studies, in spite of conflicting findings from rodent studies [180–184], have not yielded any strong evidence for efficacy in clinical trials. In phase II, positive results have been documented for the studies involving CoQ10, ethyl-EPA, and creatine, but these are likely underpowered. Other studies have been completed to assess tolerability and safety in HD [173, 185–191]. Unfortunately, most clinical studies did not incorporate pharmacodynamic or energetic measures pre- and post treatment, giving no definitive reason for the negative findings in pivotal trials. Therefore, the mechanisms purported to be modulated by these drugs remain untested.

In terms of novel potential therapeutics aimed at these mechanisms in preclinical stages, the pathways that stand out are the AMPK, PPARγ, and PGC1α pathways [69, 168, 192–198]. As PGC1α itself is not amenable to pharmacological modulation, other approaches have been explored to activate its transcriptional activity, through the use of small-molecule agonists of the nuclear hormone receptor PPARγ, such as rosiglitazone [197], or via the modulation of AMPK or sirtuin signaling [168, 192, 196, 199–201]. Finally, other approaches targeting energetic mechanisms are in early preclinical stages in HD research, including the antioxidant response system mediated by the Nrf2/Keap complex [202].

Small Molecules Targeting Lysine Deacetylases

The HDACs (histone deacetylases; herein referred to as lysine deacetylases) are composed of 11 enzymes, class I (1–3 and 8), class IIa (4, 5, 7, and 9), class IIb (6, 10), class III (sirtuins 1–6), and class IV (HDAC 11) [196]. Nonselective HDAC inhibitors such as SAHA, sodium butyrate, and TSA have been reported to ameliorate disease phenotypes in HD models. Subsequent work has identified a subset of HDACs whose modulation modifies disease progression in a variety of HD models, such as HDAC-2, HDAC-3, and HDAC-4 [56, 203–210]. Both SAHA and TSA are broad-spectrum inhibitors of class I and are weak class IIa and III inhibitors. Because of the initial encouraging findings of nonselective class I HDAC inhibitors in HD models, a large effort was undertaken to identify the HDAC subtype responsible for the beneficial effects of these inhibitors. Although SAHA and TSA administration showed promising results in HD models, their therapeutic index is low, and significant side effects were observed during dose-escalation studies, which limited their therapeutic potential.

Recently, the genetic deletion of one allele of HDAC4 was shown to ameliorate disease progression in the R6.2 model of HD, the only HDAC thus far evaluated genetically to have such a profound effect in disease progression [203, 211–213]. HDAC4 heterozygote animals crossed with R6/2 mice demonstrated a significant delay in progression in a variety of endpoints, including survival, locomotor endpoints, synaptic dysfunction in hippocampal and corticostriatal acute slice assays, HTT aggregation, and a rescue of a subset of cortical transcriptional alterations [203]. Support for the role of HDAC4 in mediating HTT toxicity has also been obtained in cellular systems of HD by other investigators [214]. Based on the genetic findings described above, CHDI developed potent, selective class IIa HDAC inhibitors with suitable PK properties for chronic dosing in HD models, which are currently being tested. Our lead molecules display 1,000-fold selectivity over class I and III enzymes, and a subset of the molecules can disrupt an HDAC4-3 complex in a cellular context. These inhibitors bind to the active site of class II HDACs as demonstrated by crystallography studies [215].

Class III lysine deacetylases of the sirtuin family have received considerable interest given their role in aging, metabolism, and DNA damage mechanisms [130, 216]. Sirtuins are NAD+ dependent lysine deacetylases, and it is the requirement for NAD+ binding for catalysis that explains their involvement in energy-sensing mechanisms, including the DNA damage response and energetic mechanisms. In the context of HD, the enzymes SIRT1 and SIRT2 have been implicated in various biochemical pathways thought to be important for disease progression; these include the p53 signaling response in the context of DNA damage and apoptosis, and the mitochondrial deficits attributed to the loss of PGC1α [130, 194, 198, 199, 216, 217] signaling. For SIRT1 modulation, both inhibition and activation have been reported to affect disease phenotypes; in the case of SIRT2, findings in genetic deletion studies contrast with the results obtained with small-molecule inhibitors of SIRT1 and SIRT2 [194, 198, 212, 218].

Selisistat (EX-527), a SIRT1,2 inhibitor developed by Elixir Pharmaceuticals and Siena Biotech, entered clinical trials for HD in 2010, on the hypothesis that SIRT1 inhibition modulates HTT levels. The outcome of this study has not yet been revealed, but results are expected soon. The development of "sirtuin activators" remains an area of active investigation, and one fraught with controversy, given the artifactual findings with resveratrol and several Sirtris molecules in their ability to modulate SIRT activity directly [219–221]. Nonetheless, new molecules are making their way through medicinal chemistry efforts, and their potential for the treatment for HD will have to be tested [221].

References

1. Zuccato C, Valenza M, Cattaneo E. Molecular mechanisms and potential therapeutical targets in Huntington's disease. Physiol Rev. 2010;90(3):905–81.
2. The Huntington's Disease Collaborative Research Group. A novel gene containing a trinucleotide repeat that is expanded and unstable on Huntington's disease chromosomes. The Huntington's Disease Collaborative Research Group. Cell. 1993;72(6):971–83.
3. Langbehn DR, Hayden MR, Paulsen JS, PREDICT-HD Investigators of the Huntington Study Group. CAG-repeat length and the age of onset in Huntington disease (HD): a review and validation study of statistical approaches. Am J Med Genet B Neuropsychiatr Genet. 2010; 153B(2):397–408.
4. Vonsattel JP, DiFiglia M. Huntington disease. J Neuropathol Exp Neurol. 1998;57(5):369–84.
5. Sanberg PR, Fibiger HC, Mark RF. Body weight and dietary factors in Huntington's disease patients compared with matched controls. Med J Aust. 1981;1(8):407–9.
6. Duff K, Paulsen JS, Beglinger LJ, Langbehn DR, Stout JC, Predict-HD Investigators of the Huntington Study Group. Psychiatric symptoms in Huntington's disease before diagnosis: the predict-HD study. Biol Psychiatry. 2007;62(12):1341–6.
7. Fischbeck KH. Polyglutamine expansion neurodegenerative disease. Brain Res Bull. 2001;56(3–4):161–3.
8. Margolis RL, Ross CA. Expansion explosion: new clues to the pathogenesis of repeat expansion neurodegenerative diseases. Trends Mol Med. 2001;7(11):479–82.
9. Hult S, Schultz K, Soylu R, Petersén A. Hypothalamic and neuroendocrine changes in Huntington's disease. Curr Drug Targets. 2010;11(10):1237–49.
10. Pouladi MA, Morton AJ, Hayden MR. Choosing an animal model for the study of Huntington's disease. Nat Rev Neurosci. 2013;14(10):708–21.
11. Andrade MA, Bork P. HEAT repeats in the Huntington's disease protein. Nat Genet. 1995;11(2):115–6.
12. Wang Z-M, Lashuel HA. Discovery of a novel aggregation domain in the huntingtin protein: implications for the mechanisms of Htt aggregation and toxicity. Angew Chem Int Ed Engl. 2013;52(2):562–7.
13. Becher MW, Kotzuk JA, Sharp AH, Davies SW, Bates GP, Price DL, et al. Intranuclear neuronal inclusions in Huntington's disease and dentatorubral and pallidoluysian atrophy: correlation between the density of inclusions and IT15 CAG triplet repeat length. Neurobiol Dis. 1998;4(6):387–97.
14. Martín-Aparicio E, Yamamoto A, Hernández F, Hen R, Avila J, Lucas JJ. Proteasomal-dependent aggregate reversal and absence of cell death in a conditional mouse model of Huntington's disease. J Neurosci. 2001;21(22):8772–81.
15. Yamamoto A, Lucas JJ, Hen R. Reversal of neuropathology and motor dysfunction in a conditional model of Huntington's disease. Cell. 2000;101(1):57–66.
16. Bhide PG, Day M, Sapp E, Schwarz C, Sheth A, Kim J, et al. Expression of normal and mutant huntingtin in the developing brain. J Neurosci. 1996;16(17):5523–35.
17. Shelbourne PF, Keller-McGandy C, Bi WL, Yoon S-R, Dubeau L, Veitch NJ, et al. Triplet repeat mutation length gains correlate with cell-type specific vulnerability in Huntington disease brain. Hum Mol Genet. 2007;16(10):1133–42.
18. Kennedy L, Evans E, Chen C-M, Craven L, Detloff PJ, Ennis M, et al. Dramatic tissue-specific mutation length increases are an early molecular event in Huntington disease pathogenesis. Hum Mol Genet. 2003;12(24):3359–67.
19. Ishiguro H, Yamada K, Sawada H, Nishii K, Ichino N, Sawada M, et al. Age-dependent and tissue-specific CAG repeat instability occurs in mouse knock-in for a mutant Huntington's disease gene. J Neurosci Res. 2001;65(4):289–97.
20. Dragileva E, Hendricks A, Teed A, Gillis T, Lopez ET, Friedberg EC, et al. Intergenerational and striatal CAG repeat instability in Huntington's disease knock-in mice involve different DNA repair genes. Neurobiol Dis. 2009;33(1):37–47.

21. Wheeler VC, Lebel L-A, Vrbanac V, Teed A, te Riele H, MacDonald ME. Mismatch repair gene Msh2 modifies the timing of early disease in Hdh(Q111) striatum. Hum Mol Genet. 2003;12(3):273–81.
22. Swami M, Hendricks AE, Gillis T, Massood T, Mysore J, Myers RH, et al. Somatic expansion of the Huntington's disease CAG repeat in the brain is associated with an earlier age of disease onset. Hum Mol Genet. 2009;18(16):3039–47.
23. Watase K, Venken KJT, Sun Y, Orr HT, Zoghbi HY. Regional differences of somatic CAG repeat instability do not account for selective neuronal vulnerability in a knock-in mouse model of SCA1. Hum Mol Genet. 2003;12(21):2789–95.
24. Gray M, Shirasaki DI, Cepeda C, André VM, Wilburn B, Lu X-H, et al. Full-length human mutant huntingtin with a stable polyglutamine repeat can elicit progressive and selective neuropathogenesis in BACHD mice. J Neurosci. 2008;28(24):6182–95.
25. Pal A, Severin F, Lommer B, Shevchenko A, Zerial M. Huntingtin-HAP40 complex is a novel Rab5 effector that regulates early endosome motility and is up-regulated in Huntington's disease. J Cell Biol. 2006;172(4):605–18.
26. Sahlender DA, Roberts RC, Arden SD, Spudich G, Taylor MJ, Luzio JP, et al. Optineurin links myosin VI to the Golgi complex and is involved in Golgi organization and exocytosis. J Cell Biol. 2005;169(2):285–95.
27. Caviston JP, Ross JL, Antony SM, Tokito M, Holzbaur ELF. Huntingtin facilitates dynein/dynactin-mediated vesicle transport. Proc Natl Acad Sci U S A. 2007;104(24): 10045–50.
28. Caviston JP, Holzbaur ELF. Huntingtin as an essential integrator of intracellular vesicular trafficking. Trends Cell Biol. 2009;19(4):147–55.
29. Colin E, Zala D, Liot G, Rangone H, Borrell-Pagès M, Li X-J, et al. Huntingtin phosphorylation acts as a molecular switch for anterograde/retrograde transport in neurons. EMBO J. 2008;27(15):2124–34.
30. Morfini G. Axonal transport. In: Siegel GJ, Albers RW, Brady ST, Price DL, editors. Basic neurochemistry: molecular, cellular, and medical aspects. San Diego: Elsevier Academic Press; 2006. p. 485–502.
31. Roy S, Zhang B, Lee VM-Y, Trojanowski JQ. Axonal transport defects: a common theme in neurodegenerative diseases. Acta Neuropathol. 2005;109(1):5–13.
32. Han I, You Y, Kordower JH, Brady ST, Morfini GA. Differential vulnerability of neurons in Huntington's disease: the role of cell type-specific features. J Neurochem. 2010;113(5): 1073–91.
33. Morfini G, Pigino G, Szebenyi G, You Y, Pollema S, Brady ST. JNK mediates pathogenic effects of polyglutamine-expanded androgen receptor on fast axonal transport. Nat Neurosci. 2006;9(7):907–16.
34. Szebenyi G, Morfini GA, Babcock A, Gould M, Selkoe K, Stenoien DL, et al. Neuropathogenic forms of huntingtin and androgen receptor inhibit fast axonal transport. Neuron. 2003; 40(1):41–52.
35. Her L-S, Goldstein LSB. Enhanced sensitivity of striatal neurons to axonal transport defects induced by mutant huntingtin. J Neurosci. 2008;28(50):13662–72.
36. Sinadinos C, Burbidge-King T, Soh D, Thompson LM, Marsh JL, Wyttenbach A, et al. Live axonal transport disruption by mutant huntingtin fragments in Drosophila motor neuron axons. Neurobiol Dis. 2009;34(2):389–95.
37. Morfini GA, You Y-M, Pollema SL, Kaminska A, Liu K, Yoshioka K, et al. Pathogenic huntingtin inhibits fast axonal transport by activating JNK3 and phosphorylating kinesin. Nat Neurosci. 2009;12(7):864–71.
38. DiProspero NA, Chen E-Y, Charles V, Plomann M, Kordower JH, Tagle DA. Early changes in Huntington's disease patient brains involve alterations in cytoskeletal and synaptic elements. J Neurocytol. 2004;33(5):517–33.
39. Apostol BL, Illes K, Pallos J, Bodai L, Wu J, Strand A, et al. Mutant huntingtin alters MAPK signaling pathways in PC12 and striatal cells: ERK1/2 protects against mutant huntingtin-associated toxicity. Hum Mol Genet. 2006;15(2):273–85.

40. Brecht S, Kirchhof R, Chromik A, Willesen M, Nicolaus T, Raivich G, et al. Specific pathophysiological functions of JNK isoforms in the brain. Eur J Neurosci. 2005; 21(2):363–77.
41. Yang DD, Kuan CY, Whitmarsh AJ, Rincón M, Zheng TS, Davis RJ, et al. Absence of excitotoxicity-induced apoptosis in the hippocampus of mice lacking the Jnk3 gene. Nature. 1997;389(6653):865–70.
42. Merienne K, Helmlinger D, Perkin GR, Devys D, Trottier Y. Polyglutamine expansion induces a protein-damaging stress connecting heat shock protein 70 to the JNK pathway. J Biol Chem. 2003;278(19):16957–67.
43. Kordasiewicz HB, Stanek LM, Wancewicz EV, Mazur C, McAlonis MM, Pytel KA, et al. Sustained therapeutic reversal of Huntington's disease by transient repression of huntingtin synthesis. Neuron. 2012;74(6):1031–44.
44. Hilditch-Maguire P, Trettel F, Passani LA, Auerbach A, Persichetti F, MacDonald ME. Huntingtin: an iron-regulated protein essential for normal nuclear and perinuclear organelles. Hum Mol Genet. 2000;9(19):2789–97.
45. Grondin R, Kaytor MD, Ai Y, Nelson PT, Thakker DR, Heisel J, et al. Six-month partial suppression of Huntingtin is well tolerated in the adult rhesus striatum. Brain. 2012;135 (Pt 4):1197–209.
46. Woda JM, Calzonetti T, Hilditch-Maguire P, Duyao MP, Conlon RA, MacDonald ME. Inactivation of the Huntington's disease gene (Hdh) impairs anterior streak formation and early patterning of the mouse embryo. BMC Dev Biol. 2005;5:17.
47. Sah DWY, Aronin N. Oligonucleotide therapeutic approaches for Huntington disease. J Clin Invest. 2011;121(2):500–7.
48. Stiles DK, Zhang Z, Ge P, Nelson B, Grondin R, Ai Y, et al. Widespread suppression of huntingtin with convection-enhanced delivery of siRNA. Exp Neurol. 2012;233(1):463–71.
49. Sathasivam K, Neueder A, Gipson TA, Landles C, Benjamin AC, Bondulich MK, et al. Aberrant splicing of HTT generates the pathogenic exon 1 protein in Huntington's disease. Proc Natl Acad Sci U S A. 2013;110(6):2366–7.
50. Galka-Marciniak P, Urbanek MO, Krzyzosiak WJ. Triplet repeats in transcripts: structural insights into RNA toxicity. Biol Chem. 2012;393(11):1299–315.
51. Li L-B, Yu Z, Teng X, Bonini NM. RNA toxicity is a component of ataxin-3 degeneration in Drosophila. Nature. 2008;453(7198):1107–11.
52. Brandt J, Folstein SE, Wong DF, Links J, Dannals RF, McDonnell-Sill A, et al. D2 receptors in Huntington's disease: positron emission tomography findings and clinical correlates. J Neuropsychiatry Clin Neurosci. 1990;2(1):20–7.
53. André VM, Cepeda C, Levine MS. Dopamine and glutamate in Huntington's disease: a balancing act. CNS Neurosci Ther. 2010;16(3):163–78.
54. Van Laere K, Casteels C, Dhollander I, Goffin K, Grachev I, Bormans G, et al. Widespread decrease of type 1 cannabinoid receptor availability in Huntington disease in vivo. J Nucl Med. 2010;51(9):1413–7.
55. Giampà C, Laurenti D, Anzilotti S, Bernardi G, Menniti FS, Fusco FR. Inhibition of the striatal specific phosphodiesterase PDE10A ameliorates striatal and cortical pathology in R6/2 mouse model of Huntington's disease. PLoS One. 2010;5(10):e13417.
56. Sadri-Vakili G, Bouzou B, Benn CL, Kim M-O, Chawla P, Overland RP, et al. Histones associated with downregulated genes are hypo-acetylated in Huntington's disease models. Hum Mol Genet. 2007;16(11):1293–306.
57. Sugars KL, Brown R, Cook LJ, Swartz J, Rubinsztein DC. Decreased cAMP response element-mediated transcription: an early event in exon 1 and full-length cell models of Huntington's disease that contributes to polyglutamine pathogenesis. J Biol Chem. 2004;279(6):4988–99.
58. Ribeiro FM, Paquet M, Ferreira LT, Cregan T, Swan P, Cregan SP, et al. Metabotropic glutamate receptor-mediated cell signaling pathways are altered in a mouse model of Huntington's disease. J Neurosci. 2010;30(1):316–24.
59. Hunter A, Bordelon Y, Cook I, Leuchter A. QEEG measures in Huntington's disease: a pilot study. PLoS Curr. 2010;2, RRN1192.

60. Eidelberg D, Surmeier DJ. Brain networks in Huntington disease. J Clin Invest. 2011; 121(2):484–92.
61. Gray MA, Egan GF, Ando A, Churchyard A, Chua P, Stout JC, et al. Prefrontal activity in Huntington's disease reflects cognitive and neuropsychiatric disturbances: the IMAGE-HD study. Exp Neurol. 2013;239:218–28.
62. Wolf RC, Thomann PA, Thomann AK, Vasic N, Wolf ND, Landwehrmeyer GB, et al. Brain structure in preclinical Huntington's disease: a multi-method approach. Neurodegener Dis. 2013;12(1):13–22.
63. Qin Z-H, Wang Y, Sapp E, Cuiffo B, Wanker E, Hayden MR, et al. Huntingtin bodies sequester vesicle-associated proteins by a polyproline-dependent interaction. J Neurosci. 2004;24(1):269–81.
64. Heikkinen T, Lehtimäki K, Vartiainen N, Puolivali J, Hendricks SJ, Glaser JR, et al. Characterization of neurophysiological and behavioral changes, MRI brain volumetry and 1H MRS in zQ175 knock-in mouse model of Huntington's disease. PLoS One. 2012;7(12):e50717.
65. Figiel M, Szlachcic WJ, Switonski PM, Gabka A, Krzyzosiak WJ. Mouse models of polyglutamine diseases: review and data table. Part I. Mol Neurobiol. 2012;46(2):393–429.
66. Switonski PM, Szlachcic WJ, Gabka A, Krzyzosiak WJ, Figiel M. Mouse models of polyglutamine diseases in therapeutic approaches: review and data table. Part II. Mol Neurobiol. 2012;46(2):430–66.
67. Cepeda C, Galvan L, Holley SM, Rao SP, André VM, Botelho EP, et al. Multiple sources of striatal inhibition are differentially affected in Huntington's disease mouse models. J Neurosci. 2013;33(17):7393–406.
68. Horne EA, Coy J, Swinney K, Fung S, Cherry AET, Marrs WR, et al. Downregulation of cannabinoid receptor 1 from neuropeptide Y interneurons in the basal ganglia of patients with Huntington's disease and mouse models. Eur J Neurosci. 2013;37(3):429–40.
69. Vlamings R, Benazzouz A, Chetrit J, Janssen MLF, Kozan R, Visser-Vandewalle V, et al. Metabolic and electrophysiological changes in the basal ganglia of transgenic Huntington's disease rats. Neurobiol Dis. 2012;48(3):488–94.
70. Pisani A, Bernardi G, Ding J, Surmeier DJ. Re-emergence of striatal cholinergic interneurons in movement disorders. Trends Neurosci. 2007;30(10):545–53.
71. Frank S. Tetrabenazine: the first approved drug for the treatment of chorea in US patients with Huntington disease. Neuropsychiatr Dis Treat. 2010;6:657–65.
72. Ross CA, Tabrizi SJ. Huntington's disease: from molecular pathogenesis to clinical treatment. Lancet Neurol. 2011;10(1):83–98.
73. Chen JJ, Ondo WG, Dashtipour K, Swope DM. Tetrabenazine for the treatment of hyperkinetic movement disorders: a review of the literature. Clin Ther. 2012;34(7):1487–504.
74. Lundin A, Dietrichs E, Haghighi S, Göller M-L, Heiberg A, Loutfi G, et al. Efficacy and safety of the dopaminergic stabilizer Pridopidine (ACR16) in patients with Huntington's disease. Clin Neuropharmacol. 2010;33(5):260–4.
75. Dyhring T, Nielsen EØ, Sonesson C, Pettersson F, Karlsson J, Svensson P, et al. The dopaminergic stabilizers pridopidine (ACR16) and (-)-OSU6162 display dopamine D(2) receptor antagonism and fast receptor dissociation properties. Eur J Pharmacol. 2010;628(1–3):19–26.
76. Kara E, Lin H, Svensson K, Johansson AM, Strange PG. Analysis of the actions of the novel dopamine receptor-directed compounds (S)-OSU6162 and ACR16 at the D2 dopamine receptor. Br J Pharmacol. 2010;161(6):1343–50.
77. Ponten H, Kullingsjö J, Lagerkvist S, Martin P, Pettersson F, Sonesson C, et al. In vivo pharmacology of the dopaminergic stabilizer pridopidine. Eur J Pharmacol. 2010;644(1–3):88–95.
78. Rung JP, Rung E, Helgeson L, Johansson AM, Svensson K, Carlsson A, et al. Effects of (-)-OSU6162 and ACR16 on motor activity in rats, indicating a unique mechanism of dopaminergic stabilization. J Neural Transm (Vienna Austria 1996). 2008;115(6):899–908.
79. De Yebenes JG, Landwehrmeyer B, Squitieri F, Reilmann R, Rosser A, Barker RA, et al. Pridopidine for the treatment of motor function in patients with Huntington's disease (MermaiHD): a phase 3, randomised, double-blind, placebo-controlled trial. Lancet Neurol. 2011;10(12):1049–57.

80. Squitieri F, Landwehrmeyer B, Reilmann R, Rosser A, de Yebenes JG, Prang A, et al. One-year safety and tolerability profile of pridopidine in patients with Huntington disease. Neurology. 2013;80(12):1086–94.
81. Munoz-Sanjuan I, Bates GP. The importance of integrating basic and clinical research toward the development of new therapies for Huntington disease. J Clin Invest. 2011;121(2): 476–83.
82. Venuto CS, McGarry A, Ma Q, Kieburtz K. Pharmacologic approaches to the treatment of Huntington's disease. Mov Disord. 2012;27(1):31–41.
83. Hassel B, Tessler S, Faull RLM, Emson PC. Glutamate uptake is reduced in prefrontal cortex in Huntington's disease. Neurochem Res. 2008;33(2):232–7.
84. Miller BR, Dorner JL, Bunner KD, Gaither TW, Klein EL, Barton SJ, et al. Up-regulation of GLT1 reverses the deficit in cortically evoked striatal ascorbate efflux in the R6/2 mouse model of Huntington's disease. J Neurochem. 2012;121(4):629–38.
85. Faideau M, Kim J, Cormier K, Gilmore R, Welch M, Auregan G, et al. In vivo expression of polyglutamine-expanded huntingtin by mouse striatal astrocytes impairs glutamate transport: a correlation with Huntington's disease subjects. Hum Mol Genet. 2010;19(15): 3053–67.
86. Sari Y, Prieto AL, Barton SJ, Miller BR, Rebec GV. Ceftriaxone-induced up-regulation of cortical and striatal GLT1 in the R6/2 model of Huntington's disease. J Biomed Sci. 2010;17:62.
87. Landwehrmeyer GB, Dubois B, de Yébenes JG, Kremer B, Gaus W, Kraus PH, et al. Riluzole in Huntington's disease: a 3-year, randomized controlled study. Ann Neurol. 2007;62(3): 262–72.
88. Arregui L, Benítez JA, Razgado LF, Vergara P, Segovia J. Adenoviral astrocyte-specific expression of BDNF in the striata of mice transgenic for Huntington's disease delays the onset of the motor phenotype. Cell Mol Neurobiol. 2011;31(8):1229–43.
89. Doria JG, Silva FR, de Souza JM, Vieira LB, Carvalho TG, Reis HJ, et al. Metabotropic glutamate receptor 5 positive allosteric modulators are neuroprotective in a mouse model of Huntington's disease. Br J Pharmacol. 2013;169(4):909–21.
90. Reiner A, Lafferty DC, Wang HB, Del Mar N, Deng YP. The group 2 metabotropic glutamate receptor agonist LY379268 rescues neuronal, neurochemical and motor abnormalities in R6/2 Huntington's disease mice. Neurobiol Dis. 2012;47(1):75–91.
91. Schiefer J, Sprünken A, Puls C, Lüesse H-G, Milkereit A, Milkereit E, et al. The metabotropic glutamate receptor 5 antagonist MPEP and the mGluR2 agonist LY379268 modify disease progression in a transgenic mouse model of Huntington's disease. Brain Res. 2004;1019(1–2):246–54.
92. Milnerwood AJ, Kaufman AM, Sepers MD, Gladding CM, Zhang L, Wang L, et al. Mitigation of augmented extrasynaptic NMDAR signaling and apoptosis in cortico-striatal co-cultures from Huntington's disease mice. Neurobiol Dis. 2012;48(1):40–51.
93. Cachope R. Functional diversity on synaptic plasticity mediated by endocannabinoids. Philos Trans R Soc Lond B Biol Sci. 2012;367(1607):3242–53.
94. Skaper SD, Di Marzo V. Endocannabinoids in nervous system health and disease: the big picture in a nutshell. Philos Trans R Soc Lond B Biol Sci. 2012;367(1607):3193–200.
95. Chiodi V, Uchigashima M, Beggiato S, Ferrante A, Armida M, Martire A, et al. Unbalance of CB1 receptors expressed in GABAergic and glutamatergic neurons in a transgenic mouse model of Huntington's disease. Neurobiol Dis. 2012;45(3):983–91.
96. Bari M, Battista N, Valenza M, Mastrangelo N, Malaponti M, Catanzaro G, et al. In vitro and in vivo models of Huntington's disease show alterations in the endocannabinoid system. FEBS J. 2013;280(14):3376–88.
97. Richfield EK, Herkenham M. Selective vulnerability in Huntington's disease: preferential loss of cannabinoid receptors in lateral globus pallidus. Ann Neurol. 1994;36(4):577–84.
98. Allen KL, Waldvogel HJ, Glass M, Faull RLM. Cannabinoid (CB(1)), GABA(A) and GABA(B) receptor subunit changes in the globus pallidus in Huntington's disease. J Chem Neuroanat. 2009;37(4):266–81.

99. Glass M, Dragunow M, Faull RL. The pattern of neurodegeneration in Huntington's disease: a comparative study of cannabinoid, dopamine, adenosine and GABA(A) receptor alterations in the human basal ganglia in Huntington's disease. Neuroscience. 2000;97(3): 505–19.

100. Glass M, Faull RL, Dragunow M. Loss of cannabinoid receptors in the substantia nigra in Huntington's disease. Neuroscience. 1993;56(3):523–7.

101. Lastres-Becker I, Berrendero F, Lucas JJ, Martín-Aparicio E, Yamamoto A, Ramos JA, et al. Loss of mRNA levels, binding and activation of GTP-binding proteins for cannabinoid CB1 receptors in the basal ganglia of a transgenic model of Huntington's disease. Brain Res. 2002;929(2):236–42.

102. Denovan-Wright EM, Robertson HA. Cannabinoid receptor messenger RNA levels decrease in a subset of neurons of the lateral striatum, cortex and hippocampus of transgenic Huntington's disease mice. Neuroscience. 2000;98(4):705–13.

103. Curtis A, Mitchell I, Patel S, Ives N, Rickards H. A pilot study using nabilone for symptomatic treatment in Huntington's disease. Mov Disord. 2009;24(15):2254–9.

104. Chou S-Y, Lee Y-C, Chen H-M, Chiang M-C, Lai H-L, Chang H-H, et al. CGS21680 attenuates symptoms of Huntington's disease in a transgenic mouse model. J Neurochem. 2005;93(2):310–20.

105. Dowie MJ, Bradshaw HB, Howard ML, Nicholson LFB, Faull RLM, Hannan AJ, et al. Altered CB1 receptor and endocannabinoid levels precede motor symptom onset in a transgenic mouse model of Huntington's disease. Neuroscience. 2009;163(1):456–65.

106. Orru M, Bakešová J, Brugarolas M, Quiroz C, Beaumont V, Goldberg SR, et al. Striatal pre- and postsynaptic profile of adenosine A(2A) receptor antagonists. PLoS One. 2011;6(1):e16088.

107. Tarditi A, Camurri A, Varani K, Borea PA, Woodman B, Bates G, et al. Early and transient alteration of adenosine A2A receptor signaling in a mouse model of Huntington disease. Neurobiol Dis. 2006;23(1):44–53.

108. Popoli P, Blum D, Domenici MR, Burnouf S, Chern Y. A critical evaluation of adenosine A2A receptors as potentially «druggable» targets in Huntington's disease. Curr Pharm Des. 2008;14(15):1500–11.

109. Orrú M, Zanoveli JM, Quiroz C, Nguyen HP, Guitart X, Ferré S. Functional changes in post-synaptic adenosine A(2A) receptors during early stages of a rat model of Huntington disease. Exp Neurol. 2011;232(1):76–80.

110. Domenici MR, Scattoni ML, Martire A, Lastoria G, Potenza RL, Borioni A, et al. Behavioral and electrophysiological effects of the adenosine A2A receptor antagonist Sch 58261 in R6/2 Huntington's disease mice. Neurobiol Dis. 2007;28(2):197–205.

111. Ferrante A, Martire A, Armida M, Chiodi V, Pézzola A, Potenza RL, et al. Influence of CGS 21680, a selective adenosine A(2A) receptor agonist, on NMDA receptor function and expression in the brain of Huntington's disease mice. Brain Res. 2010; 1323:184–91.

112. Martire A, Ferrante A, Potenza RL, Armida M, Ferretti R, Pézzola A, et al. Remodeling of striatal NMDA receptors by chronic A(2A) receptor blockade in Huntington's disease mice. Neurobiol Dis. 2010;37(1):99–105.

113. Mizuno Y, Hasegawa K, Kondo T, Kuno S, Yamamoto M, Japanese Istradefylline Study Group. Clinical efficacy of istradefylline (KW-6002) in Parkinson's disease: a randomized, controlled study. Mov Disord. 2010;25(10):1437–43.

114. Hauser RA, Cantillon M, Pourcher E, Micheli F, Mok V, Onofrj M, et al. Preladenant in patients with Parkinson's disease and motor fluctuations: a phase 2, double-blind, randomised trial. Lancet Neurol. 2011;10(3):221–9.

115. Hebb ALO, Robertson HA. Role of phosphodiesterases in neurological and psychiatric disease. Curr Opin Pharmacol. 2007;7(1):86–92.

116. Rose GM, Hopper A, De Vivo M, Tehim A. Phosphodiesterase inhibitors for cognitive enhancement. Curr Pharm Des. 2005;11(26):3329–34.

117. Jeon YH, Heo Y-S, Kim CM, Hyun Y-L, Lee TG, Ro S, et al. Phosphodiesterase: overview of protein structures, potential therapeutic applications and recent progress in drug development. Cell Mol Life Sci. 2005;62(11):1198–220.

118. Giampà C, DeMarch Z, D'Angelo V, Morello M, Martorana A, Sancesario G, et al. Striatal modulation of cAMP-response-element-binding protein (CREB) after excitotoxic lesions: implications with neuronal vulnerability in Huntington's disease. Eur J Neurosci. 2006; 23(1):11–20.

119. Giampà C, Patassini S, Borreca A, Laurenti D, Marullo F, Bernardi G, et al. Phosphodiesterase 10 inhibition reduces striatal excitotoxicity in the quinolinic acid model of Huntington's disease. Neurobiol Dis. 2009;34(3):450–6.

120. Gines S, Seong IS, Fossale E, Ivanova E, Trettel F, Gusella JF, et al. Specific progressive cAMP reduction implicates energy deficit in presymptomatic Huntington's disease knock-in mice. Hum Mol Genet. 2003;12(5):497–508.

121. Kleiman RJ, Kimmel LH, Bove SE, Lanz TA, Harms JF, Romegialli A, et al. Chronic suppression of phosphodiesterase 10A alters striatal expression of genes responsible for neurotransmitter synthesis, neurotransmission, and signaling pathways implicated in Huntington's disease. J Pharmacol Exp Ther. 2011;336(1):64–76.

122. Obrietan K, Hoyt KR. CRE-mediated transcription is increased in Huntington's disease transgenic mice. J Neurosci. 2004;24(4):791–6.

123. Sathasivam K, Lane A, Legleiter J, Warley A, Woodman B, Finkbeiner S, et al. Identical oligomeric and fibrillar structures captured from the brains of R6/2 and knock-in mouse models of Huntington's disease. Hum Mol Genet. 2010;19(1):65–78.

124. Giralt A, Saavedra A, Carretón O, Xifró X, Alberch J, Pérez-Navarro E. Increased PKA signaling disrupts recognition memory and spatial memory: role in Huntington's disease. Hum Mol Genet. 2011;20(21):4232–47.

125. Ahn HS, Bercovici A, Boykow G, Bronnenkant A, Chackalamannil S, Chow J, et al. Potent tetracyclic guanine inhibitors of PDE1 and PDE5 cyclic guanosine monophosphate phosphodiesterases with oral antihypertensive activity. J Med Chem. 1997;40(14):2196–210.

126. DeMarch Z, Giampà C, Patassini S, Bernardi G, Fusco FR. Beneficial effects of rolipram in the R6/2 mouse model of Huntington's disease. Neurobiol Dis. 2008;30(3):375–87.

127. Giampà C, Middei S, Patassini S, Borreca A, Marullo F, Laurenti D, et al. Phosphodiesterase type IV inhibition prevents sequestration of CREB binding protein, protects striatal parvalbumin interneurons and rescues motor deficits in the R6/2 mouse model of Huntington's disease. Eur J Neurosci. 2009;29(5):902–10.

128. DeMarch Z, Giampà C, Patassini S, Martorana A, Bernardi G, Fusco FR. Beneficial effects of rolipram in a quinolinic acid model of striatal excitotoxicity. Neurobiol Dis. 2007; 25(2):266–73.

129. Chandrasekaran A, Toh KY, Low SH, Tay SKH, Brenner S, Goh DLM. Identification and characterization of novel mouse PDE4D isoforms: molecular cloning, subcellular distribution and detection of isoform-specific intracellular localization signals. Cell Signal. 2008;20(1):139–53.

130. Hall JA, Dominy JE, Lee Y, Puigserver P. The sirtuin family's role in aging and age-associated pathologies. J Clin Invest. 2013;123(3):973–9.

131. Iona S, Cuomo M, Bushnik T, Naro F, Sette C, Hess M, et al. Characterization of the rolipram-sensitive, cyclic AMP-specific phosphodiesterases: identification and differential expression of immunologically distinct forms in the rat brain. Mol Pharmacol. 1998;53(1):23–32.

132. Threlfell S, Sammut S, Menniti FS, Schmidt CJ, West AR. Inhibition of phosphodiesterase 10A increases the responsiveness of striatal projection neurons to cortical stimulation. J Pharmacol Exp Ther. 2009;328(3):785–95.

133. Kotera J, Sasaki T, Kobayashi T, Fujishige K, Yamashita Y, Omori K. Subcellular localization of cyclic nucleotide phosphodiesterase type 10A variants, and alteration of the localization by cAMP-dependent protein kinase-dependent phosphorylation. J Biol Chem. 2004; 279(6):4366–75.

134. Sotty F, Montezinho LP, Steiniger-Brach B, Nielsen J. Phosphodiesterase 10A inhibition modulates the sensitivity of the mesolimbic dopaminergic system to D-amphetamine: involvement of the D1-regulated feedback control of midbrain dopamine neurons. J Neurochem. 2009;109(3):766–75.

135. Seeger TF, Bartlett B, Coskran TM, Culp JS, James LC, Krull DL, et al. Immunohistochemical localization of PDE10A in the rat brain. Brain Res. 2003;985(2):113–26.

136. Rodefer JS, Saland SK, Eckrich SJ. Selective phosphodiesterase inhibitors improve performance on the ED/ID cognitive task in rats. Neuropharmacology. 2012;62(3):1182–90.

137. Rutten K, Basile JL, Prickaerts J, Blokland A, Vivian JA. Selective PDE inhibitors rolipram and sildenafil improve object retrieval performance in adult cynomolgus macaques. Psychopharmacology (Berl). 2008;196(4):643–8.

138. Rutten K, Lieben C, Smits L, Blokland A. The PDE4 inhibitor rolipram reverses object memory impairment induced by acute tryptophan depletion in the rat. Psychopharmacology (Berl). 2007;192(2):275–82.

139. Arenas E, Akerud P, Wong V, Boylan C, Persson H, Lindsay RM, et al. Effects of BDNF and NT-4/5 on striatonigral neuropeptides or nigral GABA neurons in vivo. Eur J Neurosci. 1996;8(8):1707–17.

140. Besusso D, Geibel M, Kramer D, Schneider T, Pendolino V, Picconi B, et al. BDNF-TrkB signaling in striatopallidal neurons controls inhibition of locomotor behavior. Nat Commun. 2013;4:2031.

141. Brito V, Puigdellívol M, Giralt A, del Toro D, Alberch J, Ginés S. Imbalance of p75(NTR)/TrkB protein expression in Huntington's disease: implication for neuroprotective therapies. Cell Death Dis. 2013;4:e595.

142. Buckley NJ, Johnson R, Zuccato C, Bithell A, Cattaneo E. The role of REST in transcriptional and epigenetic dysregulation in Huntington's disease. Neurobiol Dis. 2010;39(1):28–39.

143. Canals JM, Pineda JR, Torres-Peraza JF, Bosch M, Martín-Ibañez R, Muñoz MT, et al. Brain-derived neurotrophic factor regulates the onset and severity of motor dysfunction associated with enkephalinergic neuronal degeneration in Huntington's disease. J Neurosci. 2004;24(35):7727–39.

144. Conforti P, Mas Monteys A, Zuccato C, Buckley NJ, Davidson B, Cattaneo E. In vivo delivery of DN:REST improves transcriptional changes of REST-regulated genes in HD mice. Gene Ther. 2013;20(6):678–85.

145. Conforti P, Zuccato C, Gaudenzi G, Ieraci A, Camnasio S, Buckley NJ, et al. Binding of the repressor complex REST-mSIN3b by small molecules restores neuronal gene transcription in Huntington's disease models. J Neurochem. 2013;127(1):22–35.

146. Giampà C, Montagna E, Dato C, Melone MAB, Bernardi G, Fusco FR. Systemic delivery of recombinant brain derived neurotrophic factor (BDNF) in the R6/2 mouse model of Huntington's disease. PLoS One. 2013;8(5):e64037.

147. Giralt A, Carretón O, Lao-Peregrin C, Martín ED, Alberch J. Conditional BDNF release under pathological conditions improves Huntington's disease pathology by delaying neuronal dysfunction. Mol Neurodegener. 2011;6(1):71.

148. Goggi J, Pullar IA, Carney SL, Bradford HF. Modulation of neurotransmitter release induced by brain-derived neurotrophic factor in rat brain striatal slices in vitro. Brain Res. 2002;941(1–2):34–42.

149. Ivkovic S, Ehrlich ME. Expression of the striatal DARPP-32/ARPP-21 phenotype in GABAergic neurons requires neurotrophins in vivo and in vitro. J Neurosci. 1999;19(13):5409–19.

150. Jiang M, Peng Q, Liu X, Jin J, Hou Z, Zhang J, et al. Small-molecule TrkB receptor agonists improve motor function and extend survival in a mouse model of Huntington's disease. Hum Mol Genet. 2013;22(12):2462–70.

151. Jiao Y, Zhang Z, Zhang C, Wang X, Sakata K, Lu B, et al. A key mechanism underlying sensory experience-dependent maturation of neocortical GABAergic circuits in vivo. Proc Natl Acad Sci U S A. 2011;108(29):12131–6.

152. Kells AP, Fong DM, Dragunow M, During MJ, Young D, Connor B. AAV-mediated gene delivery of BDNF or GDNF is neuroprotective in a model of Huntington disease. Mol Ther. 2004;9(5):682–8.
153. Liot G, Zala D, Pla P, Mottet G, Piel M, Saudou F. Mutant Huntingtin alters retrograde transport of TrkB receptors in striatal dendrites. J Neurosci. 2013;33(15):6298–309.
154. Martire A, Pepponi R, Domenici MR, Ferrante A, Chiodi V, Popoli P. BDNF prevents NMDA-induced toxicity in models of Huntington's disease: the effects are genotype specific and adenosine A(2A) receptor is involved. J Neurochem. 2013;125(2):225–35.
155. Massa SM, Yang T, Xie Y, Shi J, Bilgen M, Joyce JN, et al. Small molecule BDNF mimetics activate TrkB signaling and prevent neuronal degeneration in rodents. J Clin Invest. 2010;120(5):1774–85.
156. Soldati C, Bithell A, Conforti P, Cattaneo E, Buckley NJ. Rescue of gene expression by modified REST decoy oligonucleotides in a cellular model of Huntington's disease. J Neurochem. 2011;116(3):415–25.
157. Xie Y, Hayden MR, Xu B. BDNF overexpression in the forebrain rescues Huntington's disease phenotypes in YAC128 mice. J Neurosci. 2010;30(44):14708–18.
158. Zala D, Colin E, Rangone H, Liot G, Humbert S, Saudou F. Phosphorylation of mutant huntingtin at S421 restores anterograde and retrograde transport in neurons. Hum Mol Genet. 2008;17(24):3837–46.
159. Zuccato C, Marullo M, Vitali B, Tarditi A, Mariotti C, Valenza M, et al. Brain-derived neurotrophic factor in patients with Huntington's disease. PLoS One. 2011;6(8):e22966.
160. Perreault M, Feng G, Will S, Gareski T, Kubasiak D, Marquette K, et al. Activation of TrkB with TAM-163 results in opposite effects on body weight in rodents and non-human primates. PLoS One. 2013;8(5):e62616.
161. Tsao D, Thomsen HK, Chou J, Stratton J, Hagen M, Loo C, et al. TrkB agonists ameliorate obesity and associated metabolic conditions in mice. Endocrinology. 2008;149(3): 1038–48.
162. Vanevski F, Xu B. Molecular and neural bases underlying roles of BDNF in the control of body weight. Front Neurosci. 2013;7:37.
163. Waterhouse EG, Xu B. The skinny on brain-derived neurotrophic factor: evidence from animal models to GWAS. J Mol Med (Berl). 2013;91(11):1241–7.
164. Fletcher JM, Hughes RA. Modified low molecular weight cyclic peptides as mimetics of BDNF with improved potency, proteolytic stability and transmembrane passage in vitro. Bioorg Med Chem. 2009;17(7):2695–702.
165. Marongiu D, Imbrosci B, Mittmann T. Modulatory effects of the novel TrkB receptor agonist 7,8-dihydroxyflavone on synaptic transmission and intrinsic neuronal excitability in mouse visual cortex in vitro. Eur J Pharmacol. 2013;709(1–3):64–71.
166. O'Leary PD, Hughes RA. Design of potent peptide mimetics of brain-derived neurotrophic factor. J Biol Chem. 2003;278(28):25738–44.
167. Simmons DA, Belichenko NP, Yang T, Condon C, Monbureau M, Shamloo M, et al. A small molecule TrkB ligand reduces motor impairment and neuropathology in R6/2 and BACHD mouse models of Huntington's disease. J Neurosci. 2013;33(48):18712–27.
168. Mochel F, Durant B, Meng X, O'Callaghan J, Yu H, Brouillet E, et al. Early alterations of brain cellular energy homeostasis in Huntington disease models. J Biol Chem. 2012;287(2):1361–70.
169. Mochel F, Haller RG. Energy deficit in Huntington disease: why it matters. J Clin Invest. 2011;121(2):493–9.
170. Gellerich FN, Gizatullina Z, Nguyen HP, Trumbeckaite S, Vielhaber S, Seppet E, et al. Impaired regulation of brain mitochondria by extramitochondrial Ca2+ in transgenic Huntington disease rats. J Biol Chem. 2008;283(45):30715–24.
171. Reynolds NC, Prost RW, Mark LP, Joseph SA. MR-spectroscopic findings in juvenile-onset Huntington's disease. Mov Disord. 2008;23(13):1931–5.
172. Mochel F, Duteil S, Marelli C, Jauffret C, Barles A, Holm J, et al. Dietary anaplerotic therapy improves peripheral tissue energy metabolism in patients with Huntington's disease. Eur J Hum Genet. 2010;18(9):1057–60.

173. Tabrizi SJ, Blamire AM, Manners DN, Rajagopalan B, Styles P, Schapira AHV, et al. Creatine therapy for Huntington's disease: clinical and MRS findings in a 1-year pilot study. Neurology. 2003;61(1):141–2.

174. Van den Bogaard SJA, Dumas EM, Teeuwisse WM, Kan HE, Webb A, Roos RAC, et al. Exploratory 7-Tesla magnetic resonance spectroscopy in Huntington's disease provides in vivo evidence for impaired energy metabolism. J Neurol. 2011;258(12):2230–9.

175. Ciammola A, Sassone J, Sciacco M, Mencacci NE, Ripolone M, Bizzi C, et al. Low anaerobic threshold and increased skeletal muscle lactate production in subjects with Huntington's disease. Mov Disord. 2011;26(1):130–7.

176. Jenkins BG, Rosas HD, Chen YC, Makabe T, Myers R, MacDonald M, et al. 1H NMR spectroscopy studies of Huntington's disease: correlations with CAG repeat numbers. Neurology. 1998;50(5):1357–65.

177. Lin Y-S, Chen C-M, Soong B, Wu Y-R, Chen H-M, Yeh W-Y, et al. Dysregulated brain creatine kinase is associated with hearing impairment in mouse models of Huntington disease. J Clin Invest. 2011;121(4):1519–23.

178. Zhang SF, Hennessey T, Yang L, Starkova NN, Beal MF, Starkov AA. Impaired brain creatine kinase activity in Huntington's disease. Neurodegener Dis. 2011;8(4):194–201.

179. Kim J, Amante DJ, Moody JP, Edgerly CK, Bordiuk OL, Smith K, et al. Reduced creatine kinase as a central and peripheral biomarker in Huntington's disease. Biochim Biophys Acta. 2010;1802(7–8):673–81.

180. Hickey MA, Zhu C, Medvedeva V, Franich NR, Levine MS, Chesselet M-F. Evidence for behavioral benefits of early dietary supplementation with CoEnzymeQ10 in a slowly progressing mouse model of Huntington's disease. Mol Cell Neurosci. 2012;49(2):149–57.

181. Menalled LB, Patry M, Ragland N, Lowden PAS, Goodman J, Minnich J, et al. Comprehensive behavioral testing in the R6/2 mouse model of Huntington's disease shows no benefit from CoQ10 or minocycline. PLoS One. 2010;5(3):e9793.

182. Yang L, Calingasan NY, Wille EJ, Cormier K, Smith K, Ferrante RJ, et al. Combination therapy with coenzyme Q10 and creatine produces additive neuroprotective effects in models of Parkinson's and Huntington's diseases. J Neurochem. 2009;109(5):1427–39.

183. Van Raamsdonk JM, Pearson J, Rogers DA, Lu G, Barakauskas VE, Barr AM, et al. Ethyl-EPA treatment improves motor dysfunction, but not neurodegeneration in the YAC128 mouse model of Huntington disease. Exp Neurol. 2005;196(2):266–72.

184. Borrell-Pagès M, Canals JM, Cordelières FP, Parker JA, Pineda JR, Grange G, et al. Cystamine and cysteamine increase brain levels of BDNF in Huntington disease via HSJ1b and transglutaminase. J Clin Invest. 2006;116(5):1410–24.

185. Spina D. PDE4 inhibitors: current status. Br J Pharmacol. 2008;155(3):308–15.

186. Puri BK, Leavitt BR, Hayden MR, Ross CA, Rosenblatt A, Greenamyre JT, et al. Ethyl-EPA in Huntington disease: a double-blind, randomized, placebo-controlled trial. Neurology. 2005;65(2):286–92.

187. Puri BK, Bydder GM, Counsell SJ, Corridan BJ, Richardson AJ, Hajnal JV, et al. MRI and neuropsychological improvement in Huntington disease following ethyl-EPA treatment. Neuroreport. 2002;13(1):123–6.

188. Hersch SM, Gevorkian S, Marder K, Moskowitz C, Feigin A, Cox M, et al. Creatine in Huntington disease is safe, tolerable, bioavailable in brain and reduces serum 8OH2'dG. Neurology. 2006;66(2):250–2.

189. Verbessem P, Lemiere J, Eijnde BO, Swinnen S, Vanhees L, Van Leemputte M, et al. Creatine supplementation in Huntington's disease: a placebo-controlled pilot trial. Neurology. 2003;61(7):925–30.

190. Dubinsky R, Gray C. CYTE-I-HD: phase I dose finding and tolerability study of cysteamine (Cystagon) in Huntington's disease. Mov Disord. 2006;21(4):530–3.

191. Huntington Study Group Pre2CARE Investigators, Hyson HC, Kieburtz K, Shoulson I, McDermott M, Ravina B, et al. Safety and tolerability of high-dosage coenzyme Q10 in Huntington's disease and healthy subjects. Mov Disord. 2010;25(12):1924–8.

192. Weydt P, Pineda VV, Torrence AE, Libby RT, Satterfield TF, Lazarowski ER, et al. Thermoregulatory and metabolic defects in Huntington's disease transgenic mice implicate PGC-1alpha in Huntington's disease neurodegeneration. Cell Metab. 2006;4(5):349–62.
193. Johri A, Calingasan NY, Hennessey TM, Sharma A, Yang L, Wille E, et al. Pharmacologic activation of mitochondrial biogenesis exerts widespread beneficial effects in a transgenic mouse model of Huntington's disease. Hum Mol Genet. 2012;21(5):1124–37.
194. Jiang M, Wang J, Fu J, Du L, Jeong H, West T, et al. Neuroprotective role of Sirt1 in mammalian models of Huntington's disease through activation of multiple Sirt1 targets. Nat Med. 2012;18(1):153–8.
195. Chaturvedi RK, Beal MF. Mitochondria targeted therapeutic approaches in Parkinson's and Huntington's diseases. Mol Cell Neurosci. 2013;55:101–14.
196. Chaturvedi RK, Adhihetty P, Shukla S, Hennessy T, Calingasan N, Yang L, et al. Impaired PGC-1alpha function in muscle in Huntington's disease. Hum Mol Genet. 2009;18(16):3048–65.
197. Jin J, Albertz J, Guo Z, Peng Q, Rudow G, Troncoso JC, et al. Neuroprotective effects of PPAR-γ agonist rosiglitazone in N171-82Q mouse model of Huntington's disease. J Neurochem. 2013;125(3):410–9.
198. Jeong H, Cohen DE, Cui L, Supinski A, Savas JN, Mazzulli JR, et al. Sirt1 mediates neuroprotection from mutant huntingtin by activation of the TORC1 and CREB transcriptional pathway. Nat Med. 2012;18(1):159–65.
199. Ho DJ, Calingasan NY, Wille E, Dumont M, Beal MF. Resveratrol protects against peripheral deficits in a mouse model of Huntington's disease. Exp Neurol. 2010;225(1):74–84.
200. Zhao W, Kruse J-P, Tang Y, Jung SY, Qin J, Gu W. Negative regulation of the deacetylase SIRT1 by DBC1. Nature. 2008;451(7178):587–90.
201. Nin V, Escande C, Chini CC, Giri S, Camacho-Pereira J, Matalonga J, et al. Role of deleted in breast cancer 1 (DBC1) protein in SIRT1 deacetylase activation induced by protein kinase A and AMP-activated protein kinase. J Biol Chem. 2012;287(28):23489–501.
202. Johnson JA, Johnson DA, Kraft AD, Calkins MJ, Jakel RJ, Vargas MR, et al. The Nrf2-ARE pathway: an indicator and modulator of oxidative stress in neurodegeneration. Ann N Y Acad Sci. 2008;1147:61–9.
203. Mielcarek M, Landles C, Weiss A, Bradaia A, Seredenina T, Inuabasi L, et al. HDAC4 reduction: a novel therapeutic strategy to target cytoplasmic huntingtin and ameliorate neurodegeneration. PLoS Biol. 2013;11(11):e1001717.
204. Hockly E, Richon VM, Woodman B, Smith DL, Zhou X, Rosa E, et al. Suberoylanilide hydroxamic acid, a histone deacetylase inhibitor, ameliorates motor deficits in a mouse model of Huntington's disease. Proc Natl Acad Sci U S A. 2003;100(4):2041–6.
205. Jia H, Kast RJ, Steffan JS, Thomas EA. Selective histone deacetylase (HDAC) inhibition imparts beneficial effects in Huntington's disease mice: implications for the ubiquitin-proteasomal and autophagy systems. Hum Mol Genet. 2012;21(24):5280–93.
206. Jia H, Pallos J, Jacques V, Lau A, Tang B, Cooper A, et al. Histone deacetylase (HDAC) inhibitors targeting HDAC3 and HDAC1 ameliorate polyglutamine-elicited phenotypes in model systems of Huntington's disease. Neurobiol Dis. 2012;46(2):351–61.
207. McFarland KN, Das S, Sun TT, Leyfer D, Xia E, Sangrey GR, et al. Genome-wide histone acetylation is altered in a transgenic mouse model of Huntington's disease. PLoS One. 2012;7(7):e41423.
208. Quinti L, Chopra V, Rotili D, Valente S, Amore A, Franci G, et al. Evaluation of histone deacetylases as drug targets in Huntington's disease models. Study of HDACs in brain tissues from R6/2 and CAG140 knock-in HD mouse models and human patients and in a neuronal HD cell model. PLoS Curr. 2010;2. pii: RRN1172. doi: 10.1371/currents.RRN1172.
209. Steffan JS, Bodai L, Pallos J, Poelman M, McCampbell A, Apostol BL, et al. Histone deacetylase inhibitors arrest polyglutamine-dependent neurodegeneration in Drosophila. Nature. 2001;413(6857):739–43.
210. Thomas EA, Coppola G, Desplats PA, Tang B, Soragni E, Burnett R, et al. The HDAC inhibitor 4b ameliorates the disease phenotype and transcriptional abnormalities in Huntington's disease transgenic mice. Proc Natl Acad Sci U S A. 2008;105(40):15564–9.

211. Benn CL, Butler R, Mariner L, Nixon J, Moffitt H, Mielcarek M, et al. Genetic knock-down of HDAC7 does not ameliorate disease pathogenesis in the R6/2 mouse model of Huntington's disease. PLoS One. 2009;4(6):e5747.

212. Bobrowska A, Paganetti P, Matthias P, Bates GP. Hdac6 knock-out increases tubulin acetylation but does not modify disease progression in the R6/2 mouse model of Huntington's disease. PLoS One. 2011;6(6):e20696.

213. Moumné L, Campbell K, Howland D, Ouyang Y, Bates GP. Genetic knock-down of HDAC3 does not modify disease-related phenotypes in a mouse model of Huntington's disease. PLoS One. 2012;7(2):e31080.

214. Jovicic A, Zaldivar Jolissaint JF, Moser R, Mde Silva Santos F, Luthi-Carter R. MicroRNA-22 (miR-22) overexpression is neuroprotective via general anti-apoptotic effects and may also target specific Huntington's disease-related mechanisms. PLoS One. 2013;8(1):e54222.

215. Bürli RW, Luckhurst CA, Aziz O, Matthews KL, Yates D, Lyons KA, et al. Design, synthesis, and biological evaluation of potent and selective class IIa Histone Deacetylase (HDAC) inhibitors as a potential therapy for Huntington's disease. J Med Chem. 2013;56(24): 9934–54.

216. Sebastián C, Satterstrom FK, Haigis MC, Mostoslavsky R. From sirtuin biology to human diseases: an update. J Biol Chem. 2012;287(51):42444–52.

217. Shin BH, Lim Y, Oh HJ, Park SM, Lee S-K, Ahnn J, et al. Pharmacological activation of Sirt1 ameliorates polyglutamine-induced toxicity through the regulation of autophagy. PLoS One. 2013;8(6):e64953.

218. Luthi-Carter R, Taylor DM, Pallos J, Lambert E, Amore A, Parker A, et al. SIRT2 inhibition achieves neuroprotection by decreasing sterol biosynthesis. Proc Natl Acad Sci U S A. 2010;107(17):7927–32.

219. Beher D, Wu J, Cumine S, Kim KW, Lu S-C, Atangan L, et al. Resveratrol is not a direct activator of SIRT1 enzyme activity. Chem Biol Drug Des. 2009;74(6):619–24.

220. Dai H, Kustigian L, Carney D, Case A, Considine T, Hubbard BP, et al. SIRT1 activation by small molecules: kinetic and biophysical evidence for direct interaction of enzyme and activator. J Biol Chem. 2010;285(43):32695–703.

221. La Spada AR. Finding a sirtuin truth in Huntington's disease. Nat Med. 2012;18(1):24–6.

222. Liebman SW, Meredith SC. Protein folding: sticky N17 speeds huntingtin pile-up. Nat Chem Biol. 2010;6(1):7–8. doi:10.1038/nchembio.279.

223. Williamson TE, Vitalis A, Crick SL, Pappu RV. Modulation of polyglutamine conformations and dimer formation by the N-terminus of huntingtin. J Mol Biol. 2010;396(5):1295–309.

Chapter 7
Huntington Disease and Huntington Disease-Like Syndromes: An Overview

Susanne A. Schneider and Felix Gövert

Abstract The differential diagnosis of chorea syndromes may be complex. It includes inherited forms, the most important of which is autosomal dominant Huntington disease (HD). In addition there are disorders mimicking HD, the so-called HD-like syndromes, and molecular workup revealed that they account for about 1 % of suspected HD cases. The aim of this review is to summarize the main characteristics of these rare conditions in order to familiarize clinicians with them. While treatment remains symptomatic, advances have been made with genetic delineation. Hopefully with better understanding of their pathophysiology, we will also move towards mechanistic therapies.

Keywords Chorea • Huntington disease • HDL disorders • SCA17 • Differential diagnosis

Introduction

Chorea is a hyperkinetic movement disorder characterized by excessive spontaneous, involuntary movements of abrupt, irregular, unpredictable nature. Severity may range from mild focal involvement (e.g., of the hands) to severe generalized chorea affecting the limbs, trunk, head, and face. In some instances, chorea may be restricted to one side of the body (hemichorea) which is important to recognize as it may point to a secondary form due to contralateral structural lesions. As outlined in

The authors have no conflicts of interest.

S.A. Schneider (✉)
Department of Neurology, Christian-Albrechts-University Kiel, University-Hospital-Schleswig-Holstein, Kiel Campus, Schittenhelmstr. 10, Kiel, 24105, Germany
e-mail: S.Schneider@neurologie.uni-kiel.de

F. Gövert
Department of Neurology, University of Kiel, Kiel, Germany

F.E. Micheli, P.A. LeWitt (eds.), *Chorea*,
DOI 10.1007/978-1-4471-6455-5_7, © Springer-Verlag London 2014

this book, chorea may have numerous causes including acquired and inherited etiologies. Among the inherited forms, the most important cause is Huntington disease (HD), a slowly progressive autosomal dominant neurodegenerative disease characterized by impaired motor function (in particular by chorea, dystonia, and parkinsonism), cognitive impairment, and psychiatric symptoms. The condition is discussed in detail in Chaps. 6 and 7.

Since the identification of the gene underlying HD, *huntingtin* (*HTT*), in 1993, excellent progress has been made with regard to understanding the pathophysiology underlying HD. Furthermore, it has become clear that not all patients who present with a clinical phenotype and family history suggestive of HD actually turn out to have the HD mutation (i.e., an expanded CAG repeat in exon 1). Instead molecular testing may reveal mutations in other genes. These related but distinct disorders are referred to as HD-like disorders, an overall very rare group of diseases (Table 7.1).

In the following, we will summarize their main clinical features, pathophysiological and genetic underpinnings, and other key characteristics. We will begin with a brief summary of the main features in HD, but refer to other chapters in this book for further details and nongenetic differential diagnoses of chorea syndromes.

Huntington Disease

The motor phenotype of HD is mainly characterized by slowly progressive (usually generalized) chorea with onset in midlife, inherited in an autosomal dominant fashion. Interestingly, homozygous patients have also rarely been described [1]. Alleles with <27 CAG repeats are classified as normal measured by the number of uninterrupted CAG, whereas alleles with ≥36 repeats are detected in affected individuals. Intermediate range CAG repeats are incompletely penetrant and may be found in affected individuals as well as individuals without clinical symptoms [2]. The age of onset correlates with the number of trinucleotide repeats in the HD gene: longer repeats cause earlier onset of disease and vice versa. In successive generations, disease onset may develop earlier in life (correlating with longer repeat sizes; so-called genetic anticipation), particularly when the repeat expansion is inherited through the father due to unstable CAG repeat during spermatogenesis. Genetic anticipation also occurs in other repeat disorders including some of the HDL disorders (e.g., HDL2, see below). The rate of new mutations in HD is low.

The clinical presentation in HD is typically predominated by chorea as the main movement disorder, and as the disease progresses dystonia and parkinsonism may occur. However, in early-onset cases, the phenotype may be characterized by parkinsonism (Westphal variant) right from the beginning. Studies show that about 6–10 % of HD cases start before the age of 20 years with an akinetic-rigid syndrome.

In classic HD, patients demonstrate abnormal facial expression with choreic movements with inability to keep the tongue outstretched. Saccadic eye movements are impaired. With progressing disease, there is increasing postural instability and

Table 7.1 Summary of important genetic causes of chorea syndromes

Condition	Synonym	Inheritance	Position	Gene	Number of exons[a]	Triplet repeat disorder[a]	Regions of high penetrance
HD		AD	4p16	HTT	67[a]	√	Venezuela and worldwide
HDL1		AD	20p13	PRNP	2		
HDL2		AD	16q24	JPH3	5[a]	√	Black Africa
HDL3		AR	4p15	unknown			
HDL4	SCA17	AD	6q27	TBP1	8[a]	√	
DRPLA	NOD, HRS	AD	12p13	ATN1	10[a]	√	Japan
Neuroferritinopathy		AD	19p13	FTL1[b]	4		Cumbrian region of Northern England
Benign hereditary chorea	Thyroid-lung syndrome	AD	14q13	TITF-1	3		
Benign hereditary chorea 2 (c)		AD	8q21	not known			
Chorea-acanthocytosis	Levine-Critchley syndrome	AR	9q21	VPS13A	73		
McLeod syndrome		X-linked	Xp21	XK	3		
Wilson disease	Hepatolenticular degeneration	AR	13q14	ATP7B	21		

AD autosomal dominant, *AR* autosomal recessive, *HD* Huntington disease, *HDL* Huntington-like disorder, *DRPLA* dentatorubral-pallidoluysian atrophy, *SCA 17* spinocerebellar ataxia 17, *NOD* Naito-Oyanagi disease, *HRS* Haw River syndrome
[a]Disease-causing triplet repeat located in exon 1 (HD and HDL2), exon 3 (HDL4/SCA17), and exon 5 (DRPLA)
[b]Also associated with hereditary hyperferritinemia cataract syndrome
[c]based on a single family from Saudia Arabia
[d]based on two Japanese families

associated dysarthria and dysphagia. Early in the course, personality changes or psychiatric symptoms (depression, anxiety, dysphoria) develop, often preceding the motor onset. Cognitive dysfunction is characterized by affecting executive function (abstract thinking, planning, and inhibition of inappropriate behavior) and, later, memory dysfunction.

Recent research aims at more accuracy of prediction of symptom onset and understanding earliest (including presymptomatic) disease stages in order to define therapeutic windows when potential therapy may be beneficial. Thus, in large consortia (such as TRACK-HD and PREDICT-HD [3, 4]), individuals at risk (genetically confirmed asymptomatic gene mutation carriers) are longitudinally followed and regularly examined to detect subtle changes on neuroimaging, cognitive function tests, and by other parameters.

The pathophysiology of HD is not fully understood. The mutant huntingtin protein (HTT) is large and ubiquitously expressed with damaging effects on neurons. Animal models recapitulate the molecular, cellular, and clinical phenotypes and have shed light on the pathophysiological underpinnings allowing the search for mechanistic treatments [5]. However, so far treatment remains symptomatic, ideally in a multidisciplinary setting. Likewise, genetic counseling should occur in centers with expertise.

Huntington Disease-Like Syndromes

As mentioned above HDL disorders are rare. Only about 1 % of suspected HD cases emerge as phenocopy syndromes [6, 7]. Differential diagnosis includes the HD-like syndromes which will be discussed in the following.

Prion Disease: Huntington Disease-Like 1

HDL1 is a rare familial prion disease with autosomal dominant inheritance, first reported in 2001 [8]. The progressive disorder is caused by eight (sometimes six) extra octapeptide (Pro-His-Gly-Gly-Gly-Trp-Gly-Gln) repeats in the *prion protein* (*PrP*) gene (*PRNP*). Mean onset age is in early adulthood between 20 and 45 years [9]. Mean survival time after onset is only between 1 and 10 years [10], and rapid progression is suspicious of this cause. Familial prion disease may produce a diverse range of phenotypes, even within the same pedigree. It may resemble HD with prominent personality change, psychiatric symptoms and cognitive decline, chorea, rigidity, and dysarthria. Limb and truncal ataxia and seizures may be present [11, 12]. Myoclonus, rigidity, and other neurological signs, evolving to mutism and immobility may occur.

The characteristic EEG pattern of generalized bi- or triphasic periodic sharp wave complexes seen in sporadic Creutzfeldt-Jakob disease (CJD) is less frequently present in the genetic prion variant. Similarly, detection of the 14-3-3 protein in the CSF is less consistent in genetic than in sporadic CJD. Mild to moderate generalized atrophy may be demonstrated on brain MRI. FLAIR and T_2-weighted images may reveal hyperintensity of the basal ganglia.

Neuropathological examination in HDL1 revealed atrophy and prion deposition in the basal ganglia, frontal and temporal lobes, and cerebellar cortex. In comparison to other prion diseases, spongiosis is not prominent [11, 12].

Huntington Disease-Like 2

HDL2 is overall rare accounting for about 0.7–2.6 % of HD phenocopies; however, it is quite frequent in Black South Africans of sub-Saharan descent [13–16]. In this ethnic group, HDL2 is responsible for about 24–50 % of patients with an HD-like presentation and should be considered early in the diagnostic workup [17, 18].

Mutations in *junctophilin 3* (*JPH3*) cause HDL2, also inherited in an autosomal dominant manner and highly penetrant. The condition takes a progressive course and may show remarkable similarities to HD. The classic form of HDL2 presents with similar cognitive, psychiatric, and motor features. Mean age of onset is in the third or fourth decade. However, early-onset cases with a progressive akinetic-rigid syndrome, but clinically insignificant chorea, have been reported [19, 20] In contrast to juvenile-onset HD in early-onset HDL2, there is absence of seizures and mostly normal eye movements. The disease leads to death within 10–20 years [21].

Molecular analysis reveals CTG-CAG triplet repeat expansions in the *JPH3* gene [22] on chromosome 16q24.3. Normal alleles range from 6 to 28 triplets, whereas pathological repeat expansions range from 40 to 59 triplets [23]. The impact of triplets in between remains unclear [21]. Similar to HD, there is a negative correlation between age of onset and repeat length [20]. In contrast to HD, the anticipation phenomenon is more likely when the disease is maternally inherited and the repeat expansions are more unstable [10].

Neuropathologically, both HDL2 and HD show marked cortical and striatal neurodegeneration and neuronal protein aggregates staining positive for antiubiquitin antibodies and expanded polyglutamine tracts. However, there may be more brainstem involvement in HD compared to HDL2 in which is rather concentrated to cortical involvement (with prominent occipital atrophy) [20, 24]. Neuroimaging may reveal generalized brain atrophy, predominantly affecting the caudate head and putamen. The presence of a putaminal rim hyperintensity has been described (Fig. 7.1).

Gene function remains poorly understood. The encoded protein plays a role in stabilization of junctional membrane complexes and regulates neuronal calcium flux. Consequently, loss of JPH3 protein expression may lead to cellular vulnerability due to disruption of calcium flux or endoplasmic reticulum dysregulation. However, a multifactorial pathogenic mechanism is suggested with a toxic effect caused by both a toxic loss of JPH3 expression on the one hand and toxic gain of function of JPH3 RNA on the other hand [25, 26].

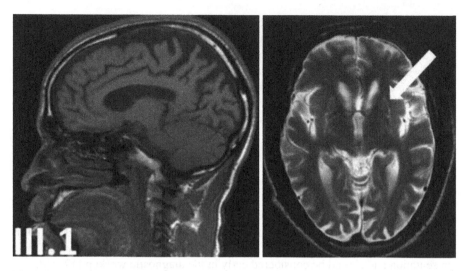

Fig. 7.1 Magnetic resonance images of a patient with a 10-year disease duration of HDL2. The number *III.1* refers to the family member of the African-American pedigree described by Schneider et al. [19] The sagittal brain MRIs demonstrate generalized brain atrophy which predominantly affects the caudate heads and putamina bilaterally in HDL2. The *arrow* emphasizes a putaminal rim hyperintensity. The brainstem and cerebellum are relatively well preserved

Huntington Disease-Like 3

In view of the recessive pattern of inheritance, HDL3 will be discussed below.

Spinocerebellar Ataxia Type 17: Huntington Disease-Like 4

Triplet repeat expansions in the *TATA box-binding protein* (TBP) gene located on chromosome 6q27 cause HDL4, also classified as spinocerebellar ataxia type 17 (SCA17). HDL4 is the most frequent identified HD phenocopy syndrome among Caucasians. However, among SCA17 patients, chorea only occurs in about 20 % of patients [14].

Inheritance is autosomal dominant. The normal allele size ranges from 25 to 40 CAG/CAA repeats, whereas 41 to 48 CAG/CAA repeats lead to reduced and 49 and greater repeats to full penetrance [27]. Similar to HD, a negative correlation between the size of repeat expansion and the age of onset and intergenerational instability with anticipation have been recognized [28, 29]. The encoded protein, TBP, functions as transcription initiation factor [30]. Interestingly, transcriptional dysregulation also plays a role in the pathophysiology of HD [31]. Animal models of SCA17 have been developed and shed further light on the underlying pathophysiology [32, 33].

The clinical phenotype of HDL4/SCA17 is markedly heterogeneous, and the age at onset ranges from age 3–75 years [34]. Cerebellar ataxia is the most common clinical feature (95 %), usually presenting with a slowly progressive course, but rapid progression resembling paraneoplastic disorders or prion disease has been reported [35]. Extrapyramidal signs (73 %), in particular dystonia and chorea, and dementia (76 %) frequently occur in SCA17. A Huntington-like presentation occurs in a subset of patients [6, 7]. Furthermore, pyramidal signs, epilepsy, and psychiatric disturbances are not uncommon [36]. Similarly, there is a broad neuropathological variation with wide participation across the CNS reflecting the broad clinical spectrum [37]. Thus, the cerebellum, cerebral neocortex, basal ganglia (in particular the caudate nucleus), and hippocampus may be involved. Neuronal intranuclear inclusions containing the abnormal protein TBP, ubiquitin, 1C2, and other proteins are widely distributed throughout the brain gray matter [38]. MR imaging demonstrated atrophy of the cerebellum and the caudate nucleus. Rim enhancement of the putamen has also been described [39, 40] but is not specific to HDL4/SCA17 as it may also occur in HDL2 [20] and other disorders. Similar to HD, presymptomatic imaging changes may be detected. Thus, MRI volumetry, as well as (11) C-raclopride and (18)F-FDG PET, reveals neuronal dysfunction and neurodegeneration even in the presymptomatic stage [40].

Notably, other forms of SCAs may also present with chorea and should be kept in mind in patients with an ataxic HDL phenotype, in particular SCA 1, 2, and 3 [41].

Dentatorubral-Pallidoluysian Atrophy

DRPLA shares many key characteristics of HD and the HDLs. It is a trinucleotide repeat disorder with autosomal dominant inheritance. The affected gene, *atrophin 1* (*ATN1*), is located on chromosome 12p13.31. In DRPLA repeat expansions in exon 5 range from 49 to 88, compared to 8 to 25 repeats in healthy individuals. The length of the CAG repeats correlates inversely with age of onset and directly with disease severity [42–44]. Like in HD and many other polyglutamine diseases, the phenomenon of marked anticipation occurs with longer stretches, particularly in the context of paternal transmission [44, 45].

The average age of onset of DRPLA is between 20 and 30 years and leads to death within 10–15 years [46]. The clinical presentation of DRPLA is, however, considerably heterogeneous and shows an age-dependent phenotype. Juvenile-onset cases develop severe progressive myoclonus epilepsy and cognitive decline. In contrast, adults tend to develop ataxia, choreoathetosis, and dementia as cardinal features which may resemble HD [47]. The common MRI findings in DRPLA are atrophy of the cerebellum and the brainstem in particular affecting the pontine tegmentum. Furthermore, in adult-onset DRPLA, diffuse high signal intensity lesions in the cerebral white matter are observed, in contrast to HD. However, rather than formation of neuronal intranuclear inclusions, diffuse accumulation of mutant DRPLA protein is found in the neuronal nuclei [46]. It has been suggested that prominent cortical involvement in juvenile-onset DRPLA patients may explain the severe cognitive deterioration and epilepsy in these patients [48].

DRPLA clusters in Japan, where the prevalence is estimated to be similar to the prevalence of HD. The condition is rare in other regions [10]: a recent review identified 183 non-Asian patients of 27 families reported with DRPLA [49]. In this study the clinicogenetic phenomenology was similar between Asian and non-Asian DRPLA patients.

Neuroferritinopathy

Neuroferritinopathy is a progressive autosomal dominant neurodegenerative disease characterized by elevated serum ferritin levels caused by mutations in the *ferritin light chain* gene (*FTL1*) located on chromosome 19q13 [50].

Like in HD, onset of neuroferritinopathy is in midlife, but early onset in teenage years and late onset in the sixth decade have also been reported [51]. Typically, the disease presents with chorea or dystonia, while other clinical features like parkinsonism, cerebellar signs, dysarthria, frontal lobe syndrome, and dementia may be variably present. Cognitive deficits and psychiatric features appear to be less prominent compared to HD. Treatment is symptomatic. The literature includes reports on iron chelation therapy in individual cases resulting in clinical deterioration in one and no change in two other patients [52].

MRI findings in neuroferritinopathy show a broad variety and may include progressive cystic degeneration of the basal ganglia, particularly cavitation of the globus pallidus and putamen, and thalamic T2 hypointense lesions reflecting iron deposits in addition to cortical atrophy [53, 54]. In rare cases, a pattern resembling the eye-of-the-tiger sign otherwise described in pantothenate kinase-associated neurodegeneration (PKAN) has been reported in neuroferritinopathy [55]. Ferritin is a ubiquitous iron storage protein, and dysfunction results in formation of iron-rich intranuclear and intracytoplasmic inclusion bodies not only within neurons and glia in the brain but also in the peripheral nerves, skin, muscles, liver, and even kidneys. Iron deposition is particularly prominent in the basal ganglia.

So far, only a handful of different mutations in the *FTL1* gene causing neuroferritinopathy have been described (in cases from England, France/French Canada, Spain/Portugal, Australia, and Japan) mainly affecting the tertiary structure of the ferritin light chain polypeptide [50, 56–61]. Notably, mutations in the 5' non-translated region of the same gene (resulting in elevated L-ferritin production irrespective of iron levels) have been associated with a distinct disorder, hereditary hyperferritinemia cataract syndrome, characterized by early-onset bilateral cataracts due to intracellular accumulation of ferritin in the lens (in the absence of neurological features). The prevalence of the latter has been estimated to a minimum of 1 in 200,000 [62–64]. In Germany, for example, about 50–100 cases have been recognized (oral communication). It yet remains unclear as to why deposits are found in the brain (iron and ferritin) in the one disease and in the lens (ferritin) in the other. From the genetic point of view, in neuroferritinopathy mutations are mainly found towards the 3' region, whereas in hereditary hyperferritinemia cataract syndrome mutations are located in the 5' non-translated region, mostly in the iron-responsive element (IRE) of the gene.

Benign Hereditary Chorea

Benign hereditary chorea (BHC) is a rare autosomal dominant disease which is characterized by nonprogressive chorea with onset in childhood and absence of dementia and caudate atrophy. Several mutations in the associated small *TITF1* (*NKX2-1*) gene but also deletions (in some cases also encompassing adjacent genes) have been described. About 30 different mutations have been reported [65]. The encoded thyroid transcription factor 1 is essential for the organogenesis of the lungs, thyroid, and basal ganglia [66, 67].

The typical clinical phenotype is infancy-onset hypotonia and chorea. Other movement disorders like myoclonus, dystonia, tics, tremor, and ataxia may be associated. Chorea may improve or resolve in adulthood; however, it may also persist as mild chorea or convert to disabling myoclonus. Notably, learning difficulties are not infrequent, and involvement of the thyroid (67 %) and lung (46 %) may also occur [65]. Symptomatic relief may be achieved with tetrabenazine, levodopa, haloperidol, chlorpromazine, or prednisone [65, 68, 69].

Neuroimaging is usually normal apart from reduced basal ganglia and thalamic uptake in single-photon emission computed tomography [70]. Pathological studies do not reveal significant abnormalities using standard methods. Using immunohistochemical staining loss of most TITF-1-mediated striatal interneurons was revealed in BHC brains [71, 72].

Shimohata et al. [73] proposed genetic heterogeneity of BHC when he reported two Japanese families with autosomal dominant adult-onset slowly progressive chorea without dementia. MRI was normal and HD, BHC, and other HDL syndromes had been excluded by genetic testing. Genetic workup showed linkage to chromosome 8q21.3-q23.3. Due to the clinical resemblance with BHC, the disease was named benign hereditary chorea type 2 [73].

Selected Autosomal Recessive Chorea Syndromes

When the family history is incomplete or no information can be retrieved, autosomal recessive chorea syndromes should also be considered in patients with choreic phenotype resembling HD as some of these may produce similar phenotypes. However, in view of word limitations, only an incomplete selection can be discussed in the following.

Huntington Disease-Like 3

HDL3 is an autosomal recessive Huntington disease-like neurodegenerative disorder described in a single Saudi Arabian family. Considering the early onset and the recessive pattern of inheritance, HDL3 clearly differs from the other HDL syndromes and is thus described in this section. The clinical phenotype was complex with childhood-onset mental deterioration, speech disturbance, dystonia, chorea, and other extrapyramidal and pyramidal features. MRI showed progressive bilateral atrophy of the caudate nucleus and the frontal cortex, and a link to HD was suggested by the authors. The causative gene still remains unclear, but the disease locus initially was mapped to chromosome 4p15.3 [74]. No similar families have been described as yet.

Chorea-Acanthocytosis and McLeod Syndrome

Both chorea-acanthocytosis and McLeod syndrome are core neuroacanthocytosis syndromes characterized by neurodegeneration of the basal ganglia and red cell acanthocytosis [75] (Fig. 7.2).

Chorea-acanthocytosis (ChAc) is a rare autosomal recessive neurodegenerative disorder due to mutations in the *VPS13A* gene on chromosome 9 encoding for chorein [77, 78]. It is estimated that about 1,000 ChAc cases exist worldwide [79]. ChAc causes movement disorders (including chorea, dystonia, parkinsonism, and tics), cognitive impairment, and psychiatric features with great similarities to HD. However, clinical characteristics like dystonia with prominent orofacial involvement with tongue protrusion, involuntary tongue- and lip biting, head thrusts, and rubberman-like appearance may indicate a diagnosis distinct from classic HD [80, 81]. Furthermore, seizures (which occur infrequently in late-onset HD) are seen in half of patients, and myopathy and axonal neuropathy are common [82]. The disease usually starts in the 20s and progresses slowly over 15–30 years [75].

Blood tests reveal elevated levels of creatine phosphokinase (CK) in most cases. The detection of acanthocytosis often remains elusive, although the probability to detect the characteristic deformed erythrocytes can be increased by using a 1:1 dilution with physiological saline and phase-contrast microscopy [83] (Fig. 7.2). However, many hematology laboratories no longer prepare wet blood films due to health and safety policies [84], and analysis of protein (chorein) levels is therefore recommended. The function of the protein is not fully understood, but a yet unknown role of chorein in regulation of secretion and aggregation of blood platelets has recently been suggested [85].

Neuroradiological findings include progressive striatal atrophy with a maximum in the caudate head [86, 87]. Postmortem examinations have shown neuronal loss and gliosis predominantly affecting the caudate nucleus, putamen, globus pallidus, thalamus, and substantia nigra. In comparison to HD, neither significant cortical

Fig. 7.2 An electron micrograph of a peripheral blood film demonstrating acanthocytes (*arrowed*) (Image courtesy of the National Institutes of Health Clinical Center and National Heart Lung and Blood Institute (McDonald Horne, Kazuyo Takeda, Zu-Xi Yu, Bill Riemenschneider and Adrian Danek) [76])

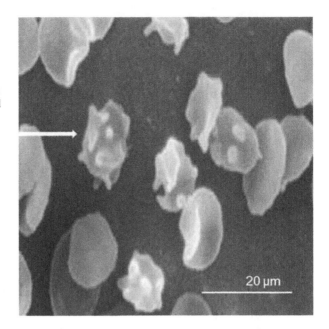

pathology nor specific neuropathological features as inclusion bodies have been detected [88, 89].

Treatment is symptomatic. When intractable to conventional treatments, such as tetrabenazine, atypical antipsychotic agents, and botulinum toxin injections, deep brain stimulation may be an option in individual cases (Table 7.2) [104, 105, [106].

McLeod syndrome is inherited in an X-linked matter and is caused by mutations in the *XK* gene commonly resulting in absence or truncation of the encoded XK protein.

Final Remarks

As outlined above, an increasing number of genetically defined choreic syndromes have been delineated which should come into the differential diagnosis, in particular when testing for HD is negative. However, there are of course also other, nongenetic causes including infectious or paraneoplastic and metabolic causes which are beyond the scope of this chapter.

Genetic testing has familial implications, and interdisciplinary counseling should be offered to patients and their families accordingly [107].

Treatment of HD and HD-like disorders remains symptomatic, and recently, evidence-based guidelines have been proposed [108]. Dopamine-depleting agents such as tetrabenazine and the glutamate antagonist riluzole have received a level B rating of evidence, while there is insufficient data to make recommendations regarding

Table 7.2 Summary of DBS-treated Huntington and chorea-acanthocytosis patients

Reference	Diagnosis	Number of patients	Age at onset (years)	Age at operation (years)	DBS target	Result[a]
Moro et al. [90]	HD	1	35	43	GPi	30–40 % motor improvement on UHDRS; 10–20 % functional improvement
Hebb et al. [91]	HD	1	28	41	Posteroventral GPi	Improvement of chorea and in overall motor functioning
Fasano et al. [92]	HD	1	55	72	GPi	Improvement of UHDRS 17→4 (at 1-year follow-up)
Biolsi et al. [93]	HD	1	50	60	GPi	Improvement sustained after 4 years
Kang et al. [94]	HD	2	45/42(?)	55/48	GPi	At 2-year follow-up improvement of chorea, however, continuation of deterioration of gait, bradykinesia, and dystonia scores
Spielberger et al. [95]	HD	1	21	30	GPi	UHDRS 79 (pre-op)→62 (post-op)→82 (at 4-year follow-up)
Garcia-Ruiz et al. [96]	HD	1	20	30	GPi	UHDRS 63 (pre-op)→44 (at 6 months)→33 (at 1-year follow-up) including improvement of abnormal vocalizations
Wihl et al. [97]	ChAc	1	35	38	GPi	No improvement
Burbaud et al. [98, 99]	ChAc	1	31	43	Thalamus (ventral oral part of motor thalamus)	Improvement (Barthel index 15→35 out of 100; Marsden and Schachter choreic score 54→34 out of 60); 2-year follow-up
Guehl et al. [100]	ChAc	1	24	32	GPi	Improvement
	McLeod syndrome	1	n.d.	n.d.		
Ruiz et al. [101]	ChAc	1	25	35	GPi	UHDRS 24→14 at 2-month follow-up
Li et al. [102]	ChAc	2	17/18	39/30	GPi	UHDRS 36→13 in patient 1; 53→26 in patient 2
Shin et al. [103]	ChAc	1	35	39	GPi	UHDRS 44→12

Kefalopoulou et al. [104]	ChAc	2	n.d./32	54/43	GPi	Improvement of UHDRS total 42/32 %; improvement of UHDRS chorea subscore 87/67 %
Lim et al. [105]	ChAc	1	26	32	GPi	50 % improvement from baseline; 8-month follow-up

ChAc Chorea-acanthocytosis, *HD* Huntington disease

[a]Not all information from original reports may be summarized

the use of typical and atypical neuroleptics (including haloperidol and olanzapine, clozapine, and quetiapine). GABAergic agents (clonazepam, gabapentin, and valproate) can be used as adjunctive therapy. Possible side effects should be discussed which include occurrence of depression/suicidality and parkinsonism with tetrabenazine and elevated liver enzymes with riluzole.

Deep brain stimulation has been used in selected cases (Table 7.2) but will also not halt the inevitable deterioration of these patients owing to the degenerative nature of these diseases. Therapy should also address accompanying symptoms (depression, etc.), and supportive treatments (physiotherapy, speech therapy, etc.) should be offered.

References

1. Tabrizi SJ, Scahill RI, Durr A, et al. Biological and clinical changes in premanifest and early stage Huntington's disease in the TRACK-HD study: the 12-month longitudinal analysis. Lancet Neurol. 2011;10:31–42.
2. Losekoot M, van Belzen MJ, Seneca S, Bauer P, Stenhouse SA, Barton DE. EMQN/CMGS best practice guidelines for the molecular genetic testing of Huntington disease. Eur J Hum Genet. 2013;21:480–6.
3. Tabrizi SJ, Reilmann R, Roos RA, et al. Potential endpoints for clinical trials in premanifest and early Huntington's disease in the TRACK-HD study: analysis of 24 month observational data. Lancet Neurol. 2012;11:42–53.
4. Wild EJ, Tabrizi SJ. Predict-HD and the future of therapeutic trials. Lancet Neurol. 2006;5:724–5.
5. Ross CA, Tabrizi SJ. Huntington's disease: from molecular pathogenesis to clinical treatment. Lancet Neurol. 2011;10:83–98.
6. Schneider SA, Walker RH, Bhatia KP. The Huntington's disease-like syndromes: what to consider in patients with a negative Huntington's disease gene test. Nat Clin Pract Neurol. 2007;3:517–25.
7. Wild EJ, Tabrizi SJ. Huntington's disease phenocopy syndromes. Curr Opin Neurol. 2007;20:681–7.
8. Moore RC, Xiang F, Monaghan J, et al. Huntington disease phenocopy is a familial prion disease. Am J Hum Genet. 2001;69:1385–8.
9. Schneider SA, Bhatia KP. Huntington's disease look-alikes. Handb Clin Neurol. 2011;100: 101–12.
10. Martino D, Stamelou M, Bhatia KP. The differential diagnosis of Huntington's disease-like syndromes: 'red flags' for the clinician. J Neurol Neurosurg Psychiatry. 2013;84(6):650–6.
11. Xiang F, Almqvist EW, Huq M, et al. A Huntington disease-like neurodegenerative disorder maps to chromosome 20p. Am J Hum Genet. 1998;63:1431–8.
12. Laplanche JL, Hachimi KH, Durieux I, et al. Prominent psychiatric features and early onset in an inherited prion disease with a new insertional mutation in the prion protein gene. Brain. 1999;122(Pt 12):2375–86.
13. Stevanin G, Camuzat A, Holmes SE, et al. CAG/CTG repeat expansions at the Huntington's disease-like 2 locus are rare in Huntington's disease patients. Neurology. 2002;58:965–7.
14. Stevanin G, Fujigasaki H, Lebre AS, et al. Huntington's disease-like phenotype due to trinucleotide repeat expansions in the TBP and JPH3 genes. Brain. 2003;126:1599–603.
15. Wild EJ, Mudanohwo EE, Sweeney MG, et al. Huntington's disease phenocopies are clinically and genetically heterogeneous. Mov Disord. 2008;23:716–20.
16. Paradisi I, Ikonomu V, Arias S. Huntington disease-like 2 (HDL2) in Venezuela: frequency and ethnic origin. J Hum Genet. 2013;58:3–6.

17. Krause A, Greenberg J. Genetic testing for Huntington's disease in South Africa. S Afr Med J. 2008;98:193–4.
18. Magazi DS, Krause A, Bonev V, et al. Huntington's disease: genetic heterogeneity in black African patients. S Afr Med J. 2008;98:200–3.
19. Schneider SA, Marshall KE, Xiao J, LeDoux MS. JPH3 repeat expansions cause a progressive akinetic-rigid syndrome with severe dementia and putaminal rim in a five-generation African-American family. Neurogenetics. 2012;13:133–40.
20. Greenstein PE, Vonsattel JP, Margolis RL, Joseph JT. Huntington's disease like-2 neuropathology. Mov Disord. 2007;22:1416–23.
21. Margolis RL, Rudnicki DD, Holmes SE. Huntington's disease like-2: review and update. Acta Neurol Taiwan. 2005;14:1–8.
22. Holmes SE, O'Hearn E, Rosenblatt A, et al. A repeat expansion in the gene encoding junctophilin-3 is associated with Huntington disease-like 2. Nat Genet. 2001;29:377–8.
23. Margolis RL, Holmes SE, Rosenblatt A, et al. Huntington's disease-like 2 (HDL2) in North America and Japan. Ann Neurol. 2004;56:670–4.
24. Rudnicki DD, Pletnikova O, Vonsattel JP, Ross CA, Margolis RL. A comparison of Huntington disease and Huntington disease-like 2 neuropathology. J Neuropathol Exp Neurol. 2008;67:366–74.
25. Seixas AI, Holmes SE, Takeshima H, et al. Loss of junctophilin-3 contributes to Huntington disease-like 2 pathogenesis. Ann Neurol. 2012;71:245–57.
26. Rudnicki DD, Holmes SE, Lin MW, Thornton CA, Ross CA, Margolis RL. Huntington's disease–like 2 is associated with CUG repeat-containing RNA foci. Ann Neurol. 2007;61:272–82.
27. Toyoshima Y, Onodera O, Yamada M, Tsuji S, Takahashi H. Spinocerebellar ataxia type 17. In: Pagon RA, Adam MP, Ardinger HH, Bird TD, Dolan CR, Fong CT, Smith RJH, Stephens K, editors. GeneReviews® [Internet]. University of Washington, Seattle, WA; 1993–2014. 2005 Mar 29 [updated 2012 May 17].
28. Gao R, Matsuura T, Coolbaugh M, et al. Instability of expanded CAG/CAA repeats in spinocerebellar ataxia type 17. Eur J Hum Genet. 2008;16:215–22.
29. Rasmussen A, De Biase I, Fragoso-Benitez M, et al. Anticipation and intergenerational repeat instability in spinocerebellar ataxia type 17. Ann Neurol. 2007;61:607–10.
30. Nakamura K, Jeong SY, Uchihara T, et al. SCA17, a novel autosomal dominant cerebellar ataxia caused by an expanded polyglutamine in TATA-binding protein. Hum Mol Genet. 2001;10:1441–8.
31. Cha JH. Transcriptional dysregulation in Huntington's disease. Trends Neurosci. 2000;23:387–92.
32. Ren J, Jegga AG, Zhang M, et al. A Drosophila model of the neurodegenerative disease SCA17 reveals a role of RBP-J/Su(H) in modulating the pathological outcome. Hum Mol Genet. 2011;20:3424–36.
33. Chang YC, Lin CY, Hsu CM, et al. Neuroprotective effects of granulocyte-colony stimulating factor in a novel transgenic mouse model of SCA17. J Neurochem. 2011;118:288–303.
34. Stevanin G, Brice A. Spinocerebellar ataxia 17 (SCA17) and Huntington's disease-like 4 (HDL4). Cerebellum. 2008;7:170–8.
35. Mehanna R, Itin I. From normal gait to loss of ambulation in 6 months: a novel presentation of SCA17. Cerebellum. 2013;12(4):568–71.
36. Craig K, Keers SM, Walls TJ, Curtis A, Chinnery PF. Minimum prevalence of spinocerebellar ataxia 17 in the North east of England. J Neurol Sci. 2005;239:105–9.
37. Seidel K, Siswanto S, Brunt ER, den Dunnen W, Korf HW, Rub U. Brain pathology of spinocerebellar ataxias. Acta Neuropathol. 2012;124:1–21.
38. Rolfs A, Koeppen AH, Bauer I, et al. Clinical features and neuropathology of autosomal dominant spinocerebellar ataxia (SCA17). Ann Neurol. 2003;54:367–75.
39. Loy CT, Sweeney MG, Davis MB, et al. Spinocerebellar ataxia type 17: extension of phenotype with putaminal rim hyperintensity on magnetic resonance imaging. Mov Disord. 2005;20:1521–3.

40. Brockmann K, Reimold M, Globas C, et al. PET and MRI reveal early evidence of neurode-generation in spinocerebellar ataxia type 17. J Nucl Med. 2012;53:1074–80.
41. Schols L, Bauer P, Schmidt T, Schulte T, Riess O. Autosomal dominant cerebellar ataxias: clinical features, genetics, and pathogenesis. Lancet Neurol. 2004;3:291–304.
42. Naito H, Oyanagi S. Familial myoclonus epilepsy and choreoathetosis: hereditary dentatorubral-pallidoluysian atrophy. Neurology. 1982;32:798–807.
43. Nagafuchi S, Yanagisawa H, Sato K, et al. Dentatorubral and pallidoluysian atrophy expansion of an unstable CAG trinucleotide on chromosome 12p. Nat Genet. 1994;6:14–8.
44. Koide R, Ikeuchi T, Onodera O, et al. Unstable expansion of CAG repeat in hereditary dentatorubral-pallidoluysian atrophy (DRPLA). Nat Genet. 1994;6:9–13.
45. Nagafuchi S, Yanagisawa H, Ohsaki E, et al. Structure and expression of the gene responsible for the triplet repeat disorder, dentatorubral and pallidoluysian atrophy (DRPLA). Nat Genet. 1994;8:177–82.
46. Tsuji S. Dentatorubral-pallidoluysian atrophy. Handb Clin Neurol. 2012;103:587–94.
47. Tsuji S. Dentatorubral-pallidoluysian atrophy (DRPLA): clinical features and molecular genetics. Adv Neurol. 1999;79:399–409.
48. Sunami Y, Koide R, Arai N, Yamada M, Mizutani T, Oyanagi K. Radiologic and neuropatho-logic findings in patients in a family with dentatorubral-pallidoluysian atrophy. AJNR Am J Neuroradiol. 2011;32:109–14.
49. Wardle M, Morris HR, Robertson NP. Clinical and genetic characteristics of non-Asian dentatorubral-pallidoluysian atrophy: a systematic review. Mov Disord. 2009;24:1636–40.
50. Curtis AR, Fey C, Morris CM, et al. Mutation in the gene encoding ferritin light polypeptide causes dominant adult-onset basal ganglia disease. Nat Genet. 2001;28:350–4.
51. Chinnery PF. Neuroferritinopathy. In: GeneReviews® [Internet]. University of Washington, Seattle, WA; 1993–2014. 2005 Apr 25 [updated 2010 Dec 23].
52. Chinnery PF, Crompton DE, Birchall D, et al. Clinical features and natural history of neurofer-ritinopathy caused by the FTL1 460InsA mutation. Brain. 2007;130:110–9.
53. Ohta E, Takiyama Y. MRI findings in neuroferritinopathy. Neurol Res Int. 2012;2012:197438.
54. McNeill A, Gorman G, Khan A, Horvath R, Blamire AM, Chinnery PF. Progressive brain iron accumulation in neuroferritinopathy measured by the thalamic T2* relaxation rate. AJNR Am J Neuroradiol. 2012;33:1810–3.
55. Shah SO, Mehta H, Fekete R. Late-onset neurodegeneration with brain iron accumulation with diffusion tensor magnetic resonance imaging. Case Rep Neurol. 2012;4:216–23.
56. Vidal R, Ghetti B, Takao M, et al. Intracellular ferritin accumulation in neural and extraneural tissue characterizes a neurodegenerative disease associated with a mutation in the ferritin light polypeptide gene. J Neuropathol Exp Neurol. 2004;63:363–80.
57. Ohta E, Nagasaka T, Shindo K, et al. Neuroferritinopathy in a Japanese family with a duplica-tion in the ferritin light chain gene. Neurology. 2008;70:1493–4.
58. Mancuso M, Davidzon G, Kurlan RM, et al. Hereditary ferritinopathy: a novel mutation, its cellular pathology, and pathogenetic insights. J Neuropathol Exp Neurol. 2005;64:280–94.
59. Maciel P, Cruz VT, Constante M, et al. Neuroferritinopathy: missense mutation in FTL causing early-onset bilateral pallidal involvement. Neurology. 2005;65:603–5.
60. Kubota A, Hida A, Ichikawa Y, et al. A novel ferritin light chain gene mutation in a Japanese family with neuroferritinopathy: description of clinical features and implications for genotype-phenotype correlations. Mov Disord. 2009;24:441–5.
61. Devos D, Tchofo PJ, Vuillaume I, et al. Clinical features and natural history of neuroferritinopathy caused by the 458dupA FTL mutation. Brain. 2009;132:e109.
62. Nonnenmacher L, Langer T, Blessing H, et al. Hereditary hyperferritinemia cataract syndrome: clinical, genetic, and laboratory findings in 5 families. Klin Padiatr. 2011;223:346–51.
63. Girelli D, Olivieri O, De Franceschi L, Corrocher R, Bergamaschi G, Cazzola M. A linkage between hereditary hyperferritinaemia not related to iron overload and autosomal dominant congenital cataract. Br J Haematol. 1995;90:931–4.
64. Craig JE, Clark JB, McLeod JL, et al. Hereditary hyperferritinemia-cataract syndrome: preva-lence, lens morphology, spectrum of mutations, and clinical presentations. Arch Ophthalmol. 2003;121:1753–61.

65. Gras D, Jonard L, Roze E, et al. Benign hereditary chorea: phenotype, prognosis, therapeutic outcome and long term follow-up in a large series with new mutations in the TITF1/NKX2-1 gene. J Neurol Neurosurg Psychiatry. 2012;83:956–62.
66. Devriendt K, Vanhole C, Matthijs G, de Zegher F. Deletion of thyroid transcription factor-1 gene in an infant with neonatal thyroid dysfunction and respiratory failure. N Engl J Med. 1998;338:1317–8.
67. Kimura S. Thyroid-specific enhancer-binding protein Role in thyroid function and organogenesis. Trends Endocrinol Metab. 1996;7:247–52.
68. Wheeler PG, Weaver DD, Dobyns WB. Benign hereditary chorea. Pediatr Neurol. 1993;9:337–40.
69. Asmus F, Horber V, Pohlenz J, et al. A novel TITF-1 mutation causes benign hereditary chorea with response to levodopa. Neurology. 2005;64:1952–4.
70. Mahajnah M, Inbar D, Steinmetz A, Heutink P, Breedveld GJ, Straussberg R. Benign hereditary chorea: clinical, neuroimaging, and genetic findings. J Child Neurol. 2007;22:1231–4.
71. Kleiner-Fisman G, Rogaeva E, Halliday W, et al. Benign hereditary chorea: clinical, genetic, and pathological findings. Ann Neurol. 2003;54:244–7.
72. Kleiner-Fisman G, Calingasan NY, Putt M, Chen J, Beal MF, Lang AE. Alterations of striatal neurons in benign hereditary chorea. Mov Disord. 2005;20:1353–7.
73. Shimohata T, Hara K, Sanpei K, et al. Novel locus for benign hereditary chorea with adult onset maps to chromosome 8q21.3 q23.3. Brain. 2007;130:2302–9.
74. Kambouris M, Bohlega S, Al-Tahan A, Meyer BF. Localization of the gene for a novel autosomal recessive neurodegenerative Huntington-like disorder to 4p15.3. Am J Hum Genet. 2000;66:445–52.
75. Walker RH, Jung HH, Dobson-Stone C, et al. Neurologic phenotypes associated with acanthocytosis. Neurology. 2007;68:92–8.
76. Sokolov E, Schneider SA, Bain PG. Chorea-acanthocytosis. Pract Neurol. 2012;12:40–3.
77. Rampoldi L, Dobson-Stone C, Rubio JP, et al. A conserved sorting-associated protein is mutant in chorea-acanthocytosis. Nat Genet. 2001;28:119–20.
78. Ueno S, Maruki Y, Nakamura M, et al. The gene encoding a newly discovered protein, chorein, is mutated in chorea-acanthocytosis. Nat Genet. 2001;28:121–2.
79. Jung HH, Danek A, Walker RH. Neuroacanthocytosis syndromes. Orphanet J Rare Dis. 2011;6:68.
80. Bader B, Walker RH, Vogel M, Prosiegel M, McIntosh J, Danek A. Tongue protrusion and feeding dystonia: a hallmark of chorea-acanthocytosis. Mov Disord. 2010;25:127–9.
81. Schneider SA, Lang AE, Moro E, Bader B, Danek A, Bhatia KP. Characteristic head drops and axial extension in advanced chorea-acanthocytosis. Mov Disord. 2010;25:1487–91.
82. Danek A, Walker RH. Neuroacanthocytosis. Curr Opin Neurol. 2005;18:386–92.
83. Storch A, Kornhass M, Schwarz J. Testing for acanthocytosis A prospective reader-blinded study in movement disorder patients. J Neurol. 2005;252:84–90.
84. Alawneh J, Baker MR, Young GR. Blood films in the investigation of chorea. Pract Neurol. 2012;12:268.
85. Schmidt EM, Schmid E, Munzer P, et al. Chorein sensitivity of cytoskeletal organization and degranulation of platelets. FASEB J. 2013;27(7):2799–806.
86. Henkel K, Danek A, Grafman J, Butman J, Kassubek J. Head of the caudate nucleus is most vulnerable in chorea-acanthocytosis: a voxel-based morphometry study. Mov Disord. 2006;21:1728–31.
87. Walterfang M, Yucel M, Walker R, et al. Adolescent obsessive compulsive disorder heralding chorea-acanthocytosis. Mov Disord. 2008;23:422–5.
88. Hardie RJ, Pullon HW, Harding AE, et al. Neuroacanthocytosis. A clinical, haematological and pathological study of 19 cases. Brain. 1991;114(Pt 1A):13–49.
89. Rinne JO, Daniel SE, Scaravilli F, Pires M, Harding AE, Marsden CD. The neuropathological features of neuroacanthocytosis. Mov Disord. 1994;9:297–304.
90. Moro E, Lang AE, Strafella AP, Poon YY, Arango PM, Dagher A, Hutchison WD, Lozano AM. Bilateral globus pallidus stimulation for Huntington's disease. Ann Neurol. 2004;56(2):290–4.

91. Hebb MO, Garcia R, Gaudet P, Mendez IM. Bilateral stimulation of the globus pallidus internus to treat choreathetosis in Huntington's disease: technical case report. Neurosurgery. 2006;58(2):E383; discussion E383.
92. Fasano A, Mazzone P, Piano C, Quaranta D, Soleti F, Bentivoglio AR. GPi-DBS in Huntington's disease: results on motor function and cognition in a 72-year-old case. Mov Disord. 2008;23(9):1289–92.
93. Biolsi B, Cif L, Fertit HE, Robles SG, Coubes P. Long-term follow-up of Huntington disease treated by bilateral deep brain stimulation of the internal globus pallidus. J Neurosurg. 2008;109(1):130–2.
94. Kang GA, Heath S, Rothlind J, Starr PA. Long-term follow-up of pallidal deep brain stimulation in two cases of Huntington's disease. J Neurol Neurosurg Psychiatry. 2011;82(3):272–7.
95. Spielberger S, Hotter A, Wolf E, Eisner W, Müller J, Poewe W, Seppi K. Deep brain stimulation in Huntington's disease: a 4-year follow-up case report. Mov Disord. 2012;27(6):806–7.
96. Garcia-Ruiz PJ, Ayerbe J, del Val J, Herranz A. Deep brain stimulation in disabling involuntary vocalization associated with Huntington's disease. Parkinsonism Relat Disord. 2012;18(6):803–4.
97. Wihl G, Volkmann J, Allert N, Lehrke R, Sturm V, Freund HJ. Deep brain stimulation of the internal pallidum did not improve chorea in a patient with neuro-acanthocytosis. Mov Disord. 2001;16(3):572–5.
98. Burbaud P, Rougier A, Ferrer X, Guehl D, Cuny E, Arne P, Gross C, Bioulac B. Improvement of severe trunk spasms by bilateral high-frequency stimulation of the motor thalamus in a patient with chorea-acanthocytosis. Mov Disord. 2002;17(1):204–7.
99. Burbaud P, Vital A, Rougier A, Bouillot S, Guehl D, Cuny E, Ferrer X, Lagueny A, Bioulac B. Minimal tissue damage after stimulation of the motor thalamus in a case of chorea-acanthocytosis. Neurology. 2002;59(12):1982–4.
100. Guehl D, Cuny E, Tison F, Benazzouz A, Bardinet E, Sibon Y, Ghorayeb I, Yelnick J, Rougier A, Bioulac B, Burbaud P. Deep brain pallidal stimulation for movement disorders in neuro-acanthocytosis. Neurology. 2007;68(2):160–1.
101. Ruiz PJ, Ayerbe J, Bader B, Danek A, Sainz MJ, Cabo I, Frech FA. Deep brain stimulation in chorea acanthocytosis. Mov Disord. 2009;24(10):1546–7.
102. Li P, Huang R, Song W, Ji J, Burgunder JM, Wang X, Zhong Q, Kaelin-Lang A, Wang W, Shang HF. Deep brain stimulation of the globus pallidus internal improves symptoms of chorea-acanthocytosis. Neurol Sci. 2012;33(2):269–74.
103. Shin H, Ki CS, Cho AR, Lee JI, Ahn JY, Lee JH, Cho JW. Globus pallidus interna deep brain stimulation improves chorea and functional status in a patient with chorea-acanthocytosis. Stereotact Funct Neurosurg. 2012;90(4):273–7.
104. Kefalopoulou Z, Zrinzo L, Aviles-Olmos I, et al. Deep brain stimulation as a treatment for chorea-acanthocytosis. J Neurol. 2013;260:303–5.
105. Lim TT, Fernandez HH, Cooper S, Wilson KM, Machado AG. Successful DBS surgery with intraoperative MRI on a difficult neuroacanthocytosis case. Neurosurgery. 2013;73(1):E184–7.
106. Miquel M, Spampinato U, Latxague C, Aviles-Olmos I, Bader B, Bertram K, Bhatia K, Burbaud P, Burghaus L, Cho JW, Cuny E, Danek A, Foltynie T, Garcia Ruiz PJ, Giménez-Roldán S, Guehl D, Guridi J, Hariz M, Jarman P, Kefalopoulou ZM, Limousin P, Lipsman N, Lozano AM, Moro E, Ngy D, Rodriguez-Oroz MC, Shang H, Shin H, Walker RH, Yokochi F, Zrinzo L, Tison F. Short and long term outcome of bilateral pallidal stimulation in chorea-acanthocytosis. PLoS One. 2013;8(11):e79241.
107. Schneider SA, Klein C. What is the role of genetic testing in movement disorders practice? Curr Neurol Neurosci Rep. 2011;11:351–61.
108. Armstrong MJ, Miyasaki JM. Evidence-based guideline: pharmacologic treatment of chorea in Huntington disease: report of the guideline development subcommittee of the American Academy of Neurology. Neurology. 2012;79:597–603.

Chapter 8
McLeod Syndrome

Hans H. Jung

Abstract Neuroacanthocytosis (NA) syndromes are genetically defined neurodegenerative disorders characterized by the association of red blood cell acanthocytosis and progressive striatal neurodegeneration. The so-called core NA syndromes include autosomal recessive chorea-acanthocytosis and X-linked McLeod syndrome. These two disorders have a Huntington disease-like phenotype consisting of a hyperkinetic, mostly choreatic, movement disorders, psychiatric manifestations, and cognitive decline with a relentlessly progressive course over several decades. In addition, they may have multisystem involvement including motor-dominant axonal neuropathy, myopathy, and cardiomyopathy. McLeod syndrome (MLS) is exceptionally rare with an estimated prevalence of less than 1–5 per 1,000,000 inhabitants. It is caused by mutations in the *XK* gene. Although the mechanism by which these mutations cause striatal neurodegeneration is not known, the association of the acanthocytic membrane abnormality with selective striatal degeneration suggests a common pathogenetic pathway. Useful diagnostic laboratory tests, besides blood smears to detect acanthocytosis, encompass determination of serum creatine kinase, since virtually all patients with McLeod syndrome reported to date have elevated levels. Cerebral magnetic resonance imaging may demonstrate striatal atrophy. Kell and Kx blood group antigens are reduced or absent, thus delivering an accurate diagnosis of McLeod syndrome, and identification of a distinct mutation in the *XK* gene is confirmatory. The course of McLeod syndrome is relentlessly progressive, and there is no curative therapy known yet. However, regular cardiologic studies and avoidance of transfusion complications are mandatory. The hyperkinetic movement disorder may be treated as in Huntington disease. Other symptoms including psychiatric manifestations should be managed in a symptom-oriented manner.

Keywords Chorea syndrome • Huntington-like disorder • XK gene

H.H. Jung
Department of Neurology, University Hospital Zürich,
Frauenklinikstrasse 26, 8091 Zürich, Switzerland
e-mail: hans.jung@usz.ch

F.E. Micheli, P.A. LeWitt (eds.), *Chorea*,
DOI 10.1007/978-1-4471-6455-5_8, © Springer-Verlag London 2014

Introduction

McLeod syndrome (MLS) was named after a Harvard dental student, Hugh McLeod, in whom an abnormal erythrocyte antigen pattern, consisting of absent or weak expression of Kell antigens was first described [1]. Initially, the McLeod blood group phenotype was thought to be of no clinical significance, apart from the requirement for matched blood transfusions. Later it was found that asymptomatic adult male carriers of the McLeod blood group phenotype have elevated serum levels of CK reflecting muscle cell pathology [2]. Subsequently it was recognized that McLeod carriers had a "neurological disorder characterized by involuntary dystonic or choreiform movements, areflexia, wasting of limb muscles, elevated CK, and congestive cardiomyopathy," thus defining MLS as a multisystem disorder with hematological, neuromuscular, and central nervous system (CNS) involvement ([3, 4] and http://www.geneclinics.org/profiles/mcleod).

Hematologically, MLS is characterized by the absence of Kx red blood cell (RBC) antigen, weak expression of Kell RBC antigens, acanthocytosis, and compensated hemolysis [5]. Asymptomatic carriers of the McLeod blood group phenotype may be accidentally recognized by blood bank testing [5]. All McLeod carriers reported up to date had elevated serum creatine kinase levels, and most develop neurological signs and symptoms with a mean onset age from 30 to 40 years [6, 7]. Neuromuscular manifestations include myopathy, sensory-motor axonal neuropathy, and cardiomyopathy [1, 8]. Central nervous system manifestations are similar as in Huntington disease, and consist of a mostly choreatic movement disorder, "subcortical" neurobehavioral deficits, psychiatric abnormalities, and generalized seizures [6, 7]. MLS is caused by mutations of the XK gene encoding the XK protein, which carries the Kx RBC antigen [9]. Although the exact function of the human XK protein is not yet known, available data suggest an important role for apoptosis regulation [10]. Thus, MLS might be a model disorder to study principal mechanisms that are not only involved in red blood cell physiology but also in neurodegeneration.

Classification

MLS belongs to the group of the so-called core NA syndromes. The other "core" NA syndrome, the autosomal recessive choreoacanthocytosis (ChAc), is caused by mutations of the *VPS13A* gene [11]. There are several other genetically defined disorders in which acanthocytosis is occasionally seen, such as PKAN [12] and Huntington disease-like 2 (HDL2; ORPHA98934) [13] (Table 8.1). Occasional rare cases or families are reported where acanthocytes are present in concert with other extrapyramidal features, such as paroxysmal dyskinesias [14] or mitochondrial disease [15].

Table 8.1 Neuroacanthocytosis syndromes

Disorder	MLS	ChAc	HDL2	PKAN
Gene	*XK*	*VPS13A*	*JPH3*	*PANK2*
Protein	XK protein	Chorein	Junctophilin-3	Pantothenate kinase 2
Inheritance	X-linked	Autosomal recessive	Autosomal dominant	Autosomal recessive
Acanthocytes	+++	+++	+/−	+/−
Serum CK (U/L)	300–3,000	300–3,000	Normal	Normal
Neuroimaging	Striatal atrophy	Striatal atrophy	Striatal and cortical atrophy	"Eye of the tiger" sign
Usual onset	25–60	20–30	20–40	Childhood
Chorea	+++	+++	+++	+++
Other movement disorders	Feeding and gait dystonia, parkinsonism in late stages	Feeding and gait dystonia, tongue and lip biting, parkinsonism	Dystonia, parkinsonism	Dystonia, parkinsonism, spasticity
Seizures	Generalized	Generalized, partial-complex	None	None
Neuromuscular manifestations	Areflexia, weakness, atrophy	Areflexia, weakness, atrophy	None	None
Cardiac involvement	Malignant arrhythmias, atrial fibrillation, dilative cardiomyopathy	None	None	None

Epidemiology

All NA disorders are all exceedingly rare, but also very likely to be underdiagnosed. Estimates suggest that there are probably around few hundred cases of MLS worldwide. MLS has been described in Europe, North and South America, and Japan without obvious clustering [3, 4, 8].

Clinical Characteristics

MLS has a Huntington disease-like phenotype with involuntary hyperkinetic, mostly choreatic, movement disorders, psychiatric manifestations, and cognitive alterations with late adult onset and a slow progression [6, 7]. However, there are several phenotypic peculiarities, in particular the neuromuscular involvement reflected in signs of myopathy and absent tendon reflexes, allowing a strong clinical suspicion of a MLS [8]. In addition, hepatosplenomegaly can be seen in MLS due to increased hemolysis [6].

The McLeod blood group phenotype is defined by the absence of the Kx antigen and by weak expression of the Kell antigens, and may be incidentally detected on routine screening [5]. Most carriers of the McLeod blood group phenotype have acanthocytosis and elevated CK levels, and develop MLS over up to several decades [5]. Onset of neurological symptoms ranges from 25 to 60 years and the disease duration may exceed 30 years, usually longer than in ChAc [6–8]. About one-third of MLS patients present with chorea indistinguishable from that observed in HD, and most patients will develop chorea during the course of the disease [6–8]. Additional involuntary movements include facial dyskinesias and vocalizations [6]. Formerly it was believed that ChAc could be distinguished from MLS by the presence of lip- or tongue-biting, dysphagia, head drop, and parkinsonism [6]. However, it is increasingly recognized that these phenotypic features may be also present in MLS patients thus demonstrating a considerable phenotypic overlap between these two disorders [16–18].

Psychiatric manifestations including depression, schizophrenia-like psychosis, and obsessive-compulsive disorder are frequent in MLS and may appear many years before onset of the movement disorder [7]. A subset of MLS patients develops cognitive decline, particularly in advanced disease stages and generalized seizures occur in about half of the patients [6, 7].

Elevated CK levels are almost always found, and about 50 % of the MLS patients develop muscle weakness and atrophy during the disease course. However, severe weakness and atrophy is only rarely observed [8]. By contrast, MLS patients may be predisposed to rhabdomyolysis, particularly, in the context of neuroleptic medication use [19]. Neuromuscular pathological work-up shows sensory-motor axonal neuropathy, neurogenic muscle alterations, and variable signs of myopathy [8]. About 60 % of MLS patients develop cardiomyopathy possibly manifesting with atrial fibrillation, malignant arrhythmias, and/or dilated cardiomyopathy [8, 20]. Cardiac complications, often leading to sudden cardiac death, are a frequent fatality [8]. Thus, MLS patients and asymptomatic carriers of the McLeod blood group phenotype should have a cardiologic evaluation.

Some female heterozygotes show CNS manifestations related to MLS [21]. Reduction of striatal glucose uptake was demonstrated in asymptomatic female heterozygotes [7]. In addition, MLS may be part of a "contiguous gene syndrome" on the X chromosome including CGD, Duchenne muscular dystrophy, or X-linked retinitis pigmentosa. This is of particular importance for boys with chronic granulomatous disease who survive into adulthood because of modern treatment modalities since they might require frequent blood transfusions. These boys must be screened for the McLeod phenotype and should be regularly monitored for its complications [4].

Pathophysiology

McLeod syndrome is caused by mutations of the *XK* gene encoding the XK protein, which carries the Kx erythrocyte antigen [9]. Most pathogenic mutations are nonsense mutations or deletions predicting an absent or shortened XK protein devoid of the Kell protein-binding site. Although the exact function of the human XK protein is not elucidated, data from a *C. elegans* analogue of the *XK* gene suggest a possible role in apoptosis regulation, since these mutated worms show alterations of the developmental apoptosis [10]. The XK protein has ten transmembrane domains and probably has membrane transport functions. In erythrocytes, it is linked to the Kell protein via disulfide bonds. This Kx-Kell-complex carries the antigens of the Kell blood group, the third most important blood group system in humans. The Kx antigen (on XK) is absent in McLeod syndrome, whereas expression of other Kell system antigens (on the Kell protein) is severely depressed [4]. In muscle, Kell and XK are not co-localized and only XK but not Kell is present in neuronal tissue indicating different physiological functions of the two proteins in different tissues [4].

Diagnosis

The determination of acanthocytosis in peripheral blood smears may be difficult in a routine laboratory examination and a negative routine hematological examination, does not exclude an NA syndrome [22]. Usually, automated blood counts usually show an elevated number of hyperchromic erythrocytes. A more sensitive and specific detection of acanthocytes is made by the use of a 1:1 dilution with physiological saline and phase contrast microscopy [22]. In contrast to the, often elusive, acanthocyte search, serum CK is elevated in almost all cases of McLeod syndrome [6, 7].

However, the diagnostic procedure of choice in MLS is the determination of absent Kx antigen and reduced Kell antigens on the erythrocytes in males and fluorescence absorbent cell sorting with Kell antigens in female heterozygotes [4]. This analysis is available in specialized blood transfusion centers. Analysis of the XK gene is confirmatory and may be provided by some academic laboratories.

Electroneurography may demonstrate findings compatible with a sensory-motor axonal neuropathy, whereas electromyography may show neurogenic as well as myopathic alterations. Electroencephalographic findings are not specific and comprise normal findings, generalized slowing, focal slowing, and epileptiform discharges [6–8]. Neuroradiologically, a progressive striatal atrophy especially affecting the head of caudate nucleus and impaired striatal glucose metabolism similar to that seen in HD may be observed in most patients [7, 23]. Neurodegeneration in MLS predominantly affects the caudate nucleus, putamen, and globus pallidus. In some very few cases, neuropathological findings consist of neuronal loss and gliosis of variable degree in these regions, but no inclusion bodies of any nature or other distinct neuropathological features have as yet been detected. In contrast to HD, however, no significant cortical pathology is evident [21, 24, 25].

Differential Diagnosis

The differential diagnosis of MLS depends upon the presenting symptoms. Initial symptoms may suggest psychiatric disease, including schizophrenia, depression, obsessive-compulsive disorder, tics, Tourette's syndrome, cognitive impairment, personality change, or may include parkinsonism, chorea, dystonia, peripheral neuropathy, myopathy, cardiomyopathy, or seizures [3, 4]. Patients harboring the McLeod blood group phenotype are sometimes identified upon blood donation, many years or even decades prior to development of neurological symptoms. McLeod testing should definitively be considered in the diagnostic work-up of boys with chronic granulomatous disease (CGD), particularly if X-linked. MLS may be detected incidentally by the elevation of CK or liver enzymes, and may give rise to an evaluation of a muscular dystrophy. Recognition of the syndrome may avoid the need for invasive and nondiagnostic tests such as muscle, bone marrow, or liver biopsies.

If a choreatic movement disorder is present, Huntington disease and the other Huntington-like disorders (HDLs) are an important differential diagnosis. Elevated CK, absent tendon reflexes, and the presence of the McLeod blood group phenotype clearly distinguish MLS. Most NA syndromes, including ChAc, MLS, and HDL2, are present in young to middle adulthood, but MLS has usually the latest onset of neurological symptoms, and it is the only X-linked NA syndrome [4]. HDL2 is only described in patients with African ancestry [26]. PKAN typically presents during childhood or adolescence, although adult onset has been reported, particularly in cases where mutations do not abolish all PANK2 enzyme activity [12].

Treatment

So far, no curative or disease-modifying treatments are available, and the management of MLS remains purely symptomatic. Recognition of treatable complications such as seizures, swallowing problems, and heart involvement is essential. Neuropsychiatric issues, particularly depression, can have a major impact, and these symptoms may be more amenable to pharmacotherapy than others. As in HD, dopamine antagonists or depleters such as tiapride, clozapine, or tetrabenazine may ameliorate the hyperkinetic movement disorders. Seizures usually respond to standard anticonvulsants, such as phenytoin and valproate. Anticonvulsants may have the benefit of multiple parallel effects upon involuntary movements, psychiatric symptoms, and seizures. Cardiac complications in MLS need to be particularly considered and heart function should be monitored regularly. No patient with MLS to our knowledge has yet received a heart transplant, which could nevertheless be a management option.

Results of deep brain stimulation (DBS) in MLS have been variable, and the optimal sites and preferred stimulation parameters remain to be determined. Benefits have been observed with stimulation of both the ventro-oral posterior (Vop) thalamic nucleus and the GPi [27, 28]. An ablative operation may be a surgical alternative if long-term implant management is considered to be problematic. In general, neurosurgical options should be considered experimental and must be adapted to individual cases.

Nonmedical therapies with a multidisciplinary approach are crucial for patient management. Evaluation by a facial-oral tract therapist may minimize problems due to dysphagia and weight loss. Dystonic tongue protrusion while eating may be present also in MLS and may respond to local botulinum toxin injections into the genioglossus muscle, although this method has to be applied with caution due to possible mechanical obstruction of the airway and inefficient swallowing secondary to paretic muscles. Placement of a feeding tube, temporarily or even continuously, including percutaneous gastrostomy, may be necessary to avoid nutritional compromise and reduce the risk of aspiration. Physical and occupational therapists can help to manage difficulties with gait, balance, and activities of daily living. Most importantly, extended and continuous multidisciplinary psychosocial support should be provided for the patients and their families.

Conclusions

MLS has to be considered in the differential diagnosis of Huntington disease (HD), in particular if HD genetic testing is negative. However, MLS has additional clinical characteristics such as epilepsy, peripheral neuropathy, myopathy, and cardiomyopathy. Paraclinical findings such as acanthocytosis and elevated CK levels may be crucial to suspect the diagnosis and therefore to initiate Kell blood group phenotyping. Molecular genetic analysis of the *XK* gene may confirm the diagnosis. Management of MLS is symptomatic, although life expectancy and quality of life may be augmented considerably by the appropriate measures.

References

1. Allen FH, Krabbe SMR, Corcoran PA. A new phenotype (McLeod) in the Kell blood-group system. Vox Sang. 1961;6:555–60.
2. Marsh WL, Marsh NJ, Moore A, Symmans WA, Johnson CL, Redman CM. Elevated serum creatine phosphokinase in subjects with McLeod syndrome. Vox Sang. 1981;40(6):403–11.
3. Walker RH, Jung HH, Dobson-Stone C, Rampoldi L, Sano A, Tison F, Danek A. Neurologic phenotypes associated with acanthocytosis. Neurology. 2007;68(2):92–8.

4. Jung HH, Danek A, Walker RH. Neuroacanthocytosis syndromes. Orphanet J Rare Dis. 2011;6:68.
5. Jung HH, Danek A, Frey BM. McLeod syndrome: a neurohaematological disorder. Vox Sang. 2007;93(2):112–21.
6. Danek A, Rubio JP, Rampoldi L, Ho M, Dobson-Stone C, Tison F, Symmans WA, Oechsner M, Kalckreuth W, Watt JM, et al. McLeod neuroacanthocytosis: genotype and phenotype. Ann Neurol. 2001;50:755–64.
7. Jung HH, Hergersberg M, Kneifel S, Alkadhi H, Schiess R, Weigell-Weber M, Daniels G, Kollias S, Hess K. McLeod syndrome: a novel mutation, predominant psychiatric manifestations, and distinct striatal imaging findings. Ann Neurol. 2001;49(3):384–92.
8. Hewer E, Danek A, Schoser BG, Miranda M, Reichard R, Castiglioni C, Oechsner M, Goebel HH, Heppner FL, Jung HH. McLeod myopathy revisited: more neurogenic and less benign. Brain. 2007;130(Pt 12):3285–96.
9. Ho M, Chelly J, Carter N, Danek A, Crocker P, Monaco AP. Isolation of the gene for McLeod syndrome that encodes a novel membrane transport protein. Cell. 1994;77(6):869–80.
10. Stanfield GM, Horvitz HR. The ced-8 gene controls the timing of programmed cell deaths in C. elegans. Mol Cell. 2000;5(3):423–33.
11. Rampoldi L, Dobson-Stone C, Rubio JP, Danek A, Chalmers RM, Wood NW, Verellen C, Ferrer X, Malandrini A, Fabrizi GM, et al. A conserved sorting-associated protein is mutant in chorea- acanthocytosis. Nat Genet. 2001;28(2):119–20.
12. Hayflick SJ, Westaway SK, Levinson B, Zhou B, Johnson MA, Ching KH, Gitschier J. Genetic, clinical, and radiographic delineation of Hallervorden-Spatz syndrome. N Engl J Med. 2003; 348(1):33–40.
13. Walker RH, Rasmussen A, Rudnicki D, Holmes SE, Alonso E, Matsuura T, Ashizawa T, Davidoff-Feldman B, Margolis RL. Huntington's disease-like 2 can present as chorea-acanthocytosis. Neurology. 2003;61(7):1002–4.
14. Tschopp L, Raina G, Salazar Z, Micheli F. Neuroacanthocytosis and carbamazepine responsive paroxysmal dyskinesias. Parkinsonism Relat Disord. 2008;14(5):440–2.
15. Mukoyama M, Kazui H, Sunohara N, Yoshida M, Nonaka I, Satoyoshi E. Mitochondrial myopathy, encephalopathy, lactic acidosis, and stroke-like episodes with acanthocytosis: a clinicopathological study of a unique case. J Neurol. 1986;233(4):228–32.
16. Chauveau M, Damon-Perriere N, Jung H, Latxague C, Spampinato U, Burbaud P, Tison F. Head drops are also observed in the McLeod syndrome. Mov Disord. 2011;26:1562–3.
17. Gantenbein A, Damon-Perrière N, Bohlender JE, Chauveau M, Latxague C, Miranda M, Jung HH, Tison F. Feeding dystonia in McLeod syndrome. Mov Disord. 2011;26:2123–6.
18. Miranda M, Jung HH, Danek A, Walker RH. The chorea of McLeod syndrome: progression to hypokinesia. Mov Disord. 2012;27:1701–2.
19. Jung HH, Brandner S. Malignant McLeod myopathy. Muscle Nerve. 2002;26:424–7.
20. Oechslin E, Kaup D, Jenni R, Jung HH. Cardiac abnormalities in McLeod syndrome. Int J Cardiol. 2009;132(1):130–2.
21. Hardie RJ, Pullon HW, Harding AE, Owen JS, Pires M, Daniels GL, Imai Y, Misra VP, King RH, Jacobs JM. Neuroacanthocytosis. A clinical, haematological and pathological study of 19 cases. Brain. 1991;114(Pt 1A):13–49.
22. Storch A, Kornhass M, Schwarz J. Testing for acanthocytosis A prospective reader-blinded study in movement disorder patients. J Neurol. 2005;252(1):84–90.
23. Valko PO, Hänggi J, Meyer M, Jung HH. Evolution of striatal degeneration in McLeod syndrome. Eur J Neurol. 2010;17:612–8 (IF: 3.765).
24. Brin MF, Hays A, Symmans WA, Marsh WL, Rowland LP. Neuropathology of McLeod phenotype is like choreaacanthocytosis (CA). Can J Neurol Sci. 1993;20(Suppl):S234.
25. Geser F, Tolnay M, Jung HH, Walker RH, Shinji S, Danek A. The neuropathology of McLeod syndrome in neuroacanthocytosis syndromes II. Berlin/Heidelberg: Springer; 2008. p. 197–203.

26. Margolis RL, Holmes SE, Rosenblatt A, Gourley L, O'Hearn E, Ross CA, Seltzer WK, Walker RH, Ashizawa T, Rasmussen A, et al. Huntington's disease-like 2 (HDL2) in North America and Japan. Ann Neurol. 2004;56(5):670–4.
27. Wihl G, Volkmann J, Allert N, Lehrke R, Sturm V, Freund HJ. Deep brain stimulation of the internal pallidum did not improve chorea in a patient with neuro-acanthocytosis. Mov Disord. 2001;16(3):572–5.
28. Burbaud P, Rougier A, Ferrer X, Guehl D, Cuny E, Arne P, Gross C, Bioulac B. Improvement of severe trunk spasms by bilateral high-frequency stimulation of the motor thalamus in a patient with chorea-acanthocytosis. Mov Disord. 2002;17(1):204–7.

Chapter 9
Neuroferritinopathy

Vanderci Borges and Roberta Arb Saba

Abstract Neuroferritinopathy is an autosomal-dominant neurodegenerative disorder caused by mutations in the ferritin light chain gene (*FTL*). The disease is clinically present during adulthood with movement disorders mainly with chorea, dystonia, and parkinsonism and progresses slowly over decades. Cognitive symptoms are often noted after motor signs. On brain magnetic resonance imaging (MRI), the findings are iron deposits in the basal ganglia and cavitation. Neuronal loss in the cerebral cortex, cerebellum, and basal ganglia has been demonstrated in neuropathological studies as well as ferritin inclusion bodies, shown within neurons and glia. As neuroferritinopathy is considered as one of several of neurodegenerative diseases with brain iron accumulation (NBIA), the main differential diagnosis of this disorder is with the other diseases found in this group and with Huntington's disease. There is no specific treatment to modify the progression of the disease.

Keywords Neuroferritinopathy • Iron deposits • Ferritin

Introduction

Neuroferritinopathy is an autosomal-dominant neurodegenerative movement disorder caused by mutations in the ferritin light chain gene (*FTL*). It is also known as hereditary neuroferritinopathy [1]. This condition was described for the first time in 2001 in a large family with autosomal movement disorder from Cumbria, a region in the north of England [2]. In that pedigree, the initial (and most common mutation, denominated as 460insA in the ferritin light chain gene (*FTL*) at chromosome

V. Borges (✉) • R.A. Saba
Movement Disorders Unit, Department of Neurology and Neurosurgery,
Universidade Federal de São Paulo, Rua Pedro de Toledo, 980 cj.33,
São Paulo CEP 04039-002, Brazil
e-mail: vanderci.borges@gmail.com

F.E. Micheli, P.A. LeWitt (eds.), *Chorea*,   155
DOI 10.1007/978-1-4471-6455-5_9, © Springer-Verlag London 2014

19q13.3) was identified. After that description was published, seven additional mutations have been identified. Neuroferritinopathy is considered to be part of a group of neurodegenerative diseases with brain iron accumulation (NBIA) including pantothenate kinase-associated neurodegeneration (PKAN), phospholipase-associated neurodegeneration (PLAN), fatty acid hydroxylase-associated neurodegeneration (FAHN), mitochondrial protein-associated neurodegeneration (MPAN), Kufor–Rakeb disease (PARK9), aceruloplasminemia, and static encephalopathy of childhood with neurodegeneration of adulthood (SENDA) [3].

Neuroferritinopathy is an extremely rare disorder, with approximately 60 cases published in the worldwide literature. The clinical presentation begins at adult midlife with chorea predominantly followed by dystonia and parkinsonism. Magnetic resonance imaging (MRI) shows high levels of iron deposits in caudate, globus pallidus, putamen, substantia nigra, and red nuclei. In addition, cystic changes can be found in the basal ganglia [4]. Neuropathological examination shows ferritin inclusion bodies in iron-rich areas within neurons and glia [5] and also in liver and muscles. Neuronal loss in the cerebral cortex, cerebellum, and basal ganglia has also been demonstrated in neuropathological studies.

There is no specific disease modifying treatment. For patients with dystonia, the use of botulinum toxin may be considered and tetrabenazine may be used to treat chorea and facial tics.

Epidemiology

As neuroferritinopathy is an extremely rare disorder; there are essentially no epidemiologic studies. The approximately 60 cases published in the medical literature are, for the most part, small groups of subjects or single case reports. There are descriptions of families affected with this disorder residing in Canada [1, 6], England [7], France [8], Japan [9, 10], Portugal [11], the United States [12], Australia [13], and Italy [14]. The largest case series, described by Chinnery et al. [7], consisted of 41 cases with the *460insA FTL1* mutation. Of these, 9 cases had already been reported earlier. Their goal in the 2007 report was to define the clinical spectrum of the disease.

Clinical Manifestations

Neuroferritinopathy is a disease that primarily presents in adults with a mean age of onset as 39 years; earlier- and later-onset cases are also known [7].

The first pedigree described (due to mutation *460insA* in the FTL gene) presented choreoathetosis, dystonia, spasticity, and rigidity, with onset between 40 and 55 years of age [2]. The main clinical feature is a movement disorder, of which the most common is chorea (50 %), followed by focal dystonia (43 %) and by

Table 9.1 Mutations in the ferritin light chain gene (FLT1) and clinical features

Mutation	Age of onset	Clinical features
460insA	40–55 years	Chorea, dystonia, parkinsonism, orofacial dyskinesias
498-499insTC	20–30 years	Cerebellar ataxia, parkinsonism, facial dyskinesia, cognitive dysfunction
646insC	63 years	Ataxia, dysarthria, dysphagia, involuntary movements in the face and limbs
458dupA	22–44 years	Chorea, dystonia, cerebellar ataxia, cognitive impairment, behavioral impairment, and rapid course
c.469_484dup16nt	Middle teens	Hand tremor, right foot dragging, generalized hypotonia, hyperextensibility, aphonia, micrographia, hyperreflexia, and cognitive impairment
641-642_4bp	40–60 years	Choreoathetosis, dystonia, tremor dyskinesia in tongue and face, mild cognitive impairment
474G>A	13 years	Gait disturbances psychosis, ataxia, parkinsonism

parkinsonism (7.5 %). The clinical course is slowly progressive; the majority of patients develop chorea and dystonia (especially in the legs, causing a gait disorder [7]). Oromandibular dyskinesia with tongue movements causing oral injury and dysarthria may be observed. Blepharospasm, writer's cramp, and ballism were not usually seen in the reported cases. Patients may present with startled facies (which is due to symmetric frontalis and platysma contractions). Cognition is usually spared at the onset despite the development of physical disability. Facial tics and stereotypies were described in a patient from North America who comes from a German-American family [12]. His father had facial chorea with later progression of gait disorder and dementia. Although chorea, dystonia, and oromandibular dyskinesia are common features in all FTL mutations, there are slight phenotypic differences depending on the mutation Table 9.1.

In the French-Canadian/Dutch family reported by Vidal et al. [1] (and whose mutation was *498-499insTC*), the phenotype seen was a postural tremor as initial manifestation at age 20. Afterwards, she developed a severe cerebellar syndrome, severe cognitive impairment, parkinsonism, and facial dyskinesia. Ory-Magne et al. [15] described a member of the family who presented with a cerebellar syndrome and parkinsonism at the age of 56.

The family described by Mancuso et al. [6] (affected with a *646insC* mutation) had a proband who presented at age 63 with unsteady gait, impaired dexterity of hands, dysarthria, dysphagia, and involuntary movements in the face and limbs. The progression was gradual and he developed depression, emotional lability, and impotence. There is no cognitive impairment or parkinsonism. The proband's sister, at 49 years of age, presented unsteady, festinating, and freezing of gait and progressed with blepharospasm, orolingual, mandibular and cervical dystonia, emotional lability, and mild proximal leg weakness. There was considerable phenotypic variability in this family.

A French family [8] (affected with a *458dupA* FTL mutation) demonstrated clinical presentations similar to the first described mutation. Some variations were evident, such as cerebellar ataxia, cognitive impairment, and behavioral impairment. This familiar also seemed to have a more rapid course of progression.

The phenotype caused by *c.469_484dup16nt* mutation, as described by Otta et al. [9], was that of a man who noticed hand tremor in his mid-teens. He developed right foot dragging at age 35 along with generalized hypotonia, hyperextensibility, aphonia, micrographia, hyperreflexia, and cognitive impairment at age 42.

This mutation was also observed in a patient lacking a positive family history. He had a history of hand tremor, behavioral changes, and fatigue since his 20s. The symptoms progressed gradually with parkinsonism, ataxia, cognitive impairment, and dystonia. At age 40, the diagnosis was established [14].

Choreoathetosis, dystonia, tremor, dyskinesia in tongue and face, mild cognitive impairment, and normal serum ferritin levels were described in a family from Japan [10] with a c.641-642 mutation in exon 4. The age of onset varied from 40 to 60 years.

The 474G>A mutation identified in a Roma family was associated with earlier onset of the disease [11]. The proband presented gait disturbances at the age of 13 and was followed by the development of acute psychosis. He developed confusion, apathy, ataxia, bradykinetic-rigid syndrome, and plantar extensor responses after treatment with valproic acid and trazodone. With the withdrawal of medication, the symptoms showed improvement. He remained with mild bradykinesia, rigidity, ataxia, and bilateral Babinski signs. His mother was an asymptomatic carrier of the mutation and displayed bilateral pallidal necrosis on MRI (similar to that of the proband). One noncarrier uncle had a diagnosis of schizophrenia, so it is not possible to confirm that the patient's psychosis is necessarily related to the gene mutation.

Unusual presentations have been described in this disorder, such as palatal tremor associated to oral and buccal dyskinesias and cognitive decline [16]. An intriguing report is a case of a man with parkinsonism who was unable to walk backward [17]. He could walk forward but when attempting to walk backward, he would exhibit a few shuffling steps before falling. Electromyography (EMG) showed bilateral synchronous 9 Hz tremor bursts of 40–80 ms while the patient stood. This fulfilled diagnostic criteria for slow orthostatic tremor (OT).

Various neuropsychiatric symptoms have been described in neuroferritinopathy as shown above, but their severity and the timing of presentation were not well defined until the report of a study by Keogh et al. [18]. This publication is a systematic review of neurocognitive phenotype, assessing the sets of patients with ferritinopathy and the mutation 460insA. They found 12 patients who underwent cognitive assessment (Addenbrooke's Cognitive Examination ACER-R or neuropsychometric testing). The nonmotor symptoms were heterogeneous and the most common features were deficits in verbal learning and impairment of executive function. Psychiatric symptoms such as emotional lability, aggressive behavior, and obsessive compulsive symptoms were observed in two patients. Neuropsychiatry symptoms appeared within 5 years of the onset of motor symptoms. Although cognition may not be the initial manifestation in most cases, its appearance may occur during the progress of disease.

Despite the disease usually manifests in adulthood, it believes that the iron deposits occur decades before clinical presentation.

The challenge is to understand why this phenotypic variation does occur.

Pathophysiology

The brain is vulnerable to oxidative stress. High concentrations of iron may catalyze the formation of reactive oxygen species (ROS). Iron probably enters the central nervous system (CNS) through the endothelial cells of blood-brain barrier and epithelial cells of the choroid plexus using many of the transporters utilized in duodenum [19]. Iron is essential for the function of the respiratory chain in mitochondria and those of myelinization and neurotransmission [20], but its concentration must be finely controlled. Brain iron deposits increase with age. Most of iron in the human brain is bound to ferritin.

Ferritin is a ubiquitous protein which consists of 24 subunits which form a soluble hollow shell form. The function of ferritin is to store cellular iron, protecting the cell from potential iron-dependent radical damage, and to allow the release of the metal according to the cellular demand [21]. Ferritin is composed of two peptides, the heavy subunit (FTH) and the light subunit (FTL). Normal ferritin is capable to store thousands of iron atoms inside the central cavity. The H subunit has a metal binding site within the bundle (ferroxidase center) where the catalytic oxidation of iron occurs and allows rapid uptake of iron. The L-subunit seems to be involved with the initiation and stabilization of the ferritin–iron core. The L-chain accelerates the transfer of the iron from the ferroxidase site to the iron core [22]. Both subunits consist of four parallel α-helices (A–D) and a shorter, carboxyterminal α-helix (E). The proportion of chains is determined by tissue and cellular development. Then the heart and brain contains more H-rich ferritins, which have higher ferroxidase activity and more pronounced antioxidants activities. L-rich ferritins are found in spleen and liver and have more pronounced iron storage function [22].

It seems that ferritin also plays a role in the protection against oxidative damage [23] and in apoptotic cell death [24].

In humans the H-chain is encoded on chromosome 11 and the L-chain on chromosome 19. FTMT is on chromosome 5, which encodes the precursor of the mitochondrial ferritin (FtMt).

The first mutation in the ferritin protein related to neuroferritinopathy was described by Curtis et al. [2]. They found a mutation in the gene encoding the ferritin light chain polypeptide (FTL1) in a family with dominantly inherited movement disorder. The mutation consisted of an adenine insertion at position 460–461 in the exon 4 of the gene (*460InsA*) which alters the reading frame and the carboxy terminus (C-terminus).

In sequence other mutations described such as *498-499insTC*, *469_484dup16nt*, *646InsC*, and *458dupA* seem to be located in the C-terminus of FTL gene. The effect of these mutations could be an inability of ferritin to bind iron.

Cellular models showed that the expression of the *460InsA* and *498insTC* mutant in HeLa and SH-SY5Y neuroblastoma cells has effects on iron homeostasis. Mutated L-subunits co-assembled with the endogenous H and L to form ferritin shells with low functionality and seemed to accelerate its degradation rate. As a consequence the labile iron excess stimulates iron-dependent ferritin synthesis and, at the same time, facilitates the production of ROS, the increase of levels of oxidized proteins, and cell death. The oxidized protein excess promotes the proteasome imbalance and ferritin aggregate formation [25]. According to this study the iron deregulation seems to be the underlying cause of the neuroferritinopathy, while the oxidative stress is the primary cause of the neurodegenerative process. At the same time ferritin and iron aggregates, which did not appear correlated to cell death, seemed to be a secondary effect to the impairment of proteasome function. It is possible these aggregates probably enhance the toxicity of iron imbalance in the long run.

A study with transgenic mouse model of neuroferritinopathy expressing 499insTC showed ferritin inclusions in the glia and neurons throughout the central nervous system and other organs. These inclusion bodies are positive for iron, ferritin, and ubiquitin. The histological, immunohistochemical, and biochemical findings of this model were similar to those found in individuals with hereditary neuroferritinopathy [26]. The transgenic presented reduced motor performance and lifespan. Besides there was an increase of brain lipid peroxidation signifying oxidative stress.

In the mouse model, it was also detected that the integrity of the mitochondrial DNA (mtDNA) in the brain was compromised due to the oxidative damage. The authors suggest that this process may compromise the mitochondrial function and lead to neuronal loss and exacerbate the neurodegeneration in neuroferritinopathy [27].

Macroscopic neuropathology in a case due to 498insTC mutation showed atrophy of the cerebral hemispheres, cerebellum, and caudate nucleus. Small cavities could be seen in putamen. Microscopically there was nerve cell loss and gliosis in the cerebral cortex, amygdale, thalamus, substantia nigra, locus coeruleus, caudate, putamen, and globus pallidus. There was the presence of intranuclear and intracytoplasmic bodies in glial cells and some neurons. Ferritin immunopositivity was seen in intranuclear bodies and in the cytoplasm of glial cells. It was present throughout the brain, especially in putamen [1].

Ferritin and iron deposition have been shown in the liver, kidney, muscle, and skin [15].

The neuropathological findings described by Mancuso et al. [6] in a patient with neuroferritinopathy due the *646insC* mutation were diffuse atrophy of cerebellum, softening of globus pallidus and putamen, and cavitation of the putamen. Small cavitations may have merged to form larger structures with progression of the disease [28].

Microscopically, there was neuronal loss of olivary nucleus, red nucleus, putamen, external globus pallidus, cerebellar vermis, and subthalamic nucleus.

Both neurons and glia displayed swollen to vacuolated nuclei containing ferritin and iron. Hyaline deposits, staining for both ferritin and iron, were particularly prominent in the external globus pallidus and in the paradentate white matter in the cerebellum. These morphologic features may be specific to ferritinopathies. The iron content in putamen was very high, nearly 40-fold over normal values. It appeared to be both in the ferrous ($Fe2+$) and ferric ($Fe3+$) forms. These changes in iron's ionic state might be a function of the underlying mechanism for neuronal and glial apoptosis. There was also evidence of lipid peroxidation and abnormal nitration of proteins in putamen and glia; these features may have been consequences of oxidative stress due to excessive iron deposition. Biochemical and immunohistochemical abnormalities in mitochondria were also demonstrated. These finding corroborate the notion that mitochondrial abnormalities may have a pathogenetic role in neuroferritinopathy.

Systemically, the patient had hepatocytic intranuclear accumulations of iron and ferritin.

The combination of the findings of cellular and mouse models and the results of neuropathological studies suggests that the iron deposition result from the inability of mutant ferritin to bind iron. In turn, this may lead to oxidative stress, mitochondrial dysfunction, and apoptosis.

Diagnosis

The diagnosis of neuroferritinopathy is based on combining information from clinical manifestations, laboratory exams, neuroimaging, muscle and nerve biopsy, and genetic testing (e.g., looking for the *460InsA* FTL1 mutation). A variety of abnormal, involuntary movements can be demonstrated among patients with the clinical presentation of neuroferritinopathy. The clinical descriptions include orofacial dyskinesia, chorea, focal dystonia, tremor, toe and finger stereotypies, and parkinsonism. Frontal lobe or subcortical dementia tends to develop after motor symptoms, and autonomic features may be observed. Although the neuroferritinopathy is an autosomal-dominant disease, phenotypic variations have been described. In contrast to other NBIA disorders, pigmentary retinopathy and acanthocytosis have not been reported. The absence of associated ophthalmologic features can be helpful in distinguishing neuroferritinopathy from other forms of NBIA.

Despite being the only NBIA syndrome with dominant inheritance, it is commonly challenging to diagnose neuroferritinopathy solely based on the clinical findings. Brain MR imaging offers quite characteristic findings and may facilitate differential diagnosis of neuroferritinopathy from other extrapyramidal disorders. Radiological findings in patients with neuroferritinopathy generally correlate with the observed pathology [29]. The findings are usually bilateral and symmetric but sometimes asymmetric. The abnormalities seen on MRI initially include iron deposition, as well as edema, gliosis, cystic changes, and cortical atrophy [1, 2]. Central nervous system iron deposition occurs with normal aging, especially in the globus pallidus, putamen, substantia nigra, and dentate. Patients affected by neuroferritinopathy demonstrate ferritin and iron deposits in excess of expected age-related accumulation, most notably within the basal ganglia [1, 6]. Cystic changes are most

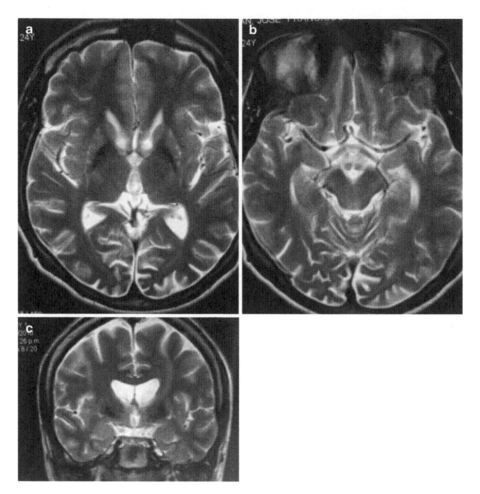

Fig. 9.1 MRI from a 24-year-old man with neuroferritinopathy shows iron deposits in globus pallidus (**a**), substantia nigra (**b**), and subthalamic region (**c**) (Courtesy by Dr. Carlos Zuñiga)

frequently pronounced within the globus pallidus and putamen [28] but may occur in the thalamus, dentate, and caudate [1, 2]. Sometimes MRI also demonstrates cortical atrophy, both within the cerebellum and cerebral cortices [2–28]. The progression of MRI changes in neuroferritinopathy can evolve from T2 hypointensities due to iron deposition to T2 hyperintensities (the latter finding likely induced by the start of cystic degeneration).

MRI imaging shows multiple stages of neuroferritinopathy. When T2* imaging is performed with a gradient echo sequence, the results are very sensitive to change in paramagnetism that is seen with iron deposition [28]. Therefore, initial MRI changes are observed with T2* imaging as hypointensities within the basal ganglia. On T2 conventional imaging, minor low-signal hypointensity is seen [30] (Fig. 9.1). As the disease progresses, the T2* signal loss and the T2 hypointensity become more pronounced. The cerebral cortex, substantia nigra, thalamus, dentate, and red nuclei may be affected [28]. The next clinical stage of neuroferritinopathy demonstrates a change to T2 hyperintensities as there will be changes in water content secondary to edema and gliosis. At advanced stages of neuroferritinopathy, there are characteristic symmetrical basal ganglia T1 hypointense and T2 hyperintense regions consistent with cystic degeneration [30]. The laboratory findings in this disorder show low serum ferritin levels [7]. Deposits of iron can be seen in muscle or nerve biopsy. Postmortem examination of the brain [5] demonstrated ferritin-positive spherical inclusions in iron-rich areas, often co-localizing in microglia, oligodendrocytes, and neurons. Neuroaxonal spheroids (immunoreactive to ubiquitin and tau) and neurofilaments have been reported [7].

Differential Diagnosis

Neurodegeneration with brain iron accumulation (NBIA), like neuroferritinopathy, encompasses a group of progressive extrapyramidal disorders characterized by iron accumulation in the brain. The main diagnoses of this disease are as follows: pantothenate kinase-associated neurodegeneration (PKAN), phospholipase A2-associated neurodegeneration (PLAN), fatty acid hydroxylase-associated neurodegeneration (FAHN), mitochondrial protein-associated neurodegeneration (MPAN), Kufor–Rakeb disease (PARK 9), aceruloplasminemia, and SENDA syndrome (static encephalopathy of childhood with neurodegeneration in adulthood) Table 9.2.

Despite not being a disease of iron accumulation, *Huntington's disease* (HD) is part of the differential diagnosis of neuroferritinopathy, due to its extrapyramidal features that generally involve chorea and dystonia. HD also shows disturbances for most patients in their behavior and cognition. HD is an autosomal-dominant neurodegenerative disorder caused by an elongated CAG repeats on short arm of chromosome 4p16.3 in the Huntingtin gene. In HD, the brain MRI shows striatal and cerebral atrophy, and the molecular genetic testing virtually always provides a confirmatory diagnosis.

Given corroborative evidence of abnormal iron deposition seen on MRI, NBIA should be considered. Perhaps the best known disorder among this entity is *pantothenate kinase-associated neurodegeneration (PKAN/NBIA1)* (formerly known as Hallervorden–Spatz disease in typical young-onset cases and Hallervorden–Spatz syndrome in atypical late-onset cases). PKAN is an autosomal recessive inborn error of coenzyme A (CoA) metabolism, caused by mutations in *PANK2*, the gene encoding pantothenate kinase 2 [31]. Typical PKAN is an early-onset movement disorder, usually beginning before 6 years of age, and rapid progression. In classic PKAN, affected children are often considered clumsy before onset of the symptoms, with impaired gait due to dystonia and usually lose the ability to ambulate by 10–15 years after disease onset. Primary clinical features include dystonia, dysarthria, and rigidity. Clinically, individuals present with progressive pigmentary retinopathy, severe dystonia, and sometimes neuropsychiatric disturbance and cognitive

Table 9.2 Overview of NBIA conditions and genes

Disease (acronym)	Synonym	Gene	Chromosomal position	Neuroimage (areas of highest iron density)
Neuroferritinopathy	–	*FTL*	*19p13*	*Caudate, GP, putamen, SN, and red nuclei*
PKAN	NBIA1	*PANK2*	20p13	GP, eye-of-the-tiger sign (central hyperintensity within a surrounding area of hypointensity)
PLAN	*NBIA2 (PARK14)*	*PLA2G6*	*22q12*	*GP/SN (in some)*
Aceruloplasminemia	–	*CP*	3q23	Basal ganglia, thalamus, dentate nuclei, and cerebral and cerebellar cortices
FAHN	*SPG35*	*FA2H*	*16q23*	*GP/often white matter changes*
Kufor–Rakeb disease	PARK9	*ATP13A2*	1q36	Putamen and caudate
WSS	–	*C2orf37*	*2q22.3-2q35*	*GP*
SENDA syndrome	–	*WDR45*	Xp11.23	GP and SN/white matter changes
MPAN	–	*C19orf12*	*19q12*	GP and SN

FA2H fatty acid 2-hydroxylase, *FTL* ferritin light chain, *GP* globus pallidus, *MPAN* mitochondrial-associated neurodegeneration, *NBIA* neurodegeneration with brain iron accumulation, *PANK2* pantothenate kinase 2, *PKAN* pantothenate kinase-associated neurodegeneration, *PLA2G6* phospholipase A2, *PLAN* phospholipase A2-associated neurodegeneration, *SENDA* static encephalopathy of childhood with neurodegeneration in adulthood, *SN* substantia nigra

decline. Corticospinal tract involvement leads to spasticity, hyperreflexia, and extensor toe signs. In the later stages of disease, patients frequently require tube feeding due to dysphagia. Gastroesophageal reflux and constipation can become chronic problems during late stage. Generally death occurs from secondary complications, including aspiration pneumonia and malnutrition. Acanthocytes may be seen in PANK2 [8]. Brain MRI findings include a hypointense outer rim with central hyperintensity of the globus pallidus on T2 and T2* sequences, described as the "eye-of-the-tiger sign" [12]. This finding is not pathognomonic for PANK2 mutation, since the "eye-of-the-tiger" sign was documented in cases of neuroferritinopathy as well as in atypical PKAN [31–33] (Fig. 9.2).

In atypical PKAN, the average age of the onset is 13–14 years and disease progression is slower. Speech difficulty is a frequent presenting sign [34]. Other symptoms include mild gait abnormalities and psychiatric symptoms, including emotional lability, impulsivity, depression, obsessive compulsive disorder, and violent outbursts [35]. Tourettism, including both motor and verbal tics, has also been observed in the early stages of atypical PKAN. Motor involvement is generally less severe than in classic PKAN and has a slower rate of progression, typically with loss of ambulation occurring within 15–40 years of onset.

Fig. 9.2 T2* bMRI scan from a 13-year-old boy with pantothenate kinase-associated neurodegeneration (PKAN). Note bilateral "eye-of-the-tiger" sign in globus pallidus

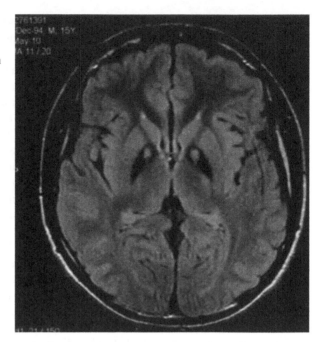

Contrary to what occurs usually in the neuroferritinopathy, the ophthalmological phenotype of PKAN has been well characterized. Pigmentary retinal degeneration has long been recognized as a prominent symptom in patients with early-onset NBIA. Ocular motility presents hypometric and slow saccadic pursuits, and neuro-ophthalmologic evaluation shows sluggish pupillary reactions with sectoral iris paralysis and patchy loss of the pupillary ruff, similar to bilateral Adie's pupils [36]. The detection of acanthocytes on peripheral blood smear supports a diagnosis of PKAN.

NBIA Type 2—PLA2G6-Associated Neurodegeneration (PLAN) has been well described and attributed to mutations in the calcium-independent phospholipase A2 gene PLA2G6 that is located on chromosome 22q and contains 17 exons. Similar to PKAN, there appears to be an age-dependent phenotype. Early-onset cases have infantile neuroaxonal dystrophy (INAD) characterized by progressive motor and mental retardation, marked truncal hypotonia, pyramidal signs, cerebellar ataxia, and early visual disturbances due to optic atrophy. Seizures may be present [37].

Late-onset cases are described as atypical neuroaxonal dystrophy, which is characterized by dystonia–parkinsonism, pyramidal signs, eye movement abnormalities, cognitive decline, and psychiatric disturbance [38]. Associated parkinsonism is characterized by the presence of tremor, rigidity, and severe bradykinesia, with a good response to levodopa. This condition has been assigned a categorization as PARK14. Neuroimaging shows cerebellar atrophy occurring in early stages of INAD, which is not a usual feature of the late-onset form. INAD patients typically develop hypointensity of the globus pallidus reflecting iron accumulation. The signal abnormality differs from the "eye-of-the-tiger" sign of PKAN in that there is no central hyperintensity. Iron deposits in the SN are present in some atypical cases [39, 40]. In some cases, brain MRI may even be completely normal; in others, there may be cortical atrophy or white matter changes [38].

Aceruloplasminemia is another NBIA disorder. The inheritance is autosomal recessive and is caused by mutations in the ceruloplasmin gene located on chromosome 3q which leads to an accumulation of iron in the brain, liver, and reticuloendothelial system [28]. Ceruloplasmin gene mutation leads to abnormal iron deposition in this disease [41]. The average age at diagnosis is 51, with a range from 16 to 71 years [42]. The clinical manifestations are blepharospasm, craniofacial dyskinesias, chorea, ataxia, and retinal degeneration [43]. The neuroradiology findings show more homogeneous lesions of the caudate, putamen, globus pallidus, thalamus, red nucleus, and dentate. The findings occur with concurrent white matter hyperintensities and cerebellar atrophy. The iron deposition in the central nervous system in aceruloplasminemia exhibits a distribution comparable to that in neuroferritinopathy, but in aceruloplasminemia, all basal ganglia nuclei and thalamus are simultaneously involved as seen in T2-weighted and T2*-weighted images. Another characteristic feature is the lack of the association of hyperintense and hypointense abnormalities that are often observed in neuroferritinopathy. The low-signal areas observed in aceruloplasminemia are homogeneous. The cystic changes of the basal ganglia observed in neuroferritinopathy are rarely seen in aceruloplasminemia [4, 7, 44].

Screening laboratories can be helpful in diagnosis, demonstrating low or undetectable ceruloplasmin levels. Decreased serum iron, microcytic hypochromic anemia, and elevated ferritin often occur. Aceruloplasminemia is the only form of NBIA that features prominent signs of peripheral organ involvement with diabetes mellitus and liver disease.

A recent addition to the NBIA group of disorders is *fatty acid hydroxylase-associated neurodegeneration (FAHN)/SPG35*, which is a result of mutations in the fatty acid-2 hydroxylase FA2H gene [45]. Patients are usually affected earlier in life compared to neuroferritinopathy patients and present with spastic quadriparesis, ataxia, and dystonia [45]. Divergent strabismus developed. Seizures may be present. MRI demonstrated bilateral globus pallidus T2 hypointensity, consistent with iron deposition, prominent pontocerebellar atrophy, mild cortical atrophy, white matter lesions, and abnormalities of corpus callosum [46].

Kufor–Rakeb disease (PARK9) is a rare autosomal recessive neurodegenerative disease. The clinical phenotype of Kufor–Rakeb includes parkinsonism, with pyramidal tract signs in some cases. There are eye movement abnormalities with incomplete supranuclear up-gaze palsy and slowing of vertical and horizontal saccades and saccadic pursuit have also been described. Oculogyric dystonic spasms, facial–faucial–finger mini-myoclonus, and autonomic dysfunction may be present. Cognitive features include visual hallucinations and dementia. Disease onset is usually in adolescence [46]. Kufor–Rakeb disease is a result of mutations in the ATP13A2 [47] gene, on chromosome 1p. The 26-kb spanning gene contains 29 exons and encodes a lysosomal 5 P-type ATPase. In the human brain, ATP13A2 is localized to pyramidal neurons within the cerebral cortex and dopaminergic neurons of the SN. A good response to levodopa has been noted; nevertheless, similar to other complicated recessive dystonia–parkinsonism variants, levodopa-induced dyskinesias tend to develop early. Brain CT and MRI may show diffuse moderate cerebral and cerebellar atrophy. Iron deposition in the basal ganglia affecting the putamen and caudate may be present. Dopamine transporter imaging shows marked bilateral symmetrical reduction of striatal activity indicative of diminished presynaptic function. Transcranial sonography of the SN is normal [48].

Another novel NBIA syndrome, *Woodhouse–Sakati syndrome (WSS)*, has been described and is the result of a single base pair deletion in the C2orf37 gene [49]. Most cases have been reported from families of Middle Eastern origin. Symptoms include diabetes mellitus, choreoathetosis, and dystonia [50]. The hallmark feature of this disease is alopecia. MRI studies demonstrate globus pallidus T2 hypointensities.

SENDA syndrome (static encephalopathy of childhood with neurodegeneration in adulthood) [7] was recently described as having clinical early-onset spastic paraplegia and mental retardation that remains static until the late 20s to early 30s; the syndrome progresses to parkinsonism and dystonia. Eye movement abnormalities, sleep disorders, dysautonomia, and frontal release signs have been included in the clinical manifestations. Imaging has shown brain iron accumulation affecting the globus pallidus and hypointensities in the substantia nigra, as well as white matter changes. It responds to L-dopa. The genetic cause is a mutation in *WDR45* (WD Repeat Domain 45), at Xp11.23 [51].

Mitochondrial protein-associated neurodegeneration (MPAN) is recently described like a childhood onset with dysarthria and gait difficulty, followed by the development of spastic paraparesis, extrapyramidal features (dystonia and parkinsonism), neuropathy, optic atrophy, and psychiatric symptoms. Iron deposition was present in the GP and SN.

Genetic workup led to identification of the new NBIA gene, *C19orf12*, at chromosome 19q12. Postmortem brain examination revealed iron-containing deposits in the GP and SN, axonal spheroids, Lewy body-like inclusions, and tau-positive inclusions [47].

Treatment

Effective treatment for neuroferritinopathy has not been established. No medication has been found to influence disease progression. Iron chelating agents like deferiprone (an oral iron chelator with good permeability across the blood-brain barrier) have been tried to attenuate the accumulation of iron, but this class of drugs has not been found to be efficacious [52]. A recent review by Lehn and colleagues collects therapies listed in case reports and reports that tetrabenazine can be beneficial as a symptomatic therapy to improve hyperkinetic movements. They also found that benzodiazepines, botulinum toxin chemodenervation, anticholinergics, and muscle relaxants, including baclofen, might be helpful in some instances as symptomatic therapy [52].

The parkinsonian features of neuroferritinopathy tend to be nonresponsive to treatment with levodopa [7, 8].

Deep brain stimulation has been shown to be effective in improving dystonia in PKAN, as in other movement disorders with dystonic manifestations, but this therapy has not demonstrated efficacy in cases of neuroferritinopathy [53, 54].

Conclusions

Neuroferritinopathy is a rare hereditary movement disorder caused by mutations in the ferritin light chain gene (*FTL*). The disease is clinically present during adulthood and progresses slowly over decades. Iron deposits and cavitation in the basal ganglia and are characteristic findings on brain MRI. Current thinking as to the disease mechanism is that the iron deposition results from the inability of mutant ferritin to bind iron, leading to oxidative stress, mitochondrial dysfunction, and apoptosis. The differential diagnosis involves mainly other entities of NBIA. There is no effective disease-modifying treatment. A few options are available as symptomatic therapy.

Acknowledgment We would like to thank Dr. Carlos Zuñiga for providing the MRI images of neuroferritinopathy.

References

1. Vidal R, Ghetti B, Takao M, et al. Intracellular ferritin accumulation in neural and extraneural tissue characterizes a neurodegenerative disease associated with a mutation in the ferritin light polypeptide gene. J Neuropathol Exp Neurol. 2004;63:363–80.
2. Curtis AR, Fey C, Morris CM, et al. Mutation in the gene encoding ferritin light polypeptide causes dominant adult onset basal ganglia disease. Nat Genet. 2001;28:350–4.
3. Schneider SA, Bhatia KP. Syndromes of neurodegeneration with brain iron accumulation. Semin Pediatr Neurol. 2012;19(2):57–66.
4. McNeill A, Birchall D, Hayflick SJ, et al. T2* and FSE MRI distinguishes four subtypes of neurodegeneration with brain iron accumulation. Neurology. 2008;70:1614–9.
5. Hautot D, Pankhurst QA, Morris CM, et al. Preliminary observation of elevated levels of nano-crystalline iron oxide in the basal ganglia of neuroferritinopathy patients. Biochim Biophys Acta. 2007;1772:21–5.
6. Mancuso M, Davidzon G, Kurlan RM, Tawil R, Bonilla E, Di Mauro S, Powers JM. Hereditary ferritinopathy: a novel mutation, its cellular pathology, and pathogenetic insights. J Neuropathol Exp Neurol. 2005;64(4):280–94.
7. Chinnery PF, Crompton DE, Birchall D, et al. The clinical features and natural history of neuroferritinopathy caused by the FTL1 460insA mutation. Brain. 2007;130:110–9.
8. Devos D, Tchofo J, Vuillaume I, et al. Clinical features and natural history of neuroferritinopathy caused by the 458dupA FTL mutation. Brain. 2009;132:e109.
9. Ohta E, Nagasaka T, Shindo K, et al. Neuroferritinopathy in a Japanese family with a duplication in the ferritin light chain gene. Neurology. 2008;70:1493–4.
10. Kubota A, Hida A, Ichikawa Y, et al. A novel ferritin light chain gene mutation in a Japanese family with neuroferritinopathy: description of clinical features and implications for genotype-phenotype correlations. Mov Disord. 2009;24:441–5.

11. Maciel P, Cruz VT, Constante M, et al. Neuroferritinopathy: missense mutation in FTL causing early-onset bilateral pallidal involvement. Neurology. 2005;65:603–5.
12. Ondo WG, Adam OR, Jankovic J, Chinnery PF. Dramatic response of facial stereotype/tic to tetrabenazine in the first reported cases of neuroferritinopathy in the United States. Mov Disord. 2010;25:2470–2.
13. Lehn A, Mellick G, Boyle R. Teaching neuroimages: neuroferritinopathy. Neurology. 2011; 77(18):e107.
14. Storti E, Cortese F, Di Fabio R, Fiorillo C, Pierallini A, Tessa A, Valleriani A, Pierelli F, Santorelli FM, Casali C. De novo FTL mutation: a clinical, neuroimaging, and molecular study. Mov Disord. 2013;28(2):252–3.
15. Ory-Magne F, Brefel-Courbon C, Payoux P, et al. Clinical phenotype and neuroimaging findings in a French family with hereditary ferritinopathy (FTL498-499insTC). Mov Disord. 2009;24:1676–83.
16. Wills AJ, Sawle GV, Guilbert PR, Curtis ARJ. Palatal tremor and cognitive decline in neuroferritinopathy. J Neurol Neurosurg Psychiatry. 2002;73:86–95.
17. Cassidy AJ, Williams ER, Goldsmith P, Baker SN, Baker MR. The man who could not walk backward: an unusual presentation of neuroferritinopathy. Mov Disord. 2011;26:362–4.
18. Keogh MJ, Singh B, Chinnery PF. Early neuropsychiatry features in neuroferritinopathy. Mov Disord. 2013;28:1310–3.
19. Rouault TA. Iron metabolism in the CNS: implications for neurodegenerative diseases. Nat Rev. 2013;14:551–64.
20. Moos T, Morgan EH. The metabolism of neuronal iron and its pathogenic role in neurologic disease. Ann N Y Acad Sci. 2004;1012:14–26.
21. Friedman A, Arosio P, Finazzi D, Koziorowski D, Galazka-Friedman J. Ferritin as an important player in neurodegeneration. Parkinsonism Relat Disord. 2011;17(6):423–30.
22. Santambrogio P, Levi S, Cozzi A, Rovida E, Albertini A, Arosio P. Production and characterization of recombinant heteropolymers of human ferritin H and L-chains. J Biol Chem. 1993;268(17):12744–8.
23. Cozzi A, Levi S, Corsi B, Santambrogio P, Albertini A, Arosio P. Overexpression of wild type and mutated human ferritin H-chain in HeLa cells: in vivo role of ferritin ferroxidase activity. J Biol Chem. 2000;275:25122–9.
24. Cozzi A, Levi S, Corsi B, Santambrogio P, Campanella A, Gerardi G, Arosio P. Role of iron and ferritin in TNF alpha-induced apoptosis in HeLa cells. FEBS Lett. 2003;537:187–92.
25. Cozzi A, Rovelli E, Frizzale G, Campanella A, Amendola M, Arosio P, Levi S. Oxidative stress and cell death in cells expressing L-ferritin variants causing neuroferritinopathy. Neurobiol Dis. 2010;37(1):77–85.
26. Vidal R, Miravalle L, Gao X, Barbeito AG, Baraibar MA, Hekmatyar SK, Widel M, Bansal N, Delisle MB, Ghetti B. Expression of a mutant form of the ferritin light chain gene induces neurodegeneration and iron overload in transgenic mice. J Neurosci. 2008;28(1):60–7.
27. Deng X, Vidal R, Englander EW. Accumulation of oxidative DNA damage in brain mitochondria in mouse model of hereditary ferritinopathy. Neurosci Lett. 2010;479:44–8.
28. Ohta E, Takiyama Y. MRI findings in neuroferritinopathy. Neurol Res Int. 2012;2012: 19743–8.
29. McNeill A, Gorman G, Khan A, Horvath R, Blamire AM, Chinnery PF. Progressive brain iron accumulation in neuroferritinopathy measured by the thalamic T2* relaxation rate. AJNR Am J Neuroradiol. 2012;33(9):1810–3.
30. Crompton DE, Chinnery PF, Bates D, et al. Spectrum of movement disorders in neuroferritinopathy. Mov Disord. 2005;20:95–9.
31. Morphy MA, Feldman JA, Kilburn G. Hallervorden-Spatz disease in a psychiatric setting. J Clin Psychiatry. 1989;50:66–8.
32. Szanto J, Gallyas F. A study of iron metabolism in neuropsychiatric patients. Hallervorden-Spatz disease. Arch Neurol. 1966;14:438–42.
33. Williamson K, Sima AA, Curry B, Ludwin SK. Neuroaxonal dystrophy in young adults: a clinicopathological study of two unrelated cases. Ann Neurol. 1982;11:335–43.

34. Hayflick SJ, Westaway SK, Levinson B, Zhou B, Johnson MA, Ching KH, Gitschier J. Genetic, clinical, and radiographic delineation of Hallervorden-Spatz syndrome. N Engl J Med. 2003;348:33–40.
35. Muthane UB, Shetty R, Panda K, Yasha TC, Jayakumar PN, Taly AB. Hallervordern Spatz disease and acanthocytes. Neurology. 1999;53:32A.
36. Dooling EC, Schoene WC, Richardson Jr EP. Hallervorden-Spatz syndrome. Arch Neurol. 1974;30:70–83.
37. Morgan NV, Westaway SK, Morton JE, et al. PLA2G6, encoding a phospholipase A2, is mutated in neurodegenerative disorders with high brain iron. Nat Genet. 2006;38:752–4.
38. Paisan-Ruiz C, Bhatia KP, Li A, et al. Characterization of PLA2G6 as a locus for dystonia-Parkinsonism. Ann Neurol. 2009;65(1):19–23.
39. Gregory A, Polster BJ, Hayflick SJ. Clinical and genetic delineation of neurodegeneration with brain iron accumulation. J Med Genet. 2009;46:73–80.
40. Paisan-Ruiz C, Li A, Schneider SA, et al. Widespread Lewy body and tau accumulation in childhood and adult onset dystonia-parkinsonism cases with PLA2G6 mutations. Neurobiol Aging. 2012;33:814–23.
41. Yoshida K, Furihata K, Takeda S, Nakamura A, Yamamoto K, Morita H, Hiyamuta S, Ikeda S, Shimizu N, Yanagisawa N. A mutation in the ceruloplasmin gene is associated with systemic hemosiderosis in humans. Nat Genet. 1995;9:267–72.
42. McNeill A, Pandolfo M, Kuhn J, et al. The neurological presentation of ceruloplasmin gene mutations. Eur Neurol. 2008;60:200–5.
43. Miyajima H, Takahashi Y, Kono S, et al. An inherited disorder of iron metabolism. Biol Met. 2003;16:205–13.
44. Morita H, Ikeda S, Yamamoto K, et al. Hereditary ceruloplasmin deficiency with hemosiderosis: a clinicopathological study of a Japanese family. Ann Neurol. 1995;37:646–56.
45. Newell FW, Johnson 2nd RO, Huttenlocher PR. Pigmentary degeneration of the retina in the Hallervorden-Spatz syndrome. Am J Ophthalmol. 1979;88(3 Pt 1):467–71.
46. Schneider AS, Hardy J, Bhatia KP. Syndromes of neurodegeneration with brain iron accumulation (NBIA): an update on clinical presentations, histological and genetic underpinnings, and treatment considerations. Mov Disord. 2012;27:42–53.
47. Ramirez A, Heimbach A, Grundemann J, et al. Hereditary parkinsonism with dementia is caused by mutations in ATP13A2, encoding a lysosomal type 5 P-type ATPase. Nat Genet. 2006;38:1184–91.
48. Brüggemann N, Hagenah J, Reetz K, et al. Recessively inherited parkinsonism: effect of ATP13A2 mutations on the clinical and neuroimaging phenotype. Arch Neurol. 2010;67: 1357–63.
49. Alazami AM, Al-Saif A, Al-Semari A, et al. Mutations in C2orf37, encoding a nucleolar protein, cause hypogonadism, alopecia, diabetes mellitus, mental retardation, and extrapyramidal syndrome. Am J Hum Genet. 2008;83:684–91.
50. Woodhouse NJ, Sakati NA. A syndrome of hypogonadism, alopecia, diabetes mellitus, mental retardation, deafness, and ECG abnormalities. J Med Genet. 1983;20:216–9.
51. Saitsu H, et al. De novo mutations in the autophagy gene encoding WDR45 cause static encephalopathy of childhood with neurodegeneration in adulthood. Nat Genet. 2013;45(4): 445–9.
52. Lehn A, Boyle R, Brown H, Airey C, Mellick G. Neuroferritinopathy. Parkinsonism Relat Disord. 2012;18:909–15.
53. Timmermann L, Pauls KA, Wieland K, et al. Dystonia in neurodegeneration with brain iron accumulation: Outcome of bilateral pallidal stimulation. Brain. 2010;133:701–12.
54. Ge M, Zhang K, Ma Y, et al. Bilateral subthalamic nucleus stimulation in the treatment of neurodegeneration with brain iron accumulation type 1. Stereotact Funct Neurosurg. 2011;89: 162–6.

Chapter 10
Neurodegeneration with Brain Iron Accumulation

Nardocci Nardo, Vanessa Cavallera, Luisa Chiapparini, and Giovanna Zorzi

Abstract The spectrum of phenotypes included in the group of neurodegeneration with brain iron accumulation (NBIA) has been recently growing. It includes the genetically defined neurodegeneration associated with mutation in genes PANK2 (PKAN), PLA2G6 (PLAN), FA2H (FAHN), ATP13A2 (Kufor–Rakeb disease), CP (aceruloplasminemia), FTL1 (neuroferritinopathy) and the more recently identified C19orf12 (MPAN), and WDR45 (BPAN). Other genetic causes are still to be identified. The clinical manifestations are highly heterogenic and often overlap among the NBIA group. The core features are characterized by progressive extrapyramidal deterioration and iron accumulation in the basal ganglia. Current therapeutic options include various symptomatic approaches, and new therapies are under consideration such as surgical approaches and chelating agents.

Keywords NBIA • Neurodegeneration with brain iron accumulation • Childhood • Diagnosis • Treatment

N. Nardo (✉) • V. Cavallera • G. Zorzi
Department of Pediatric Neurology, Fondazione IRCCS Istituto Neurologico
Carlo Besta, Via Celoria 11, Milan 20133, Italy

Department of Pediatric Neuroscience, Istituto Neurologico Carlo Besta,
Milan, Italy
e-mail: nnardocci@istituto-besta.it; cavallera.v@istituto-besta.it;
gzorzi@istituto-besta.it

L. Chiapparini
Fondazione IRCCS Istituto Neurologico
Carlo Besta, Via Celoria 11, Milan 20133, Italy

Department of Neuroradiology, Istituto Neurologico Carlo Besta, Milan, Italy
e-mail: chiapparini.l@istituto-besta.it

F.E. Micheli, P.A. LeWitt (eds.), *Chorea*,
DOI 10.1007/978-1-4471-6455-5_10, © Springer-Verlag London 2014

Introduction

Brain iron deposition, although part of the normal aging process, can be implicated in different neurologic disorders such as Parkinson's disease, Alzheimer's disease, and numerous genetic neurodegenerative disorders. NBIA includes a heterogeneous group of neurodegenerative disorders (Table 10.1) presenting with a progressive extrapyramidal syndrome and excessive iron accumulation in the basal ganglia, particularly the globus pallidus. Most of the mutations responsible of the syndrome are inherited in an autosomal recessive fashion. The most common genetic defects occur in PANK2 gene leading to pantothenate kinase-associated neurodegeneration (PANK), followed by mutations in PLA2G6 gene leading to phospholipase-associated neurodegeneration (PLAN); both genes encode mitochondrial proteins [1]. Several additional genes have recently been identified including FA2H (fatty acid hydroxylase-associated neurodegeneration, FAHN), ATP13A2 (Kufor–Rakeb disease), C19orf12 (mitochondrial membrane protein-associated neurodegeneration, MPAN), and WDR45 (beta-propeller-associated neurodegeneration, BPAN). In a large number of cases, the underlying genetic bases of NBIA are yet to be defined (idiopathic NBIA). Onset of symptoms is usually in childhood. Adult onset is more common for aceruloplasminemia and neuroferritinopathy which are responsible for only a minority of cases. There is a clinical overlap between the phenotypes; however, additional features such as pyramidal involvement, retinopathy or optic atrophy, and neuroimaging may be of diagnostic value orienting toward etiology.

Table 10.1 Genetic defects in NBIA

Syndrome	Synonym	Gene	Chromosomal position
PKAN		PANK2	20p13
PLAN	PARK14	PLA2G6	22q12
KRS	PARK9	ATP13A2	1p36
FAHN		FA2H	16q23
MPAN		C19orf12	19q12
BPAN	SENDA	WDR45	Xp11.23
Neuroferritinopathy		FTL1	19q13
Aceruloplasminemia		CP	3q23
Idiopathic		–	–

Modified from Schneider et al. [2]

PKAN pantothenate kinase-associated neurodegeneration, *PANK2* pantothenate kinase 2, *PLAN* phospholipase A2-associated neurodegeneration, *PLA2G6* phospholipase A2, *KRS* Kufor–Rakeb syndrome, *FAHN* fatty acid hydroxylase-associated neurodegeneration, *FA2H* fatty acid 2-hydroxylase, *MPAN* mitochondrial-associated neurodegeneration, *BPAN* beta-propeller-associated neurodegeneration, *SENDA* static encephalopathy of childhood with neurodegeneration in adulthood, *FTL* ferritin light chain, *CP* ceruloplasmin

PKAN

Pantothenate kinase-associated neurodegeneration accounts for about half of NBIA cases and is caused by mutations in the pantothenate kinase 2 gene (PANK2) on chromosome 20p [3]. The gene codes for a mitochondrial enzyme catalyzing the first step in the synthesis of coenzyme A [4, 5]. The majority of patients present with an early onset which defines the so-called classic type, whereas patients with later onset may show atypical features.

Classic PKAN

The classic form is characterized by a relatively homogeneous phenotype [6, 7]. Age at onset is around 3–4 years, and before age 6 in almost 90 % (range 6 months to 12 years) [2]. A common presentation is clumsiness and gait abnormalities followed by dystonia affecting gait and posture. Dystonia is the prominent extrapyramidal feature; it is usually asymmetric and has a characteristic involvement of the oromandibular region muscles which can lead to dysarthria and tube feeding [7]. Status dystonicus may occur during the course of the disease [8]. Other movement disorders such as chorea or parkinsonism are rarely described later on. Patients may show pyramidal tract involvement with spasticity, hyperreflexia, and extensor toe sign. Oculomotor abnormalities including impaired saccadic pursuits, hypometric or slowed vertical saccades, and supranuclear gaze vertical palsy have been reported [2]. Clinical signs of pigmentary retinopathy are frequent, and around 70 % of patients show abnormal electroretinograms [9]. Behavioral changes and dementia may occur, but neuropsychological abnormalities have been studied only in few cases [10]. The course of the disease may alternate a rapid neurological deterioration in the absence of precipitating factors, with periods of relative stability [11]. The majority of children (85 %) lose ability to walk within 10–15 years [12]. Death usually occurs because of cardiorespiratory complications, the secondary effects of malnutrition, and, rarely, status dystonicus [11].

Atypical PKAN

Atypical PKAN shows a heterogeneous phenotype. It has a later onset, in the second or third decade (mean age 14, range 1–28 years), a slower course of disease, and more slowly progressive neuropsychiatric features in addition to movement disorders [13]. Patients frequently present with speech abnormalities (such as palilalia or dysarthria) or parkinsonism. Other initial signs may be unilateral dystonic tremor,

Fig. 10.1 PKAN. Axial
PD-weighted MR image
reveals the typical appearance
of the "eye-of-the-tiger":
marked hypointensity in the
globi pallidi with high signal
intensity foci in the
anteromedial region in which
is also visible a central spot
of low signal intensity

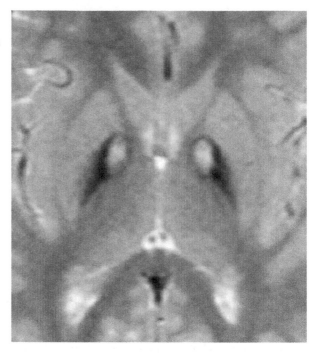

focal arm dystonia or cognitive decline, and psychiatric symptoms (such as depression, emotional lability, impulsivity, obsessive compulsive disorder, and violent outbursts). Motor involvement tends to be milder than the classic form with loss of ambulation after 15–40 years [3, 14]. Clinical or electroretinographic evidence of retinopathy is less frequent in these patients. Psychiatric symptoms, behavioral disturbances, and frontotemporal-like dementia are common in the early course of the disease [11]. Tics and obsessions suggestive of Tourette syndrome have also been described [15]. Rare presentations of PKAN include pure akinesia, a motor neuron disease-like phenotype, and early onset parkinsonism [16].

Neuroimaging

MRI characteristics are specific for PKAN. Classical and atypical forms of the disease share the same MRI findings.

The typical "eye-of-the-tiger" sign (see Fig. 10.1) is characterized by a T2-weighted central hyperintense signal surrounded by a rim of hypointensity in the globus pallidus. The central bright spot is probably due to fluid accumulation or edema, while the hypointensity indicates iron deposits [17]. Some individuals with PKAN may also show hypointensity of the substantia nigra [11]. Although the majority of patients with PANK2 mutations have this radiological sign, it is not pathognomonic of PKAN. Not all patients show the radiological sign of the "eye-of-the-tiger," particularly during the early stages of the disease [11].

Pathophysiology

PANK2 encodes the mitochondrial form of pantothenate kinase 2 which phosphory-lates pantothenate – vitamin B5 – to phosphopantothenate. This is the initial and rate-limiting step in the biosynthesis of coenzyme A [13]. The PANK2 enzyme has a mitochondrial targeting sequence and is localized in mitochondria in the human brain. Therefore, mitochondrial dysfunction due to PANK2 mutations is proposed to lead to neurodegeneration. Moreover, coenzyme A is involved in numerous bio-chemical reactions and is essential for energy metabolism as well as lipid synthesis and degradation. Mutations in this gene is thought to alter the normal cellular lipid metabolism [11]. Accumulation of cysteine, which is a potent iron chelator and a substrate for PANK2, is neurotoxic and could lead to iron deposits; neuronal injury may be increased by oxidative stress due to iron accumulation [11].

The central role of PANK2 activity in PKAN neuropathogenesis is confirmed by studies which demonstrate earlier onset of disease in homozygous null muta-tions compared with point mutations, and studies show that the greater the resid-ual enzyme activity, the later is the onset of the disease. This correlation does not include progression of the disease [7, 16]. The majority of PANK2 variants are missense mutations; deletions and splice-site variants are also described. The amount of residual enzyme activity determines the severity of the clinical mani-festations [11].

Pathological features include rusty brown discoloration of the globus pallidus and substantia nigra on gross examination indicating iron deposits. Microscopic examinations show destructive lesions characterized by loss of myelinated fibers and gliosis [13]. Moreover, dystrophic axons are also commonly observed: axonal spheroids are found throughout the central nervous system. Ceroid–lipofuscin and neuromelanin also accumulate within cells [13, 16].

PLAN (PLA2G6-Associated Neurodegeneration)

PLA2G6 mutations are the second cause of NBIA. They have been associated with a variety of diverse phenotypes that include psychomotor regression in infancy with death in early childhood at one end of the spectrum and adult-onset parkinsonism dystonia at the other [8]. PLA2G6 mutations have been identified in patients with neuroaxonal dystrophy (NAD), PANK2 negative NBIA, Karak syndrome, and dystonia–parkinsonism demonstrating that the phenotypic spec-trum of PLAN is wide and needs to be more characterized [18]. Not all PLA2G6-related conditions belong to the category of NBIA, as iron on MRI may be totally absent [3, 19, 20].

In the NBIA group, PLAN includes three different conditions: classic infantile neuroaxonal dystrophy (classic INAD), atypical neuroaxonal dystrophy (atypical NAD), and adult-onset dystonia–parkinsonism. Classic INAD and atypical NAD were not initially well distinguished leading to some confusion in the classification of single patients. After the identification of gene mutations, it became evident that the same gene can be responsible for all of the three phenotypes and that other genes are involved [21].

Classic INAD

This is the first described syndrome associated with PLA2G6 mutations. The clinicopathological diagnostic criteria have been defined by Aicardi and Castelein: onset of the disease before 3 years of age; diffuse disorder of the central nervous system with psychomotor deterioration and increasing neurological involvement comprising symmetrical pyramidal tract signs and marked hypotonia; progressive course; and the presence of spheroids at biopsy of central or peripheral tissue. Onset ranges between 6 months and 3 years, but most patients show symptoms by the end of the first year [22]. The first manifestation is often a delay in psychomotor development or loss of previously acquired milestones. Some patients may show an acute onset after a febrile illness or severe vomiting [22–24]. Patients may show pyramidal signs, marked truncal hypotonia, and visual disturbances due to optic atrophy. Visual symptoms such as nystagmus, strabismus, or visual deficit are common at early stage of the disease [22]. Later in the course of the disease, complex partial or generalized seizures occasionally appear and are usually well controlled by antiepileptic drugs. The disease evolves in dementia, dystonia, spastic tetraparesis with areflexia, and distal limb contractures associated to bulbar signs such as impaired swallowing and dyspnea [18]. The average age at death is 9.4 years [16].

Abnormal electroencephalogram (EEG), electromyogram (EMG), and visual evoked potentials (VEP) may facilitate early diagnosis [20, 21]. EEG shows high-voltage fast rhythms both in the sleeping and waking states. These abnormalities usually appear after age 2 and persist over time. Other EEG features include a progressive slowing of background activity and various epileptiform activity [18]. EMG reveals signs of chronic denervation detectable as early as 2 years. Nerve conduction velocity (NCV) studies may show abnormalities consistent with a distal axonal-type sensorimotor peripheral neuropathy [22]. VEP are markedly reduced in amplitude or extinct, while electroretinogram (ERG) is normal [18].

The most characteristic finding on MRI is cerebellar atrophy with T2-weighted hyperintensity of the cerebellar cortex (see Fig. 10.2). Additional MRI findings are thinning of the optic chiasma, signal hyperintensity of the dentate nuclei, cerebral cortical atrophy, and hyperintensity of the white matter [22]. MRI variably shows evidence of iron accumulation in the globus pallidus and substantia nigra (see Fig. 10.2d–f) [8, 21]. Half of the patients may lack the iron deposits early in the disease, but they usually develop globus pallidus hypointensities as the disease progresses [19, 21].

Mutations in PLA2G6 are found in the majority of patients with classic INAD. Large intragenic deletions or duplications, which are not detectable with direct sequencing of the gene, are probably responsible for mutation-negative patients, predicting that the mutation detection rate of PLA2G6 in typical INAD cases will

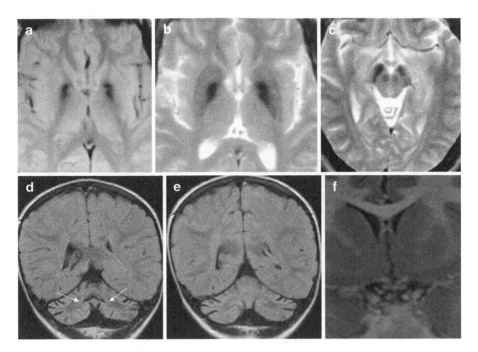

Fig. 10.2 INAD. On the top axial T2-weighted MR images (**a**–**c**) show marked hypointensity in the nucleus pallidum (**a**, **b**) and in the substantia nigra (**c**) bilaterally, due to iron accumulation. At the bottom, in another case, coronal FLAIR (**d**, **e**) and IR (**f**) images demonstrate global cerebellar atrophy with hyperintensity in the cerebellar cortex and in the dentate nuclei (*arrows*). Thin optic chiasma is also visible (**f**)

exceed 95 % [3, 18]. In the presence of a classic INAD phenotype, peripheral biopsy may be no more necessary and should be performed only if genetic testing for PLA2G6 mutations results negative [3].

Atypical NAD

These patients show a heterogeneous phenotype including a later onset ranging from 1.5 years to late teens [16, 18]. When the onset is in the early years (before age 3), the progression of the disease is slow with a late neurological deterioration [18]. The main presentation is gait instability or ataxia, delay in development of speech, and abnormal social interactions. Additional features may include spasticity, rigidity, and dystonia associated with mental impairment. The patients show optic atrophy, nystagmus, and epilepsy with the same frequency as classic INAD [16, 18].

Atypical NAD with a later onset may not show any MRI abnormalities, lacking signs of iron accumulation. Others may show cortical atrophy or white matter changes [25].

Dystonia–Parkinsonism

Another atypical phenotype characterized by transient levodopa-responsive dystonia–parkinsonism associated with psychiatric features, cognitive decline, and pyramidal signs, adult onset (12–15 years), and absence of brain iron accumulation was described as a result of PLA2G6 mutations [20].

Pathophysiology

PLA2G6 encodes a calcium-independent phospholipase A2 that catalyzes the hydrolysis of glycerophospholipids. The protein has proposed roles in phospholipid remodeling, arachidonic acid release, leukotriene and prostaglandin synthesis, cell membrane homeostasis, and apoptosis. All of the mutations identified are likely to be inactivating and could cause structural membrane abnormalities responsible for the axonal pathology observed in NAD [21]. The variability in iron accumulation is perhaps related to differential effect of mutations in the enzyme's catalytic activity toward phospholipids and lysophospholipids [8, 26]. Ubiquitin has been detected in small spheroids in some cases indicating that the impairment of axonal membrane homeostasis and protein degradation are involved in the pathogenesis of NAD [27, 28].

The neuropathological features are the hallmark of the disease in all three phenotypes, that is, the presence of neuroaxonal dystrophy throughout the central and peripheral nervous system. Spheroids can be detected on conjunctiva, skin, rectum, muscle, and nerve biopsy. Spheroids contain membranes, synaptic vesicles, degenerated mitochondria, myelin figures, neurofilaments, and membranous dense bodies [29]. Additional neuropathological findings in PLAN are widespread neuronal loss and gliosis in the cerebral cortex and basal ganglia, iron deposits in a perivascular distribution within the globus pallidus and substantia nigra, and extracellular deposits. Lewy bodies in the substantia nigra and neurofibrillary tangles have been demonstrated in a few patients [16]. These last findings could possibly reflect oxidative damage of the cytoskeleton.

KRS (Kufor–Rakeb Syndrome: PARK9)

The disease shows an autosomal recessive inheritance. It is a form of juvenile parkinsonism due to mutations in a lysosomal type 5 P-type ATPase (ATP13A2 gene) [30]. Only a few families are described; therefore, the clinical spectrum is still to be determined. It includes a transitory levodopa-responsive parkinsonism, pyramidal tract signs, eye movement abnormalities with incomplete supranuclear upward gaze palsy and saccadic pursuit, facial–finger minimyoclonus and autonomic dysfunction, psychiatric manifestations, and cognitive deterioration [30, 31]. Follow-up in a few patients showed visual hallucinations and oculogyric dystonic spasms [30].

Prominent cerebellar symptoms and signs have been reported in a Greenlandic Inuit family. These patients also showed an axonal neuropathy at electrophysiological examination [32]. Onset of disease is usually during late childhood (mean age 13 years). The course of the disease has a rapid progression, especially in the first years, and is more insidious thereafter [31, 32]. Evaluation of heterozygous carriers revealed the presence of very subtle neurological signs, and no signs in the youngest siblings, suggesting the age-related progression of the disease [30].

MRI may show pallidopyramidal degeneration followed by generalized brain atrophy and inconstant iron deposits in the basal ganglia [2, 30, 32]. There are no demonstrated phenotype–genotype correlations; however, missense mutations have been suggested to produce a phenotype without iron accumulation [33].

Pathophysiology

The function of the ATP13A2 protein remains largely unknown, but it is supposed that it might participate in autophagic protein degradation via the lysosomal pathway [34]. It has been shown that the wild-type protein is located in the perimeter lysosomal membrane, while the truncated mutants are retained in the endoplasmic reticulum and degraded by the proteasome [35, 36]. The experimental mouse model provided by Schultheis is likely due to lysosomal dysfunction resulting from the loss of ATP13A2 function with little contribution from secondary effects of a mutant protein. Nevertheless, the observation that the mutant ATP13A2 is degraded by the proteasomal pathway raises the possibility that mutant species might overload the proteasomal ways and disturb its normal function to degrade other proteins, particularly if the lysosomal pathways are not working efficiently [35, 37]. Moreover, pathogenic mutations may lead to a failure of the protein in reaching the lysosome with an impaired lysosomal function. This effect may lead to cellular accumulation of toxic proteins such as alpha-synuclein, a protein known to accumulate and contribute to the etiology of Parkinson's disease [35]. Radi et al. demonstrated the presence of high apoptosis levels in KRS patients; however, the presence of similar abnormalities in the parents suggests that this alteration is probably not pathogenetic by itself [38].

Mutations in PARK9 are associated with KRS and of particular interest is a recent study showing that a juvenile-onset form of neuronal ceroid lipofuscinosis (NCL) was caused by an ATP13A2 missense mutation. NCL is an inherited progressive degenerative disease that affects the brain and the retina [37]. Affected individuals developed typical NCL pathology but also presented with several clinical aspects commonly seen in KRS such as parkinsonism and cognitive decline in NCL families [39]. An experimental model of ATP13A2-knockdown cells resulted in an abnormal aggregation of lysosomes at perinuclear site; the vesicles were enlarged. Moreover, cathepsin D activity was reduced indicating lysosomal dysfunction; cathepsin D deficiency causes NCL [34].

No neuropathological examination has been reported [34]. However, biopsy analysis of a sural nerve from a patient revealed cytoplasmic inclusions in Schwann cells and smooth muscle cells, which may be interpreted as reminiscent of NCL inclusions [36, 40]. Moreover, accumulation of membrane-bound storage material in the muscle and skin cells of KRS patients has been demonstrated [39]. Such observations support the hypothesis that mutations in ATP13A2 may produce a clinical continuum ranging from NCL to KRS [39].

FAHN (Fatty Acid Hydroxylase-Associated Neurodegeneration)

Mutation in the FA2H gene is another autosomal recessive cause of NBIA [41]. Mutations in the same gene were known to cause leukodystrophy and SPG35, a form of hereditary spastic paraplegia (HSP), [42, 43] and overlapping syndromes [44, 45]. Onset is variable but typically during the first decade; the clinical phenotype is characterized by gait impairment, spastic paraparesis, ataxia, and dystonia. Additional features are seizures, optic atrophy, and acquired strabismus [41].

MRI shows the typical features of NBIA with iron deposition (globus pallidus hypointensity) late in the course of the disease. Additional features have been described overlapping with HSP (thin corpus callosum) and leukodystrophy (white matter changes) [2]. In the allelic variants, iron deposition appears to be variable [42, 43]. Pontocerebellar atrophy has also been described [41].

Pathophysiology

FA2H is strictly linked to the lipid and ceramide metabolism (such as PANK2 and PLA2G6). More in detail, the product of the gene participates in the synthesis of galactosylceramides and sulfatides which are important components of myelin sheaths [41, 42]. Therefore, it is not surprising that formation of abnormal myelin gives rise to leukodystrophy and HSP. Moreover, FA2H has a role in intracellular ceramide pool composition and consequent cellular protein and lipid turnover. Mutations may have effects on lipid signal transduction and lead to premature apoptosis [41]. No postmortem tissue has yet been available for pathological studies. Mouse models showed marked demyelination, profound axonal loss, and abnormally enlarged axons in the central nervous system but mostly unaffected peripheral nerves (that may be related to the presence of a second fatty acid hydroxylase in peripheral tissue) [41, 46, 47].

MPAN (Mitochondrial Membrane Protein-Associated Neurodegeneration)

Mutations on gene C19orf12 at chromosome 19q12 have been recently identified as causative in a recessive form of NBIA [1]. The age of onset is between 4 and 30 years [48]. Clinical phenotype is varied and characterized by a slow progression of extrapyramidal signs (generalized dystonia and parkinsonism), pyramidal tract involvement

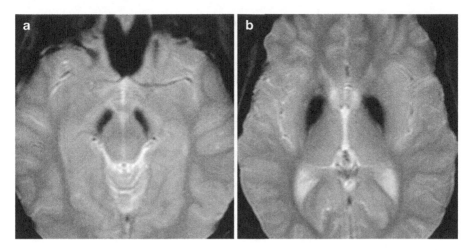

Fig. 10.3 C19orf12. Axial T2-GRE images (**a**, **b**) demonstrate a marked bilateral and symmetric hypointensity in the substantia nigra and in the globus pallidus of both sides due to abnormal iron deposition

(spastic paraparesis), and cognitive impairment [49, 50]. Most of the patients show additional optic nerve atrophy and motor axonal neuropathy [50, 51]. Three patients presented with a predominant upper and lower motor neuron dysfunction mimicking amyotrophic lateral sclerosis with juvenile onset associated with behavioral abnormalities (disinhibited or impulsive behavior); extrapyramidal signs were absent [49]. Compared with PKAN, these patients have a later onset and a slower progression (loss of ambulance 8 to over 24 years from onset) [1]. A mild phenotype initially diagnosed as idiopathic Parkinson's disease has been described [1].

MRI shows T2 hypointensities in the globus pallidus and substantia nigra (see Fig. 10.3). It is worth noting that one patient of the Polish cohort showed the eye-of-the-tiger sign [1].

Pathophysiology

The first mutation identified in the Polish cohort was a single homozygous mutation that predicted a truncation of the protein sequence through a frameshift and a premature stop codon [1]. The protein encoded by C19orf12 is highly conserved in evolution. Two alternative first exons give birth to two isoforms of the protein. The gene is mostly expressed in the brain, blood cells, and adipocytes. Genes coregulated by C19orf12 appeared to be mainly involved in the mitochondrial pathway of CoA metabolism (fatty acid biogenesis and valine, leucine, and isoleucine degradation). More is to be discovered [1].

Neuropathological finding documented widespread iron deposits, neuroaxonal spheroids, alpha-synuclein-positive Lewy bodies, and tau-positive inclusions [1, 48].

BPAN (Beta-Propeller-Associated Neurodegeneration)

SENDA (Static Encephalopathy of Childhood with Neurodegeneration in Adulthood)

SENDA is a distinctive phenotype among NBIA, and only single cases have been reported. It is an X-linked dominant form due to de novo or dominant mutation in WDR45, at Xp11.23 [52, 53]. The disease is characterized by early childhood psychomotor retardation with a static course until adulthood. During the second or third decade, patients show rapidly progressive dystonia–parkinsonism and cognitive decline [53]. Additional neurological signs include spasticity, eye movement abnormalities, sleep disorders, frontal release signs, and dysautonomia [2]. A transient good response of motor signs during levodopa treatment has been frequently observed [54].

MRI after deterioration shows the classic iron deposition in the globus pallidus and substantia nigra. A deeply suggestive feature of SENDA is the T1 hyperintensity surrounding a central band of hypointensity within the substantia nigra and cerebral peduncles (see Fig. 10.4) [8, 52, 55]. In addition, cerebral atrophy may be observed [52, 55].

Pathophysiology

WDR45 encodes for proteins involved in the autophagy pathway (the major intracellular degradation system). Mutant proteins are structurally unstable and undergo degradation. Evidence suggests that mutations incompletely block the flux of the autophagic flux at an intermediate step of the formation of the autophagosome. Studies in mice showed that impairment of autophagy in the central nervous system

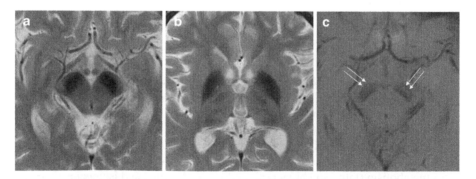

Fig. 10.4 SENDA. Axial T2-weighted images (**a, b**) and T1-weighted image (**c**) show a marked hypointensity in the substantia nigra and in the globus pallidus due to iron deposition. Note the increased volume of the substantia nigra. In (**c**) a slight hyperintensity is visible in the central portion of the substantia nigra (*arrows*)

caused neurodegeneration and neurological symptoms (motor and behavioral deficits) [54]. Saitsu et al. hypothesized that the partial autophagy activity left in mutations of WDR45 might be a possible explanation why the course of the disease is initially static. Only three male patients have been described so far with the exact same phenotype of females [52, 53] They had somatic mosaic mutation supporting the idea that male germ line mutations are nonviable [56]. The clinical phenotype is most likely determined by the stage of development at the time of the somatic mutation; therefore, it is possible that milder and less severe phenotypes exist [52]. De novo mutations could occur also in the somatic tissues in females.

Neuroferritinopathy

Neuroferritinopathy is an adult-onset autosomal dominant disease due to mutation in the ferritin light chain (FLT1) gene on chromosome 19q13.3. Onset of symptoms is typically in the fourth to sixth decades with a mean age at onset of 39 years [16, 57]. It is often reported a family history of a misdiagnosed movement disorder [58]. The clinical manifestations are variable with prominent asymmetric movement disorders (chorea, focal dystonia, and parkinsonism) that overlap with Huntington's disease, torsion dystonia, and idiopathic Parkinson's disease [57, 58]. Facial and orolingual dyskinesia is a common finding. Less common presentations include acute onset ballism, blepharospasm, and writer's cramp [16]. A single extrapyramidal disorder at onset remains the major phenotype, but most of the patients eventually develop a combination of movement disorders. The progression of symptoms is slow and may lead to aphonia, dysphagia, and severe motor disability, although most patients remain ambulant two decades after onset [16]. Cognitive deficits are usually subtle, mainly involving verbal fluency; subcortical or frontal cognitive dysfunction has been occasionally described, later in the course of the disease [58]. In isolated cases have been described additional neuropsychiatric symptoms ranging from acute psychosis to depression. Keogh et al. described a significant but heterogeneous nonmotor component which begins within 5 years of the onset of motor symptoms [59]. Specific phenotypes have been related to less common mutations: early-onset tremor, gait disturbances, and cognitive decline have been described in a Japanese family and ataxia, bradykinetic–rigid syndrome, and episodic psychosis in adolescence in a different one [60].

Serum ferritin level is a useful screening test: it is low in the majority of male and postmenopausal patients [58]. Hemoglobin and serum iron are within normal range.

Abnormal MRI characterized by hypointensity of the globus pallidus and substantia nigra can be found in presymptomatic carriers decades before onset of symptoms [58, 61]. Early in the disease, neuroimaging shows hypointensity of the red nucleus, caudate, globus pallidus, putamen, thalamus, substantia nigra, and cerebral cortex [16]. When the disease proceeds, MRI demonstrates iron accumulation predominantly in the basal ganglia that evolves into cystic degeneration associated with generalized involutional changes and an excess of small vessel ischemic lesions [57,

58, 63]. Tissue edema shows as T2-weighted hyperintensity of the globus pallidus and caudate heads [16].

Limited muscle biopsy studies showed respiratory chain defects; they are associated to brain immunocytochemical abnormalities which point to a secondary defect of the respiratory chain [62].

Pathophysiology

Ferritin is a polymer of ferritin light chains (FTL) and ferritin heavy chains (FTH). Mutations in the FLT1 gene result in an elongated peptide that alters the structure causing accumulation of ferritin and iron in central neurons, in particular within the basal ganglia, leading to oxidative stress, causing mitochondrial damage, and ultimately resulting in apoptotic cell death and neurodegeneration [61, 63]. Seven mutations have been described [59]. The most commonly found mutations result in disruption of the terminal portion of the ferritin light chain (last portion of helix D and the whole length of helix E). Helix D is assumed to facilitate incorporation of mutant ferritin light chain into ferritin; therefore, the rare mutations that leave helix D intact seem to lead to a severe phenotype [60]. A mouse model suggested the involvement of the pathway of ubiquitin and proteasome as well in the pathogenesis of neuroferritinopathy [64] Neuropathological studies showed neuronal loss and gliosis in the cerebral cortex, thalamus, substantia nigra, caudate, putamen, and globus pallidus. Moreover, intranuclear and intracytoplasmic inclusion bodies were found in the neurons, glia, and endothelial cells which stained positive for iron deposition and abnormal aggregates of ferritin [16, 57].

Aceruloplasminemia

Aceruloplasminemia is a neurodegenerative disorder resulting from autosomal recessive mutations of the ceruloplasmin (CP) gene. Clinical presentations of the homozygotes include psychiatric features in late middle age (mean age of 51), but are usually preceded by microcytic anemia with diabetes mellitus and retinal degeneration in association with a low serum ceruloplasmin and a high ferritin [58]. The most common neurological features are dementia, craniofacial dyskinesia, and cerebellar ataxia. Movement disorders such as tremor, parkinsonism, and chorea have been reported [65]. The disease progresses over months to years with additional neurological deficits [65]. Death is often due to aspiration pneumonia [16]. Only a few heterozygous cases have been described in literature. They showed either cerebellar ataxia, postural tremor, or choreoathetosis. The patients were not diabetic [66, 67].

Serum investigations of homozygous patients show absent ceruloplasmin and elevated ferritin. Most of the patients have a microcytic anemia. Heterozygous patients have halved ceruloplasmin level [65].

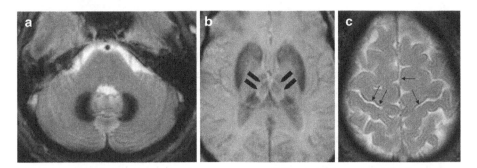

Fig. 10.5 Aceruloplasminemia. Axial T2-weighted images (GRE) of the cerebellum (**a**) and of the supratentorial show marked hypointensity in the dentate nuclei (**a**), neostriatum, and thalamic nuclei (**b**). A slight hypointensity is also visible in the superficial cortical layers of the cerebral hemispheres (*arrows* in **c**). Note the relative hyperintensity of the internal medullary lamina of the thalamus (*arrowheads* in **b**)

MRI demonstrates hypointensity of the cerebral and cerebellar cortex, globus pallidus, caudate nucleus, putamen, thalamus, red nucleus, and substantia nigra (see Fig. 10.5). There are no hyperintense lesions [16]. The few cases with cerebellar ataxia only showed cerebellar atrophy [66].

Pathophysiology

The disease is characterized by absent serum ceruloplasmin or mutant ceruloplasmin lacking ferroxidase activity [57]. Ceruloplasmin is an alpha2-glycoprotein and carries 95 % of the plasma copper. Moreover, it plays a crucial role in the mobilization of iron by interacting with ferritin through its ferroxidase activity. Therefore, ceruloplasmin deficiency inhibits iron efflux from the tissues and causes excessive accumulation in various tissues including the liver, pancreas, retina, and basal ganglia. Iron deposition is proposed to generate free radicals and thus cause tissue injury [57, 65, 68]. Nonetheless, mouse models recently indicated that ceruloplasmin plays an important neuroprotective role, even without apparent misregulation of iron homeostasis [69]. CP is not only a ferroxidase but also a scavenger of reactive oxygen species. Moreover, Texel et al. suggested that redox injury is limited in these patients and that the defect in aceruloplasminemia is associated with inefficient iron efflux from macrophages; astrocyte and microglial iron overload damages these cells and therefore neuronal survival is compromised [70].

Pathologically, there is evidence of iron deposition within astrocytes and neurons in the basal ganglia (especially the putamen and globus pallidus), thalamus, and cerebral and cerebellar cortices. Neuronal loss is mild and can be due to both direct effects of free radicals on neurons and astrocyte dysfunction [71]. The most characteristic histopathological findings are abnormal deformed astrocytes and globular structures especially in the striatum; this finding may be related to iron-induced tissue damage [71]. Moreover, assays of complex I and IV function revealed mitochondrial dysfunction [72].

Idiopathic NBIA

There remains a significant group of patients with radiologically discernible NBIA without a known genetic diagnosis. NBIA appears to be both clinically and genetically heterogeneous, and it describes all cases where imaging or autopsy shows iron accumulation in the brain. This group of patients shows a great variability of age at presentation: onset in infancy often has a rapid disease progression, while presentation in the third to fifth decade is often associated with slower disease progression [11]. Clinical symptoms lack specificity to guide additional subcategorizations and include pyramidal (spastic paraparesis), extrapyramidal (parkinsonism, dystonia, dysarthria), and ophthalmological features (optic nerve pallor or atrophy). The phenotypic spectrum is broad, although some findings are common: progressive dystonia, rigidity, and dysarthria are observed in most cases [11]. Developmental delay and cognitive impairment are described more frequently in this group than in PKAN [3]. The wide clinical spectrum in this group of patients suggests that more additional genes remain to be found. MRI demonstrates iron accumulation in the globus pallidus and substantia nigra.

Diagnosis

The clinical presentation of NBIA is nonspecific with a considerable phenotypic heterogeneity, making the diagnosis of these rare diseases challenging. Nonetheless, brain iron accumulation is radiologically recognizable and, therefore, brain MRI is a highly sensitive diagnostic tool. It is common for NBIA to be suspected only after brain MRI has been performed. MRI can also contribute to distinguishing the various forms of NBIA if combined with clinical features. A correct subtype characterization is essential for family counseling, to guide prognostic and treatment decisions, and may have relevance for clinical trials [8].

After iron deposition is identified on MRI, the pattern of accumulation should be evaluated. Further investigations include complete blood count, serum copper, serum ceruloplasmin, and serum iron indexes, fundoscopic examination (to identify pigmentary retinopathy or optic atrophy), and VEP (subclinical retinal degeneration or optic neuropathy) [8].

Statistically, neurological disorders associated with brain iron deposits are most likely due to PKAN or INAD in childhood and neuroferritinopathy or aceruloplasminemia in adults. Neuroferritinopathy and aceruloplasminemia show widespread iron in the basal ganglia and cerebral cortex. MRI in neuroferritinopathy also often demonstrates the presence of hyperintense lesions, probably due to cavitation. Laboratory findings such as elevation of serum ferritin and absent serum ceruloplasmin point toward aceruloplasminemia, while a low serum ferritin and normal ceruloplasmin level suggest neuroferritinopathy. Moreover, subjects with aceruloplasminemia may show systemic abnormalities of iron metabolism (diabetes mellitus and microcytic anemia) prior to the onset of neurological symptoms [73].

In PKAN brain iron accumulation is restricted to the globus pallidus and substantia nigra showing the classic "eye-of-the-tiger sign." However, iron accumulation may be seen also in non-NBIA disorders such as organic acidurias and acquired dystonias. NAD patients (classic INAD and atypical NAD) show cerebellar atrophy with T2-weighted hyperintensity of the cerebellar cortex or may show high globus pallidus iron not associated with the "eye-of-the-tiger sign." When the clinical and radiological suggestions for INAD is strong, molecular testing for PLA2G6 should precede invasive peripheral biopsy. For those cases in which molecular analysis is negative, biopsy should be performed to search for spheroids [3].

Differential Diagnosis

The differential diagnosis in pediatric cases is complicated because of the wide range of recessive neurological and metabolic disorders which can present in similar ways.

Consanguinity among parents strongly suggests a recessive disorder but is not always present. Clinical clues which help to differentiate pediatric NBIA from neurometabolic disorders include the absence of dysmorphisms, normal head size, and normal perinatal history. Serum lactate, ammonia, urinary organic acids, amino acids, and oligosaccharides are normal [16]. Moreover, NBIA disorders produce a characteristic hypointensity of the basal ganglia, whereas other disorders such as mitochondrial encephalopathies, organic acidurias, and abnormalities of cofactor metabolism feature T2 hyperintensity [8]. There are other disorders such as lysosomal disorders that show T2 hypointensity (storage material); in these cases specific sequences (gradient echo T2 weighted and susceptibility weighted) help distinguish iron deposition [8]. Association of hypointense basal ganglia with the lack of normal myelination should point toward hypomyelinating diseases. Friedreich's ataxia may show brain iron deposition that accumulates mainly in the dentate nucleus, and there is no involvement of the basal ganglia; extrapyramidal symptoms are usually absent [8]. Diverse acquired condition can show iron deposits in the basal ganglia such as multiple sclerosis and inflammatory disorders of the central nervous system, and these etiologies should be taken into consideration, although the clinical features are usually quite different and show no involvement of the basal ganglia. Spinocerebellar ataxia 17, Huntington's disease, and the different etiologies of cerebellar syndromes enter in the differential diagnosis of adult-onset NBIA phenotypes presenting with chorea or ataxia (neuroferritinopathy and aceruloplasminemia) [16].

Treatment

Treating NBIA disorders is a challenging issue, which requires expertise and a comprehensive approach to the patient. Current medical options remain largely unsatisfactory and do not prevent the disease from progressing to a severe and disabling state. Conventional pharmacotherapy and, in selected cases, surgical treatment are primarily symptomatic, aimed to reduce abnormal movements and spasticity which represent the most disabling symptoms. The advances in understanding pathophysiological mechanism have led to the availability of innovative and more rationale treatments that are under consideration. Among innovative therapy, the use of iron chelators is emerging, but other strategies are under investigation that will possibly modify the clinical course of such severe conditions [74].

Moreover, during the acute and prolonged episodes of dystonia is especially important to evaluate for other causes of pain such as occult bone fractures; osteopenia in non-ambulatory patients associated with marked stress on the long bones from dystonia increases the risk for fractures without apparent trauma [75].

Pharmacotherapy

Pharmacotherapy is used for the symptomatic treatment of the movement disorders, spasticities, seizures, and psychiatric disturbances in patients with NBIA.

Dystonia, to the degree of status dystonicus, is often the most severe symptom which is a potentially life-threatening condition. Several drugs may be efficacious including anticholinergics, baclofen; typical and atypical neuroleptics, benzodiazepines; and levodopa. Trihexyphenidyl, an anticholinergic drug, is used in the first line of the treatment of generalized and segmental dystonia [76]. Children usually tolerate higher dose than adults, and some patients may need up to 60–80 mg per day of trihexyphenidyl. The starting dose has to be low and increased very slowly so the accurate evaluation of efficacy requires some months. Side effects include dose-related drowsiness, confusion, memory difficulty, and hallucinations. Benzodiazepines (diazepam, lorazepam, or clonazepam) can provide additional benefit for patients whose response to anticholinergic drugs is unsatisfactory.

Dopamine-depleting drugs, such as tetrabenazine, have been proven useful in some patients with dystonia. Tetrabenazine is an inhibitor of vesicular monoamine transporter 2 and has the advantage, compared to other anti-dopaminergic drugs, not to cause tardive dyskinesia [77]. The potential clinical benefit of anti-dopaminergic drugs is usually limited by the development of side effects, especially sedation.

Improvements with levodopa appear to be mild but have been reported in patients with secondary and degenerative dystonias [78] and in those patients with associated parkinsonism. A good response to this treatment, at short- and long-term follow-up, shows in some patients with late-onset PLAN and BPAN [2, 54].

Spasticity is a common feature in most patients with NBIA and may vary in severity from a subtle neurological sign to severe spasticity causing pain and orthopedic deformities. The most used antispasticity drugs are those acting on a GABAergic system such as baclofen, gabapentin, and benzodiazepines; on alpha-2 adrenergic system such as tizanidine; and on those that block calcium release into the muscles such as dantrolene [74]. It is important to note that there are no evidence-based guidelines for the choice and titration of these drugs which are based on clinical experience [79].

Psychiatric symptoms including aggressive behavior, depression, nervousness, and irritability have been described in PKAN patients, especially in atypical cases [6], and may require specific intervention; drugs such as benzodiazepine, selective serotonin receptor inhibitor (SSRI), or antidepressant may be useful. Clear psychotic symptoms (visual and auditory hallucination, ideation, psychomotor agitation) have been described in few cases of NBIA [80–82]. In the reported cases, symptoms resolved with olanzapine, clozapine, aripiprazole, and risperidone, and in these patients the use of atypical neuroleptic is recommended [83].

Psychiatric treatment should be considered for those with a later-onset, more protracted course accompanied by neuropsychiatric symptoms. In particular, the obsessive–compulsive disorder and complex tics may be prominent symptoms especially in patients with late-onset PKAN, resembling Tourette syndrome [15, 84, 85].

Botulinum Toxin Injections

Botulinum toxin injections are the treatment of choice in adult-onset focal and segmentary dystonia [78]. This local treatment may be helpful in reducing the oromandibular or cranial involvement that can be frequently seen in NBIA, particularly in patients with PKAN.

Surgical Treatment Options in NBIA

Intrathecal Baclofen Infusion (IT)

Baclofen is a gamma-aminobutyric acid receptor agonist. It has been reported to be effective in treating dystonia in NBIA both orally and intrathecally. The administration via an intrathecal pump has been reported as effective in patients with dystonia associated with spasticity or pain, both of which have been observed in PKAN; it has in fact been reported to be effective also in patients with NBIA [86, 87].

Recently Albright and Ferson reported favorable outcomes from a new technique used to deliver intraventricular baclofen in nine children and one adult with drug-refractory secondary dystonia, including one child with PKAN. The infusion was significantly beneficial on dystonia. Baclofen delivery at this site may better treat upper body and facial dystonia, such as blepharospasm and oromandibular dystonia, which are particularly disabling in NBIA patients and may result in higher baclofen concentrations over the cortex [74, 75, 88].

Functional Neurosurgery

Surgical interventions with stereotactic procedures in NBIA is currently seen as a promising treatment option. Deep brain stimulation (DBS) has become a common treatment for primary dystonia, and it is also being used more frequently to attempt to treat secondary dystonia, in particular classic PKAN. Since early 1990s, *thalamotomy and pallidotomy* had been performed to treat some cases of PKAN presenting with status dystonicus, and the ablative procedure resolved the life-threatening dystonia [89–92]. An amelioration of cognitive performances was shown in children with PKAN under pallidal deep brain stimulation [103].

After the initial observation by Umemura et al. [93], several other patients have been reported, demonstrating the efficacy of pallidal DBS on dystonia in PKAN. A large cohort studied at the same center included six individuals with PKAN. Those treated with DBS showed overall improvements in writing, speech, walking, and global measures of motor skills [94]. Additional reports showed either long-term effects or short-time benefits due to disease progression [95–101]. A recent multi-national study reported outcomes of 23 patients with NBIA treated with pallidal DBS. The primary outcome measured dystonia severity using the Burke–Fahn–Marsden Dystonia Rating Scale (BFMDRS) and the Barry–Albright Dystonia (BAD) Scale. A mean improvement of 28.5 % in dystonia severity was seen at 2–6 months, and 25.7 % improvement was seen at 9–15 months. Two-thirds of the patients treated experienced an improvement in their dystonia of 20 % or more. Although overall improvements were modest, global quality of life ratings showed a median improvement of 83.3 % at 9–15 months, suggesting that even this relatively small effect led to clinically meaningful improvements [100]. However, in contrast to experience in adults with NBIA, a study in children with NBIA showed poor results, and infection risk was highest in the youngest patients [102].

A sustained improvement (3 years of follow-up) of dystonia in PKAN was also seen in one patient with DBS in the subthalamic nucleus [104].

Innovative Therapeutic Strategies

It is not clear yet whether iron accumulation is the effective cause or whether it is simply an epiphenomenon. Nevertheless, iron remains a clinical relevant clue of these disorders.

The identification of new chelating agents with specific characteristic, the advances in radiological techniques that have allowed a quantitative assessment of iron by magnetic resonance together with the preliminary experience of successful treatment with chelators in patients with other conditions characterized by regional iron accumulation, has very recently brought an increasing interest about the possible efficacy of chelating agents in NBIA [74].

Chelating Treatment

Until recently, trials were limited by the development of systemic iron deficiency before any clinical neurological benefits were evident. The active deferiprone (3-hydroxy-1,2-dimethylpyridin-4-one, DFP) is a chelating agent that was first used in patients with thalassemia receiving blood transfusions. It has properties of iron relocator and therefore is most suitable for treating conditions characterized by regional iron accumulation [105]. It has the ability to donate iron to physiologic acceptor, which means that it can transfer iron from cellular pools to circulating transferrin. It has high permeability across cell membranes, gaining access to mitochondria, reducing intracellular iron, and therefore reducing the formation of iron free radicals. Moreover, it has been shown to cross the blood-brain barrier with obvious therapeutic implication for neurodegenerative disorders [106].

Despite the reduction of brain iron accumulation as measured by T2* MRI, case reports on chelating treatment showed only occasional clinical improvement in adults with idiopathic NBIA [107, 108]. More recently, a young patient with classic PKAN treated with oral deferiprone (20 mg/kg/die) was reported with a sustained improvement of dystonia after 1 year of chelating therapy without remarkable side effects [109].

Besides reports of single cases, two studies on larger series of patients have been published. The first Phase II pilot open trial assessed the clinical and radiological effects of deferiprone at a dose of 25 mg/kg/day over a 6-month period. Among the 9 patients who completed the study, 6 had classic and 3 had atypical disease. There were no serious adverse events; side effects included nausea and gastralgia (44 %), but no serious adverse event occurred. The authors observed a significant (median, 30 %) reduction in globus pallidus iron content, ranging from 15 to 61 %. However, over the 6-month treatment period, there were no significant changes in clinical status suggesting perhaps the need of a longer duration of treatment. Other possible explanations indicated by the authors were long disease duration or neuronal damage too advanced to allow for a rescue of function [110]. A similar study was conducted for a longer period on a series of 6 patients with different forms of NBIA (4 PKAN and 2 idiopathic NBIA) treated with deferiprone 15/mg/kg/die and assessed at 6 and 12 months. Mean age at enrollment was 36.5 and mean disease duration was 7.5 (range: 3–13). Clinical rating scales (UPDRS/III, ICARS, and UDRS) and blinded video rating documented a slight improvement in 3 patients (2 PKAN and 1 NBIA), and no change in the remaining 3 patients. The improvement was observed after 6 months of treatment and persisted at 12 months. Quantitative analysis of brain iron through T2* relaxometry was possible only in 3 patients and demonstrated a reduction of iron content in the globus pallidus [111].

A large multicenter placebo-control study in PKAN patients is currently enrolling and will hopefully give further insight on the potential efficacy of chelating treatment in NBIA.

Other Innovative Therapeutic Strategies Under Investigation

In PKAN patients, the CoA precursors calcium pantothenate and pantethine are discussed as potential therapeutic agents [50]. Most of the data still refers to PANK2 deficiency, which is the most common form of NBIA; little is known about the functioning of the other known genes associated to NBIA but mechanism related to lipid metabolism seems to be implicated.

Lipid Metabolism

Pantothenate kinase is a regulator in the synthesis of lipid metabolism and bile acids, pathways that both require coenzyme A. Moreover, lactate is elevated in PKAN patients, suggesting dysfunctional mitochondrial metabolism. In a recent study, global metabolic profiling of plasma from patients with PANK2 deficiency together with follow-up studies in patient-derived fibroblasts reveals defects in bile acid conjugation and lipid metabolism, pathways that require coenzyme A. The authors suggested that these findings may raise a novel therapeutic hypothesis with dietary fats and bile acid supplementation [112].

Pantothenate

Pantothenate is the substrate of the defective enzyme PANK2. The existence of residual enzyme activity in some individuals with PKAN, mainly patients with late-onset PKAN, has raised the possibility of treatment using high-dose pantothenate. Moreover, pantothenate has no known toxicity in humans. However, the efficacy of pantothenate supplementation has not been systematically studied; some individuals have anecdotally reported improvement in their symptoms under treatment [54].

Pantethine

Pantethine is a vitamin B5 analog that could potentially bypass the enzymatic defect in PKAN and restore the ability to synthesize adequate quantities of coenzyme A. Pantethine was administered to a PKAN drosophila model with a neurodegenerative phenotype. The experimental observation showed restored CoA levels, improved mitochondrial function, rescued brain degeneration, enhanced locomotor abilities, and increased lifespan. The authors showed evidence for the presence of a de novo CoA biosynthesis pathway in which pantethine is used as a precursor compound [113]. Although this therapy has significant potential as a rational therapeutic, several limitations exist. These include incomplete data regarding bioavailability, pharmacokinetics, and ability of this compound to cross the blood-brain barrier in humans. Furthermore, there are important differences between drosophila and human physiology, including a single PANK in drosophila but four distinct isoforms in humans, with *PANK2* being specifically targeted to mitochondria. Despite these limitations, the identification of pantethine as a potential therapeutic agent has generated much interest within the NBIA community.

Bibliography

1. Hartig MB, Iuso A, Haack T, Kmiec T, Jurkiewicz E, Heim K, Roeber S, Tarabin V, Dusi S, Krajewska-Walasek M, Jozwiak S, Hempel M, Winkelmann J, Elstner M, Oexle K, Klopstock T, Mueller-Felber W, Gasser T, Trenkwalder C, Tiranti V, Kretzschmar H, Schmitz G, Strom TM, Meitinger T, Prokisch H. Absence of an orphan mitochondrial protein, c19orf12, causes a distinct clinical subtype of neurodegeneration with brain iron accumulation. Am J Hum Genet. 2011;89(4):543–50.

2. Schneider SA, Hardy J, Bhatia KP. Syndromes of neurodegeneration with brain iron accumulation (NBIA): an update on clinical presentations, histological and genetic underpinnings, and treatment considerations. Mov Disord. 2012;27(1):42–53.
3. Gregory A, Westaway SK, Holm IE, Kotzbauer PT, Hogarth P, Sonek S, Coryell JC, Nguyen TM, Nardocci N, Zorzi G, Rodriguez D, Desguerre I, Bertini E, Simonati A, Levinson B, Dias C, Barbot C, Carrilho I, Santos M, Malik I, Gitschier J, Hayflick SJ. Neurodegeneration associated with genetic defects in phospholipase A(2). Neurology. 2008;71(18):1402–9.
4. Zhou B, Westaway SK, Levinson B, Johnson MA, Gitschier J, Hayflick SJ. A novel pantothenate kinase gene (PANK2) is defective in Hallervorden-Spatz syndrome. Nat Genet. 2001;28(4):345–9.
5. Leonardi R, Rock CO, Jackowski S, Zhang YM. Activation of human mitochondrial pantothenate kinase 2 by palmitoylcarnitine. Proc Natl Acad Sci U S A. 2007;104(5):1494–9.
6. Hayflick SJ, Westaway SK, Levinson B, Zhou B, Johnson MA, Ching KH, Gitschier J. Genetic, clinical, and radiographic delineation of Hallervorden-Spatz syndrome. N Engl J Med. 2003;348(1):33–40.
7. Hartig MB, Hörtnagel K, Garavaglia B, Zorzi G, Kmiec T, Klopstock T, Rostasy K, Svetel M, Kostic VS, Schuelke M, Botz E, Weindl A, Novakovic I, Nardocci N, Prokisch H, Meitinger T. Genotypic and phenotypic spectrum of PANK2 mutations in patients with neurodegeneration with brain iron accumulation. Ann Neurol. 2006;59(2):248–56.
8. Kruer MC, Boddaert N. Neurodegeneration with brain iron accumulation: a diagnostic algorithm. Semin Pediatr Neurol. 2012;19(2):67–74.
9. Egan RA, Weleber RG, Hogarth P, Gregory A, Coryell J, Westaway SK, Gitschier J, Das S, Hayflick SJ. Neuro-ophthalmologic and electroretinographic findings in pantothenate kinase-associated neurodegeneration (formerly Hallervorden-Spatz syndrome). Am J Ophthalmol. 2005;140(2):267–74.
10. Marelli C, Piacentini S, Garavaglia B, Girotti F, Albanese A. Clinical and neuropsychological correlates in two brothers with pantothenate kinase-associated neurodegeneration. Mov Disord. 2005;20(2):208–12.
11. Kurian MA, McNeill A, Lin JP, Maher ER. Childhood disorders of neurodegeneration with brain iron accumulation (NBIA). Dev Med Child Neurol. 2011;53(5):394–404.
12. Keogh MJ, Chinnery PF. Current concepts and controversies in neurodegeneration with brain iron accumulation. Semin Pediatr Neurol. 2012;19(2):51–6.
13. Hayflick SJ. Neurodegeneration with brain iron accumulation: from genes to pathogenesis. Semin Pediatr Neurol. 2006;13(3):182–5.
14. Diaz N. Late onset atypical pantothenate-kinase-associated neurodegeneration. Case Rep Neurol Med. 2013;2013:860201.
15. Pellecchia MT, Valente EM, Cif L, Salvi S, Albanese A, Scarano V, Bonuccelli U, Bentivoglio AR, D'Amico A, Marelli C, Di Giorgio A, Coubes P, Barone P, Dallapiccola B. The diverse phenotype and genotype of pantothenate kinase-associated neurodegeneration. Neurology. 2005;64(10):1810–2.
16. McNeill A, Chinnery PF. Neurodegeneration with brain iron accumulation. Handb Clin Neurol. 2011;100:161–72.
17. Hayflick SJ, Hartman M, Coryell J, Gitschier J, Rowley H. Brain MRI in neurodegeneration with brain iron accumulation with and without PANK2 mutations. AJNR Am J Neuroradiol. 2006;27(6):1230–3.
18. Nardocci N, Zorzi G. Axonal dystrophies. Handb Clin Neurol. 2013;113:1919–24.
19. Kurian MA, Morgan NV, MacPherson L, Foster K, Peake D, Gupta R, Philip SG, Hendriksz C, Morton JE, Kingston HM, Rosser EM, Wassmer E, Gissen P, Maher ER. Phenotypic spectrum of neurodegeneration associated with mutations in the PLA2G6 gene (PLAN). Neurology. 2008;70(18):1623–9.
20. Paisan-Ruiz C, Bhatia KP, Li A, Hernandez D, Davis M, Wood NW, Hardy J, Houlden H, Singleton A, Schneider SA. Characterization of PLA2G6 as a locus for dystonia-parkinsonism. Ann Neurol. 2009;65(1):19–23.
21. Morgan NV, Westaway SK, Morton JE, Gregory A, Gissen P, Sonek S, Cangul H, Coryell J, Canham N, Nardocci N, Zorzi G, Pasha S, Rodriguez D, Desguerre I, Mubaidin A, Bertini E, Trembath RC, Simonati A, Schanen C, Johnson CA, Levinson B, Woods CG, Wilmot B, Kramer P, Gitschier J, Maher ER, Hayflick SJ. PLA2G6, encoding a phospholipase A2, is mutated in neurodegenerative disorders with high brain iron. Nat Genet. 2006;38(7):752–4.

22. Nardocci N, Zorzi G, Farina L, Binelli S, Scaioli W, Ciano C, Verga L, Angelini L, Savoiardo M, Bugiani O. Infantile neuroaxonal dystrophy: clinical spectrum and diagnostic criteria. Neurology. 1999;52(7):1472-8.

23. Aicardi J, Castelein P. Infantile neuroaxonal dystrophy. Brain. 1979;102(4):727-48.

24. Ramaekers VT, Lake BD, Harding B. Diagnostic difficulties in infantile neuroaxonal dystrophy. A clinico-pathological study of eight cases. Neuropediatrics. 1978;18:170-5.

25. Schneider SA, Bhatia KP. Excess iron harms the brain: the syndromes of neurodegeneration with brain iron accumulation (NBIA). J Neural Transm. 2013;120(4):695-703.

26. Engel LA, Jing Z, O'Brien DE, Sun M, Kotzbauer PT. Catalytic function of PLA2G6 is impaired by mutations associated with infantile neuroaxonal dystrophy but not dystonia-parkinsonism. PLoS One. 2010;5(9):e12897.

27. Moretto G, Sparaco M, Monaco S, Bonetti B, Rizzuto N. Cytoskeletal changes and ubiquitin expression in dystrophic axons of Seitelberger's disease. Clin Neuropathol. 1993;12(1): 34-7.

28. Malik I, Turk J, Mancuso DJ, Montier L, Wohltmann M, Wozniak DF, Schmidt RE, Gross RW, Kotzbauer PT. Disrupted membrane homeostasis and accumulation of ubiquitinated proteins in a mouse model of infantile neuroaxonal dystrophy caused by PLA2G6 mutations. Am J Pathol. 2008;172(2):406-16.

29. Seitelberger F. Neuroaxonal dystrophy: its relation to aging and neurological diseases. Handb Clin Neurol. 1986;49:391-416.

30. Behrens MI, Brüggemann N, Chana P, Venegas P, Kägi M, Parrao T, Orellana P, Garrido C, Rojas CV, Hauke J, Hahnen E, González R, Seleme N, Fernández V, Schmidt A, Binkofski F, Kömpf D, Kubisch C, Hagenah J, Klein C, Ramirez A. Clinical spectrum of Kufor-Rakeb syndrome in the Chilean kindred with ATP13A2 mutations. Mov Disord. 2010;25(12): 1929-37.

31. Williams DR, Hadeed A, al-Din AS, Wreikat AL, Lees AJ. Kufor Rakeb disease: autosomal recessive, levodopa-responsive parkinsonism with pyramidal degeneration, supranuclear gaze palsy, and dementia. Mov Disord. 2005;20(10):1264-71.

32. Eiberg H, Hansen L, Korbo L, Nielsen IM, Svenstrup K, Bech S, Pinborg LH, Friberg L, Hjermind LE, Olsen OR, Nielsen JE. Novel mutation in ATP13A2 widens the spectrum of Kufor-Rakeb syndrome (PARK9). Clin Genet. 2012;82(3):256-63.

33. Chien HF, Bonifati V, Barbosa ER. ATP13A2-related neurodegeneration (PARK9) without evidence of brain iron accumulation. Mov Disord. 2011;26(7):1364-5.

34. Matsui H, Sato F, Sato S, Koike M, Taruno Y, Saiki S, Funayama M, Ito H, Taniguchi Y, Uemura N, Toyoda A, Sakaki Y, Takeda S, Uchiyama Y, Hattori N, Takahashi R. ATP13A2 deficiency induces a decrease in cathepsin D activity, fingerprint-like inclusion body formation, and selective degeneration of dopaminergic neurons. FEBS Lett. 2013;587(9): 1316-25.

35. Park JS, Mehta P, Cooper AA, Veivers D, Heimbach A, Stiller B, Kubisch C, Fung VS, Krainc D, Mackay-Sim A, Sue CM. Pathogenic effects of novel mutations in the P-type ATPase ATP13A2 (PARK9) causing Kufor-Rakeb syndrome, a form of early-onset parkinsonism. Hum Mutat. 2011;32(8):956-64.

36. Bras J, Verloes A, Schneider SA, Mole SE, Guerreiro RJ. Mutation of the parkinsonism gene ATP13A2 causes neuronal ceroid-lipofuscinosis. Hum Mol Genet. 2012;21(12):2646-50.

37. Schultheis PJ, Fleming SM, Clippinger AK, Lewis J, Tsunemi T, Giasson B, Dickson DW, Mazzulli JR, Bardgett ME, Haik KL, Ekhator O, Chava AK, Howard J, Gannon M, Hoffman E, Chen Y, Prasad V, Linn SC, Tamargo RJ, Westbroek W, Sidransky E, Krainc D, Shull GE. Atp13a2-deficient mice exhibit neuronal ceroid lipofuscinosis, limited α-synuclein accumulation and age-dependent sensorimotor deficits. Hum Mol Genet. 2013;22(10):2067-82.

38. Radi E, Formichi P, Di Maio G, Battisti C, Federico A. Altered apoptosis regulation in Kufor-Rakeb syndrome patients with mutations in the ATP13A2 gene. J Cell Mol Med. 2012;16(8):1916-23.

39. Malandrini A, Rubegni A, Battisti C, Berti G, Federico A. Electron-dense lamellated inclusions in 2 siblings with Kufor-Rakeb syndrome. Mov Disord. 2013;28(12):1751-2.

40. Paisan-Ruiz C, Guevara R, Federoff M, Hanagasi H, Sina F, Elahi E, Schneider SA, Schwingenschuh P, Bajaj N, Emre M, Singleton AB, Hardy J, Bhatia KP, Brandner S, Lees AJ, Houlden H. Early-onset L-dopa-responsive parkinsonism with pyramidal signs due to ATP13A2, PLA2G6, FBXO7 and spatacsin mutations. Mov Disord. 2010;25(12):1791–800.
41. Kruer MC, Paisán-Ruiz C, Boddaert N, Yoon MY, Hama H, Gregory A, Malandrini A, Woltjer RL, Munnich A, Gobin S, Polster BJ, Palmeri S, Edvardson S, Hardy J, Houlden H, Hayflick SJ. Defective FA2H leads to a novel form of neurodegeneration with brain iron accumulation (NBIA). Ann Neurol. 2010;68(5):611–8.
42. Edvardson S, Hama H, Shaag A, Gomori JM, Berger I, Soffer D, Korman SH, Taustein I, Saada A, Elpeleg O. Mutations in the fatty acid 2-hydroxylase gene are associated with leukodystrophy with spastic paraparesis and dystonia. Am J Hum Genet. 2008;83(5):643–8.
43. Dick KJ, Eckhardt M, Paisán-Ruiz C, Alshehhi AA, Proukakis C, Sibtain NA, Maier H, Sharifi R, Patton MA, Bashir W, Koul R, Raeburn S, Gieselmann V, Houlden H, Crosby AH. Mutation of FA2H underlies a complicated form of hereditary spastic paraplegia (SPG35). Hum Mutat. 2010;31(4):E1251–60.
44. Garone C, Pippucci T, Cordelli DM, Zuntini R, Castegnaro G, Marconi C, Graziano C, Marchiani V, Verrotti A, Seri M, Franzoni E. FA2H-related disorders: a novel c.270+3A>T splice-site mutation leads to a complex neurodegenerative phenotype. Dev Med Child Neurol. 2011;53(10):958–61.
45. Pierson TM, Simeonov DR, Sincan M, Adams DA, Markello T, Golas G, Fuentes-Fajardo K, Hansen NF, Cherukuri PF, Cruz P, Mullikin JC, Blackstone C, Tifft C, Boerkoel CF, Gahl WA, NISC Comparative Sequencing Program. Exome sequencing and SNP analysis detect novel compound heterozygosity in fatty acid hydroxylase-associated neurodegeneration. Eur J Hum Genet. 2012;20(4):476–9.
46. Zöller I, Meixner M, Hartmann D, Büssow H, Meyer R, Gieselmann V, Eckhardt M. Absence of 2-hydroxylated sphingolipids is compatible with normal neural development but causes late-onset axon and myelin sheath degeneration. J Neurosci. 2008;28(39):9741–54.
47. Potter KA, Kern MJ, Fullbright G, Bielawski J, Scherer SS, Yum SW, Li JJ, Cheng H, Han X, Venkata JK, Khan PA, Rohrer B, Hama H. Central nervous system dysfunction in a mouse model of FA2H deficiency. Glia. 2011;59(7):1009–21.
48. Hogarth P, Gregory A, Kruer MC, Sanford L, Wagoner W, Natowicz MR, Egel RT, Subramony SH, Goldman JG, Berry-Kravis E, Foulds NC, Hammans SR, Desguerre I, Rodriguez D, Wilson C, Diedrich A, Green S, Tran H, Reese L, Woltjer RL, Hayflick SJ. New NBIA subtype: genetic, clinical, pathologic, and radiographic features of MPAN. Neurology. 2013;80(3):268–75.
49. Deschauer M, Gaul C, Behrmann C, Prokisch H, Zierz S, Haack TB. C19orf12 mutations in neurodegeneration with brain iron accumulation mimicking juvenile amyotrophic lateral sclerosis. J Neurol. 2012;259(11):2434–9.
50. Schulte EC, Claussen MC, Jochim A, Haack T, Hartig M, Hempel M, Prokisch H, Haun-Jünger U, Winkelmann J, Hemmer B, Förschler A, Ilg R. Mitochondrial membrane protein associated neurodegeneration: a novel variant of neurodegeneration with brain iron accumulation. Mov Disord. 2013;28(2):224–7.
51. Horvath R, Holinski-Feder E, Neeve VC, Pyle A, Griffin H, Ashok D, Foley C, Hudson G, Rautenstrauss B, Nürnberg G, Nürnberg P, Kortler J, Neitzel B, Bässmann I, Rahman T, Keavney B, Loughlin J, Hambleton S, Schoser B, Lochmüller H, Santibanez-Koref M, Chinnery PF. A new phenotype of brain iron accumulation with dystonia, optic atrophy, and peripheral neuropathy. Mov Disord. 2012;27(6):789–93.
52. Haack TB, Hogarth P, Kruer MC, Gregory A, Wieland T, Schwarzmayr T, Graf E, Sanford L, Meyer E, Kara E, Cuno SM, Harik SI, Dandu VH, Nardocci N, Zorzi G, Dunaway T, Tarnopolsky M, Skinner S, Frucht S, Hanspal E, Schrander-Stumpel C, Héron D, Mignot C, Garavaglia B, Bhatia K, Hardy J, Strom TM, Boddaert N, Houlden HH, Kurian MA, Meitinger T, Prokisch H, Hayflick SJ. Exome sequencing reveals de novo WDR45 mutations causing a phenotypically distinct, X-linked dominant form of NBIA. Am J Hum Genet. 2012;91(6):1144–9.

53. Saitsu H, Nishimura T, Muramatsu K, Kodera H, Kumada S, Sugai K, Kasai-Yoshida E, Sawaura N, Nishida H, Hoshino A, Ryujin F, Yoshioka S, Nishiyama K, Kondo Y, Tsurusaki Y, Nakashima M, Miyake N, Arakawa H, Kato M, Mizushima N, Matsumoto N. De novo mutations in the autophagy gene WDR45 cause static encephalopathy of childhood with neurodegeneration in adulthood. Nat Genet. 2013;45(4):445–9.
54. Gregory A, Polster BJ, Hayflick SJ. Clinical and genetic delineation of neurodegeneration with brain iron accumulation. J Med Genet. 2009;46(2):73–80.
55. Kimura Y, Sato N, Sugai K, Maruyama S, Ota M, Kamiya K, Ito K, Nakata Y, Sasaki M, Sugimoto H. MRI, MR spectroscopy, and diffusion tensor imaging findings in patient with static encephalopathy of childhood with neurodegeneration in adulthood (SENDA). Brain Dev. 2013;35(5):458–61.
56. Haack TB, Hogarth P, Kruer MC, Gregory A, Wieland T, Schwarzmayr T, Graf E, Sanford L, Meyer E, Kara E, Cuno SM, Harik SI, Dandu VH, Nardocci N, Zorzi G, Dunaway T, Tarnopolsky M, Skinner S, Frucht S, Hanspal E, Schrander-Stumpel C, Héron D, Mignot C, Garavaglia B, Bhatia K, Hardy J, Strom TM, Boddaert N, Houlden HH, Kurian MA, Meitinger T, Prokisch H, Hayflick SJ. "Exome sequencing reveals de novo WDR45 mutations causing a phenotypically distinct, X-linked dominant form of NBIA". Am J Hum Genet. 2012;91(6):1144–9. doi: 10.1016/j.ajhg.2012.10.019. Epub 2012 Nov 21.
57. Crompton DE, Chinnery PF, Fey C, Curtis AR, Morris CM, Kierstan J, Burt A, Young F, Coulthard A, Curtis A, Ince PG, Bates D, Jackson MJ, Burn J. Neuroferritinopathy: a window on the role of iron in neurodegeneration. Blood Cells Mol Dis. 2002;29(3):522–31.
58. Chinnery PF, Crompton DE, Birchall D, Jackson MJ, Coulthard A, Lombès A, Quinn N, Wills A, Fletcher N, Mottershead JP, Cooper P, Kellett M, Bates D, Burn J. Clinical features and natural history of neuroferritinopathy caused by the FTL1 460InsA mutation. Brain. 2007;130(Pt 1):110–9. Epub 2006 Dec 2.
59. Keogh MJ, Singh B, Chinnery PF. Early neuropsychiatry features in neuroferritinopathy. Mov Disord. 2013;28(9):1310–3.
60. Kubota A, Hida A, Ichikawa Y, Momose Y, Goto J, Igeta Y, Hashida H, Yoshida K, Ikeda S, Kanazawa I, Tsuji S. A novel ferritin light chain gene mutation in a Japanese family with neuroferritinopathy: description of clinical features and implications for genotype-phenotype correlations. Mov Disord. 2009;24(3):441–5.
61. Burn J, Chinnery PF. Neuroferritinopathy. Semin Pediatr Neurol. 2006;13(3):176–81.
62. Mancuso M, Davidzon G, Kurlan RM, Tawil R, Bonilla E, Di Mauro S, Powers JM. Hereditary ferritinopathy: a novel mutation, its cellular pathology, and pathogenetic insights. J Neuropathol Exp Neurol. 2005;64(4):280–94.
63. Keogh MJ, Jonas P, Coulthard A, Chinnery PF, Burn J. Neuroferritinopathy: a new inborn error of iron metabolism. Neurogenetics. 2012;13(1):93–6.
64. Vidal R, Miravalle L, Gao X, Barbeito AG, Baraibar MA, Hekmatyar SK, Widel M, Bansal N, Delisle MB, Ghetti B. Expression of a mutant form of the ferritin light chain gene induces neurodegeneration and iron overload in transgenic mice. J Neurosci. 2008;28(1):60–7.
65. McNeill A, Pandolfo M, Kuhn J, Shang H, Miyajima H. The neurological presentation of ceruloplasmin gene mutations. Eur Neurol. 2008;60(4):200–5.
66. Miyajima H, Kono S, Takahashi Y, Sugimoto M, Sakamoto M, Sakai N. Cerebellar ataxia associated with heteroallelic ceruloplasmin gene mutation. Neurology. 2001;57(12):2205–10.
67. Kuhn J, Bewermeyer H, Miyajima H, Takahashi Y, Kuhn KF, Hoogenraad TU. Treatment of symptomatic heterozygous aceruloplasminemia with oral zinc sulphate. Brain Dev. 2007;29(7):450–3.
68. Patel BN, Dunn RJ, Jeong SY, Zhu Q, Julien JP, David S. Ceruloplasmin regulates iron levels in the CNS and prevents free radical injury. J Neurosci. 2002;22(15):6578–86.
69. Hineno A, Kaneko K, Yoshida K, Ikeda S. Ceruloplasmin protects against rotenone-induced oxidative stress and neurotoxicity. Neurochem Res. 2011;36(11):2127–35.
70. Texel SJ, Xu X, Harris ZL. Ceruloplasmin in neurodegenerative diseases. Biochem Soc Trans. 2008;36(Pt 6):1277–81.
71. Kono S, Miyajima H. Molecular and pathological basis of aceruloplasminemia. Biol Res. 2006;39(1):15–23.
72. Kohno S, Miyajima H, Takahashi Y, Suzuki H, Hishida A. Defective electron transfer in complexes I and IV in patients with aceruloplasminemia. J Neurol Sci. 2000;182(1):57–60.

73. Schipper HM. Neurodegeneration with brain iron accumulation - clinical syndromes and neuroimaging. Biochim Biophys Acta. 2012;1822(3):350–60.
74. Zorzi G, Zibordi F, Chiapparini L, Nardocci N. Therapeutic advances in neurodegeneration with brain iron accumulation. Semin Pediatr Neurol. 2012;19(2):82–6.
75. Gregory A, Hayflick S. Neurodegeneration with brain iron accumulation disorders overview. In: Pagon RA, Adam MP, Bird TD, Dolan CR, Fong CT, Stephens K, editors. GeneReviews™ [Internet]. Seattle: University of Washington; 1993–2013.
76. Hoon Jr AH, Freese PO, Reinhardt EM, Wilson MA, Lawrie Jr WT, Harryman SE, Pidcock FS, Johnston MV. Age-dependent effects of trihexyphenidyl in extrapyramidal cerebral palsy. Pediatr Neurol. 2001;25(1):55–8.
77. Kenney C, Jankovic J. Tetrabenazine in the treatment of hyperkinetic movement disorders. Expert Rev Neurother. 2006;6(1):7–17.
78. Jankovic J. Dystonia: medical therapy and botulinum toxin. In: Fahn S, Hallett M, DeLong DR, editors. Dystonia 4: advances in neurology volume 94. Philadelphia: Lippincott, Williams & Wilkins; 2004. p. 275–86.
79. Shakespeare D, Boggild M, Young CA. Antispasticity agents for multiple sclerosis: a systematic review. Cochrane Database of Sys Rev. 2003;(4):CD001332.
80. Morphy MA, Feldman JA, Kilburn G. Hallervorden-Spatz disease in a psychiatric setting. J Clin Psychiatry. 1989;50(2):66–8.
81. Oner O, Oner P, Deda G, Içağasioğlu D. Psychotic disorder in a case with Hallervorden-Spatz disease. Acta Psychiatr Scand. 2003;108(5):394–7.
82. Sunwoo YK, Lee JS, Kim WH, Shin YB, Lee MJ, Cho IH, Ock SM. Psychiatric disorder in two siblings with hallervorden-spatz disease. Psychiatry Investig. 2009;6(3):226–9.
83. del Valle-López P, Pérez-García R, Sanguino-Andrés R, González-Pablos E. Adult onset Hallervorden-Spatz disease with psychotic symptoms. Actas Esp Psiquiatr. 2011;39(4):260–2.
84. Nardocci N, Rumi V, Combi ML, Angelini L, Mirabile D, Bruzzone MG. Complex tics, stereotypies, and compulsive behavior as clinical presentation of a juvenile progressive dystonia suggestive of Hallervorden-Spatz disease. Mov Disord. 1994;9(3):369–71.
85. Scarano V, Pellecchia MT, Filla A, Barone P. Hallervorden-Spatz syndrome resembling a typical Tourette syndrome. Mov Disord. 2002;17(3):618–20.
86. Albright AL, Barry MJ, Fasick P, Barron W, Shultz B. Continuous intrathecal baclofen infusion for symptomatic generalized dystonia. Neurosurgery. 1996;38(5):934–8; discussion 938–9.
87. Hou JG, Ondo W, Jankovic J. Intrathecal baclofen for dystonia. Mov Disord. 2001;16(6):1201–2.
88. Albright AL, Ferson SS. Intraventricular baclofen for dystonia: techniques and outcomes. Clinical article. J Neurosurg Pediatr. 2009;3:11–4.
89. Tsukamoto H, Inui K, Taniike M, Nishimoto J, Midorikawa M, Yoshimine T, Kato A, Ikeda T, Hayakawa T, Okada S. A case of Hallervorden-Spatz disease: progressive and intractable dystonia controlled by bilateral thalamotomy. Brain Dev. 1992;14(4):269–72.
90. Justesen CR, Penn RD, Kroin JS, Egel RT. Stereotactic pallidotomy in a child with Hallervorden-Spatz disease. Case report. J Neurosurg. 1999;90(3):551–4.
91. Kyriagis M, Grattan-Smith P, Scheinberg A, Teo C, Nakaji N, Waugh M. Status dystonicus and Hallervorden-Spatz disease: treatment with intrathecal baclofen and pallidotomy. J Paediatr Child Health. 2004;40(5–6):322–5.
92. Balas I, Kovacs N, Hollody K. Staged bilateral stereotactic pallidothalamotomy for life-threatening dystonia in a child with Hallervorden-Spatz disease. Mov Disord. 2006;21(1):82–5.
93. Umemura A, Jaggi JL, Dolinskas CA, Stern MB, Baltuch GH. Pallidal deep brain stimulation for longstanding severe generalized dystonia in Hallervorden-Spatz syndrome: case report. J Neurosurg. 2004;100(4):706–9.
94. Castelnau P, Cif L, Valente EM, Vayssiere N, Hemm S, Gannau A, Digiorgio A, Coubes P. Pallidal stimulation improves pantothenate kinase-associated neurodegeneration. Ann Neurol. 2005;57:738–41.
95. Krause M, Fogel W, Tronnier V, Pohle S, Hörtnagel K, Thyen U, Volkmann J. Long-term benefit to pallidal deep brain stimulation in a case of dystonia secondary to pantothenate kinase-associated neurodegeneration. Mov Disord. 2006;21:2255–7.
96. Shields DC, Sharma N, Gale JT, Eskandar EN. Pallidal stimulation for dystonia in pantothenate kinase-associated neurodegeneration. Pediatr Neurol. 2007;37:442–5.

97. Isaac C, Wright I, Bhattacharyya D, Baxter P, Rowe J. Pallidal stimulation for pantothenate kinase-associated neurodegeneration dystonia. Arch Dis Child. 2008;93:239–40.
98. Mikati MA, Yehya A, Darwish H, Karam P, Comair Y. Deep brain stimulation as a mode of treatment of early onset pantothenate kinase-associated neurodegeneration. Eur J Paediatr Neurol. 2009;13:61–4.
99. Grandas F, Fernandez-Carballal C, Guzman-de-Villoria J, Ampuero I. Treatment of a dystonic storm with pallidal stimulation in a patient with PANK2 mutation. Mov Disord. 2011;26(5):921–2.
100. Timmermann L, Pauls KA, Wieland K, Jech R, Kurlemann G, Sharma N, Gill SS, Haenggeli CA, Hayflick SJ, Hogarth P, Leenders KL, Limousin P, Malanga CJ, Moro E, Ostrem JL, Revilla FJ, Santens P, Schnitzler A, Tisch S, Valldeoriola F, Vesper J, Volkmann J, Woitalla D, Peker S. Dystonia in neurodegeneration with brain iron accumulation: outcome of bilateral pallidal stimulation. Brain. 2010;133(Pt 3):701–12.
101. Lim BC, Ki CS, Cho A, Hwang H, Kim KJ, Hwang YS, Kim YE, Yun JY, Jeon BS, Lim YH, Paek SH, Chae JH. Pantothenate kinase-associated neurodegeneration in Korea: recurrent R440P mutation in PANK2 and outcome of deep brain stimulation. Eur J Neurol. 2012;19:556–61.
102. Air EL, Ostrem JL, Sanger TD, Starr PA. Deep brain stimulation in children: experience and technical pearls. J Neurosurg Pediatr. 2011;8(6):566–74.
103. Mahoney R, Selway R, Lin JP. Cognitive functioning in children with pantothenate-kinase-associated neurodegeneration undergoing deep brain stimulation. Dev Med Child Neurol. 2011;53(3):275–9.
104. Ge M, Zhang K, Ma Y, Meng FG, Hu WH, Yang AC, Zhang JG. Bilateral subthalamic nucleus stimulation in the treatment of neurodegeneration with brain iron accumulation type 1. Stereotact Funct Neurosurg. 2011;89(3):162–6.
105. Kakhlon O, Breuer W, Munnich A, Cabantchik ZI. Iron redistribution as a therapeutic strategy for treating diseases of localized iron accumulation. Can J Physiol Pharmacol. 2010;88(3):187–96.
106. Boddaert N, Le Quan Sang KH, Rötig A, Leroy-Willig A, Gallet S, Brunelle F, Sidi D, Thalabard JC, Munnich A, Cabantchik ZI. Selective iron chelation in Friedreich ataxia: biologic and clinical implications. Blood. 2007;110(1):401–8.
107. Forni GL, Balocco M, Cremonesi L, Abbruzzese G, Parodi RC, Marchese R. Regression of symptoms after selective iron chelation therapy in a case of neurodegeneration with brain iron accumulation. Mov Disord. 2008;23:904–7.
108. Kwiatkowski A, Ryckewaert G, Jissendi Tchofo P, Moreau C, Vuillaume I, Chinnery PF, Destée A, Defebvre L, Devos D. Long-term improvement under deferiprone in a case of neurodegeneration with brain iron accumulation. Parkinsonism Relat Disord. 2012;18:110–2.
109. Pratini NR, Sweeters N, Vichinsky E, Neufeld JA. Treatment of classic pantothenate kinase-associated neurodegeneration with deferiprone and intrathecal baclofen. Am J Phys Med Rehabil. 2013;92(8):728–33.
110. Zorzi G, Zibordi F, Chiapparini L, Bertini E, Russo L, Piga A, Longo F, Garavaglia B, Aquino D, Savoiardo M, Solari A, Nardocci N. Iron-related MRI images in patients with pantothenate kinase-associated neurodegeneration (PKAN) treated with deferiprone: results of a phase II pilot trial. Mov Disord. 2011;26:1756–9.
111. Abbruzzese G, Cossu G, Balocco M, Marchese R, Murgia D, Melis M, Galanello R, Barella S, Matta G, Ruffinengo U, Bonuccelli U, Forni GL. A pilot trial of deferiprone for neurodegeneration with brain iron accumulation. Haematologica. 2011;96(11):1708–11.
112. Leoni V, Strittmatter L, Zorzi G, Zibordi F, Dusi S, Garavaglia B, Venco P, Caccia C, Souza AL, Deik A, Clish CB, Rimoldi M, Ciusani E, Bertini E, Nardocci N, Mootha VK, Tiranti V. Metabolic consequences of mitochondrial coenzyme A deficiency in patients with PANK2 mutations. Mol Genet Metab. 2012;105(3):463–71.
113. Rana A, Seinen E, Siudeja K, Muntendam R, Srinivasan B, van der Want JJ, Hayflick S, Reijngoud DJ, Kayser O, Sibon OC. Pantethine rescues a Drosophila model for pantothenate kinase-associated neurodegeneration. Proc Natl Acad Sci U S A. 2010;107(15):6988–93.

Chapter 11
Aceruloplasminemia

Hiroaki Miyajima

Abstract Aceruloplasminemia is an autosomal recessive neurodegenerative disorder associated with iron accumulation in the brain. The disorder is caused by a complete absence of ceruloplasmin ferroxidase activity due to a homozygous mutation of the ceruloplasmin gene. There is no apparent genotype-phenotype association. The diagnosis is based on the absence of serum ceruloplasmin, low serum iron concentration and high serum ferritin concentration, as well as hepatic iron overload. Unique among iron-overload syndromes, aceruloplasminemia involves both systemic and brain iron trafficking. To date, 71 cases have been reported, and the prevalence has been estimated at about 1/2,000,000 in non-consanguineous marriages. The clinical triad of retinal degeneration, diabetes mellitus, and neurologic disease is seen in patients ranging from age 16 years to older than 70 years. The neurological findings of ataxia, involuntary movement (dystonia, chorea, and tremors), Parkinsonism, and cognitive dysfunction correspond to the regions of iron deposition in the brain. The neuropathological process extends beyond the basal ganglia to the cerebral cortex with time. Patients with aceruloplasminemia usually present with iron-refractory anemia prior to the onset of diabetes or obvious neurological problems. The diagnosis is strongly supported by characteristic MRI findings of abnormal low intensities reflecting iron accumulation in the brain and liver on both T1- and T2-weighted images. Genetic testing can confirm the diagnosis. The differential diagnoses include hereditary hemochromatosis, Wilson's disease, Huntington's disease, and spinocerebellar ataxias, as well as other forms of neurodegeneration with brain iron accumulation, atypical pantothenate kinase-associated neurodegeneration, and neuroferritinopathy. The treatment of aceruloplasminemia is based on intravenous and oral iron chelators, which have been associated with improvements in the diabetes and neurological symptoms. The oral administration of zinc may prevent tissue damage.

H. Miyajima
First Department of Medicine, Hamamatsu University School of Medicine,
1-20-1 Handayama, Higashi-ku, Hamamatsu 431-3192, Japan
e-mail: miyajima@hama-med.ac.jp

F.E. Micheli, P.A. LeWitt (eds.), *Chorea*,
DOI 10.1007/978-1-4471-6455-5_11, © Springer-Verlag London 2014

Keywords Ceruloplasmin • Iron • Ferroxidase • Neurodegeneration of brain iron accumulation (NBIA) • Free radical injury • Iron chelator • Zinc

Introduction

Iron is a bioactive metal essential for normal brain functions. It participates in a variety of cellular functions, including the biosynthesis of neurotransmitters, cytokinesis, myelin formation, and electron transport for sustaining brain energy metabolism. However, excessive iron in the brain causes neuronal injury and cell death, because ferrous iron (Fe^{2+}) enhances oxidative stress due to the generation of the highly cytotoxic hydroxyl radical. Thus, the brain iron content is critically regulated by several iron metabolic molecules [1].

In 1987, we described the first case of aceruloplasminemia in a 52-year-old Japanese female suffering from blepharospasm, retinal degeneration, and diabetes mellitus (DM) [2]. Subsequent evaluations revealed the presence of mild anemia, low serum iron concentrations, elevated serum ferritin levels, and significant iron accumulation in the basal ganglia and liver on T2-weighted MRI, as well as a complete absence of serum ceruloplasmin (Cp). A genetic analysis of the Cp gene revealed that this patient was homozygous for a five-base insertion in exon 7, thus resulting in a truncated mutation [3]. The novel disorder was termed aceruloplasminemia (MIM 604290).

This disease revealed an essential role for Cp in brain iron homeostasis. We now know that (1) Cp regulates the efficiency of iron efflux; (2) Cp functions as a ferroxidase and regulates the oxidation of ferrous iron (Fe^{2+}) to ferric iron (Fe^{3+}); (3) Cp does not bind to transferrin directly; (4) Cp stabilizes the cell surface iron transporter, ferroportin; and (5) glycophosphosinositide-linked Cp (GPI-Cp) is the predominant form expressed in the brain [4]. In the brain, serum transferrin-bound iron is endocytosed by brain endothelial cells in a manner dependent on transferrin receptor 1, and iron is released into the brain interstitial fluid through ferroportin. Extracellular iron is oxidized by GPI-Cp, which is found in the foot processes of astrocytes, and then the iron binds to the transferrin synthesized by oligodendrocytes and is transported into neurons. β-amyloid precursor protein (APP) was found to possess ferroxidase activity like Cp and to interact with neuron ferroportin [5]. The brain needs several times the concentration of iron obtained from the blood to maintain its normal function. Taken together, the known functions of the iron metabolic molecules suggest the presence of a cycle of iron storage and reutilization within the brain (see Fig. 11.1).

Classification

Recent molecular investigations have revealed the existence of inherited neurodegenerative disorders termed "neurodegeneration with brain iron accumulation" (NBIA) due to genetic defects associated with iron metabolism. NBIA is a heterogeneous group of disorders including pantothenate kinase-associated neurodegeneration

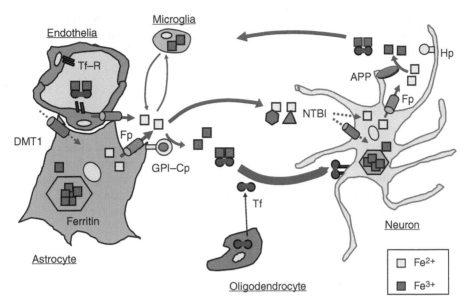

Fig. 11.1 A model for the iron metabolic cycle in the brain. In the normal brain, iron may be continuously recycled between astrocytes and neurons, with transferrin acting as a shuttle to deliver iron from astrocytes to neurons. The role of Cp, expressed as GPI-Cp on astrocytes, is a ferroxidase-mediating ferrous iron oxidation and the subsequent transfer to transferrin. In aceruloplasminemia, neurons take up the iron from alternative sources of NTBI complexed to molecules, such as citrate and ascorbate, because astrocytes without GPI-Cp cannot transport iron to transferrin. Neuronal iron starvation associated with astrocyte and microglial iron overload may result in neuronal cell loss due to iron-mediated oxidative stress and the disrupted neuronal cell protection offered by astrocytes and microglia. Hp, hephaestin, a Cp homologue with ferroxidase activity, is expressed on neurons. *Tf* transferrin, *Tf-R* transferrin receptor 1, *Fp* ferroportin, *DMT1* divalent ion transporter, *APP* β-amyloid precursor protein

(PKAN), neuroferritinopathy, and infantile neuroaxonal dystrophy. Autosomal recessive mutations in *PANK2* cause PCAN, which comprises about 50 % of NBIA cases [6]. Neuroferritinopathy and infantile neuroaxonal dystrophy result from mutations in the genes encoding the light chain of the iron storage protein, ferritin, and the A2 phospholipase, PLA2G6, a class of enzyme that catalyzes the release of fatty acids from phospholipids. So far, four additional genes (*FA2H, ATP13A2, C2orf37,* and *C19orf12*) have been linked to NBIA [7]. The most striking feature of these diseases is the accumulation of iron limited within the brain.

Unique among iron-overload syndromes, aceruloplasminemia involves both the systemic and brain iron metabolism. A marked accumulation of iron in the affected parenchymal tissues, such as the liver, pancreas, heart, and thyroid, results in DM, cardiac failure, and hypothyroidism. The iron accumulation in systemic tissues is caused by the disrupted tissue iron release, with subsequent incorporation into circulating transferrin. In a major systemic iron-overload disorder, hereditary hemochromatosis, tissue iron accumulation arises from the increased serum ferrous iron, but brain iron is not increased despite an elevation in the serum ferrous iron. Aceruloplasminemia is classified as an inherited neurodegenerative disorder with systemic iron-overload syndrome.

Epidemiology

The serum Cp concentrations of about 5,000 adults undergoing medical examination were screened in Japan. The prevalence of aceruloplasminemia was estimated to be approximately 1 per 2,000,000 in non-consanguineous marriages [8], and subsequent studies have now identified more than 40 affected families around the world. Heterozygotes for the Cp gene were estimated to account for 0.1 % of individuals with diabetes in Japan [9].

Clinical Manifestations

The clinical manifestations of aceruloplasminemia are a triad of retinal degeneration, DM, and neurological signs/symptoms [10]. A summary of the clinical manifestations in the 71 patients is shown in Table 11.1. The neurological manifestations (in order of frequency) include ataxia, involuntary movement, cognitive dysfunction, and Parkinsonism corresponding to the specific regions of brain iron accumulation. These symptoms generally appear in the fourth or fifth decade of life. More than 40 % of the involuntary movement is dystonia, and approximately 30 % of cases exhibit chorea and choreoathetosis. The cognitive dysfunction includes forgetfulness, mental slowing, and apathy. The phenotypic expression varies even within families. Ophthalmological examinations usually reveal evidence of peripheral retinal degeneration secondary to iron accumulation. Some patients have been recognized prior to the onset of neurological symptoms due to biochemical abnormalities indicating iron metabolism [11]. These laboratory findings include microcytic anemia, a decreased serum iron content, and an increased serum ferritin concentration.

Case Report of a Patient Presenting with Chorea

A 53-year-old Japanese male had been suffering from excessive, spontaneous movements in the left hand for the last 3 years. He had gradually developed facial spasms, as well as ataxia. At age 51, the patient developed insulin-dependent DM. His parents were second-degree cousins, but there was no family history of anemia or neurological diseases. He had mild anemia (hemoglobin 9.1 g/L) accompanied by a low serum iron concentration of 22 μg/dL (reference range; 60–180) and a high level of serum ferritin at 1,175 μg/dL (reference range; 45–200). Serum Cp was not detected by a Western blot analysis, and the serum copper level was decreased to 6 μg/dL (reference range; 65–125).

Table 11.1 The clinical characteristics of patients with aceruloplasminemia

Clinical manifestations in 71 patients with aceruloplasminemia
Anemia (80 %)
Retinal degeneration (76 %)
Diabetes mellitus (70 %)
Neurological symptoms (68 %)
1. Ataxia (71 %): dysarthria > gait ataxia > limb ataxia
2. Involuntary movement (64 %): dystonia (blepharospasm, grimacing, neck dystonia) > chorea > tremors
3. Parkinsonism (20 %): rigidity > akinesia
4. Cognitive dysfunction (60 %): apathy > forgetfulness
Onset of clinical manifestation
Diabetes mellitus: under 30 years old, 18 %; 30–39 years old, 35 %; 40–49 years old, 31 %; over 50 years old, 16 %
Neurological symptoms: under 40 years old, 7 %; 40–49 years old, 38 %; 50–59 years old, 42 %; over 60 years old, 13 %
Laboratory findings
Undetectable serum ceruloplasmin
Elevated serum ferritin
Decreased serum iron, iron-refractory microcytic anemia
Low serum copper and normal urinary copper levels
MRI (magnetic resonance imaging) findings
Low intensity on both T1- and T2-weighted MRI in the liver and the basal ganglia, including the caudate nucleus, putamen and pallidum, and the thalamus
Liver biopsy results
Excess iron accumulation (>1,000 μg/g dry weight) within hepatocytes and reticuloendothelial cells
Normal hepatic architecture and histology without cirrhosis or fibrosis
Normal copper accumulation

His neurological examination showed choreic movement of the muscles in the upper extremities, and facial spasms appeared during speech or voluntary movement of the extremities. His deep tendon reflexes were hyperactive in all four limbs. His sensation was normal, except for a decrease in the vibratory sense in the feet. He had slurred speech and slight gait ataxia, but did not require assistance walking. His visual acuity was not disturbed, but an opthalmoscopic examination showed retinal degeneration (Fig. 11.2). The findings differed from those of diabetic retinopathy. MRI showed an abnormal hypointensity on the T2-weighted images of the liver, as well as in the basal ganglia, thalamus, and the dentate nucleus in the brain (Fig. 11.3). There was a homogenous mutation in the Cp gene of c.2482delG in exon 14.

Pathophysiology

Cp knockout mice developed deficits in motor coordination and showed increased iron deposition in several regions of the central nervous system [12]. Increased lipid peroxidation due to iron-mediated cellular radical injury was also seen in these regions. These results indicate that Cp plays an important role in maintaining

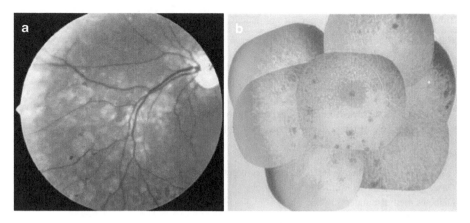

Fig. 11.2 Ocular fundus findings of an aceruloplasminemia patient. An ophthalmoscopic examination showed retinal degeneration with several small yellowish opacities, which were scattered over grayish atrophy of the retinal pigment epithelium (**a**). Fluorescein angiography demonstrated window defects corresponding to the yellowish opacities (**b**)

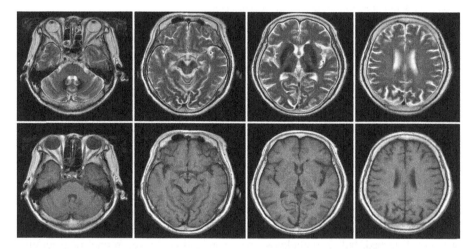

Fig. 11.3 T2- and T1-weighted magnetic resonance images of an aceruloplasminemia patient. The upper and bottom rows indicate respective T2- and T1-weighted axial images of the brain, which showed signal attenuation of the dentate nucleus of the cerebellum, striatum, caudate nucleus, and thalamus

iron homeostasis in the brain and in protecting the brain from iron-mediated free radical injury.

The mechanism underlying the neurodegeneration in aceruloplasminemia has not been clarified. However, the antioxidant activity of Cp can be mainly ascribed to its ferroxidase activity, which effectively inhibits ferrous ion-stimulated lipid peroxidation and ferrous ion-dependent formation of hydroxyl radicals in the Fenton reaction. A direct role for iron in the oxidant-mediated neuronal injury is supported by findings of increased lipid peroxidation, mitochondrial dysfunction, and oxidatively modified glial fibrillary acidic protein in the brain tissues of aceruloplasminemia patients [13–16]. The pathological findings in the brain showed severe iron deposition in both the astrocytes and neurons and neuronal loss [17]. The iron accumulation observed in neurons indicates that the neurons take up significant amounts of iron due to alternative sources of non-transferrin-bound iron (NTBI), because astrocytes without any expression of Cp cannot transport iron to the transferrin that binds to transferrin receptor 1 on neurons. Such neuronal cell loss suggests that neuronal cell injury may result from iron-mediated oxidation and from iron deficiency in specific regions. The excess iron in astrocytes could also result in oxidative damage to these cells, with the subsequent loss of glial-derived growth factors critical for neurons.

Diagnosis

Clinically, aceruloplasminemia consists of hepatic iron overload, anemia, retinal degeneration, DM, and adult-onset neurological symptoms. The diagnosis of aceruloplasminemia in a symptomatic individual relies upon the demonstration of the complete absence of serum Cp and abnormal laboratory findings, as well as neuroimages suggesting iron overload. The neuroimaging studies in aceruloplasminemia patients are strongly supported by the characteristic MRI findings of abnormal low intensities reflecting iron accumulation in the liver and brain. A T2*-weighted MRI study can be used to distinguish these patients from those with other NBIA [18]. Genetic testing can thereafter confirm the diagnosis. The genetic analyses of aceruloplasminemia patients worldwide have identified more than 40 distinct mutations in the Cp gene [4] (Fig. 11.4).

Cp is a multi-copper oxidase. A low serum concentration of Cp is not specific for aceruloplasminemia. Cp deficiency is also a characteristic feature in various copper metabolic disorders, including Wilson's disease and Menkes disease. In Wilson's disease, an inability to transfer copper onto the Cp precursor protein, apoceruloplasmin (apo-Cp), and a decrease in biliary copper excretion result in serum Cp deficiency and excess copper accumulation. In Menkes disease, copper absorption from the intestine is decreased, leading to copper and Cp deficiencies in the body. In contrast, aceruloplasminemia is an iron metabolic disorder in which Cp deficiency is caused by a lack of apo-Cp biosynthesis, and the copper metabolism is not disturbed (Fig. 11.5). Patients with typical clinical symptoms who were homozygous for the p.Gly969Ser mutation showed a slight decrease in the serum Cp concentration by an immune-nephelometry method because of the secretion of apo-Cp alone, without any ferroxidase activity [19]. These potentially confounding findings should be kept in mind when making the diagnosis based on Cp concentrations alone.

The differential diagnoses for the neurological manifestations of aceruloplasminemia include Wilson's disease, Huntington's disease, and spinocerebellar ataxias.

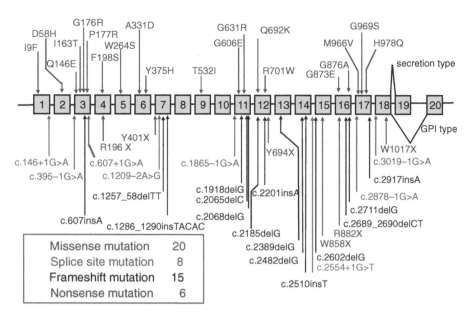

Fig. 11.4 Genetic mutations in patients with aceruloplasminemia. The structure of the human ceruloplasmin gene consists of 20 exons. Alternative splicing at exon 18 allows for the production of two Cp isoforms, secreted Cp or GPI-Cp

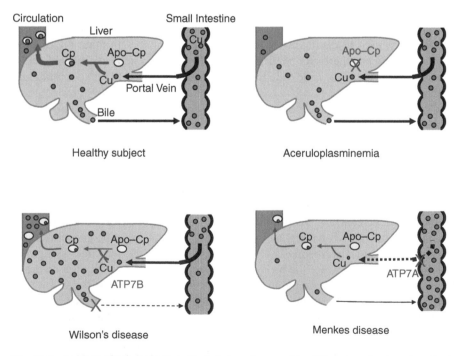

Fig. 11.5 Cp biosynthesis in the liver. Cp deficiency is caused by different mechanisms in patients with aceruloplasminemia, Wilson's disease, and Menkes disease. Wilson's disease and Menkes disease are copper transport disorders caused by the lack of ATP7B and ATP7A proteins, respectively. *Cu* copper, *Cp* ceruloplasmin, *Apo-Cp* ceruloplasmin precursor protein

Treatment

Aceruloplasminemia is a slowly progressive neurodegenerative disorder, and its early diagnosis and the early treatment of patients are important issues. Iron-mediated lipid peroxidation and oxidative stress are considered to be the main causes of the neuronal degeneration in aceruloplasminemia patients. The prognosis may involve heart failure due to cardiac iron overload. To reduce the iron accumulation, systemic iron chelation therapy has been introduced in some patients. The intravenous administration of deferoxamine was effective for reducing the hepatic iron overload and leading to a partial improvement of the neurological symptoms and brain iron accumulation [20, 21]. However, subsequent studies showed little effect of deferoxamine on neurological symptoms, despite the improvement in the brain iron levels on T2*-weighted MRI [22, 23]. Combination therapy with fresh frozen plasma to replenish the blood Cp levels and, thereafter, the administration of deferoxamine to deplete ferric iron stores showed an unprecedented improvement in the neurological symptoms [24]. Iron chelation therapy with deferasirox, an oral iron-chelating agent, led to a mild improvement in clinical symptoms, including the cognitive performance, gait, and balance, in an aceruloplasminemia patient who had no response to both deferoxamine and fresh frozen plasma therapy [25]. A marked neurological improvement was found after 15 months of treatment with oral zinc sulfate therapy (200 mg/day) in a patient with extrapyramidal and cerebellar symptoms caused by a heterozygous Cp gene mutation [26]. Zinc therapy could be useful as an alternative treatment when iron chelation therapy is discontinued due to side effects or progression of the symptoms, because zinc therapy has minimal side effects and may ameliorate the neurological symptoms.

Conclusions

Cp plays an important role in maintaining iron homeostasis in the brain and in protecting the brain from iron-mediated free radical injury. Therefore, the antioxidant effects of Cp could have important implications for various neurodegenerative disorders such as Parkinson's disease and Alzheimer's disease, in which iron depositions are known to occur.

References

1. Sipe JC, Lee P, Beutler E. Brain iron metabolism and neurodegenerative disorders. Dev Neurosci. 2002;24:188–96.
2. Miyajima H, Nishimura Y, Mizoguchi K, Sakamoto M, Shimizu T, Honda N. Familial apoceruloplasmin deficiency associated with blepharospasm and retinal degeneration. Neurology. 1987;37:761–7.

3. Harris ZL, Takahashi Y, Miyajima H, Serizawa M, MacGillivray RT, Gitlin JD. Aceruloplasminemia: molecular characterization of this disorder of iron metabolism. Proc Natl Acad Sci U S A. 1995;92:2539–43.
4. Kono S. Aceruloplasminemia. Curr Drug Targets. 2012;13:1190–9.
5. Duce JA, Tsatsanis A, Cater MA, James SA, Robb E, Wikhe K, Leong SL, Perez K, Johanssen T, Greenough MA, Cho H-H, Galatis D, Moir RD, Masters CL, McLean C, Tanzi RE, Cappai R, Barnham KJ, Ciccotosto GD, Rogers RT, Bush AI. An iron-export ferroxidase activity of beta-amyloid protein precursor is inhibited by zinc in Alzheimer's Disease. Cell. 2010;142:857–67.
6. Ke Y, Ming Qian Z. Iron misregulation in the brain: a primary cause of neurodegenerative disorders. Lancet Neurol. 2003;2:246–53.
7. Zecca L, Youdim MB, Riederer P, Connor JR, Crichton RR. Iron, brain ageing and neurodegenerative disorders. Nat Rev Neurosci. 2004;5:863–73.
8. Miyajima H, Kohno S, Takahashi Y, Yonekawa O, Kanno T. Estimation of the gene frequency of aceruloplasminemia in Japan. Neurology. 1999;53:617–9.
9. Daimon M, Yamatani K, Tominaga M, Manaka H, Kato T, Sasaki H. NIDDM with a ceruloplasmin gene mutation. Diabetes Care. 1997;20:678.
10. Miyajima H. Aceruloplasminemia, an iron metabolic disorder. Neuropathology. 2003;23:345–50.
11. Hellman NE, Schaefer M, Gehrke S, Stegen P, Hoffman WJ, Gitlin JD, Stremmel W. Hepatic iron overload in aceruloplasminaemia. Gut. 2000;47:858–60.
12. Patel BN, Dunn RJ, Jeong SY, Zhu Q, Julien JP, David S. Ceruloplasmin regulates iron levels in the CNS and prevents free radical injury. J Neurosci. 2002;22:6578–86.
13. Miyajima H, Takahashi Y, Serizawa M, Kaneko E, Gitlin JD. Increased plasma lipid peroxidation in patients with aceruloplasminemia. Free Radic Biol Med. 1996;20:757–60.
14. Kohno S, Miyajima H, Takahashi Y, Suzuki H, Hishida A. Defective electron transfer in complexes I and IV in patients with aceruloplasminemia. J Neurol Sci. 2000;182:57–60.
15. Miyajima H, Kono S, Takahashi Y, Sugimoto M. Increased lipid peroxidation and mitochondrial dysfunction in aceruloplasminemia brains. Blood Cells Mol Dis. 2002;29:433–8.
16. Kaneko K, Nakamura A, Yoshida K, Kametani F, Higuchi K, Ikeda S. Glial fibrillary acidic protein is greatly modified by oxidative stress in aceruloplasminemia brain. Free Radic Res. 2002;36:303–6.
17. Morita H, Ikeda S, Yamamoto K, Morita S, Yoshida K, Nomoto S, Kato M, Yanagisawa N. Hereditary ceruloplasmin deficiency with hemosiderosis: a clinicopathological study of a Japanese family. Ann Neurol. 1995;37:646–56.
18. McNeill A, Birchall D, Hayflick SJ, Gregory A, Schenk JF, Zimmerman EA, Shang H, Miyajima H, Chinnery PF. T2* and FSE MRI distinguishes four subtypes of neurodegeneration with brain iron accumulation. Neurology. 2008;70:1614–9.
19. Kono S, Suzuki H, Takahashi K, Takahashi Y, Shirakawa K, Murakawa Y, Yamaguchi S, Miyajima H. Hepatic iron overload associated with a decreased serum ceruloplasmin level in a novel clinical type of aceruloplasminemia. Gastroenterology. 2006;131:240–5.
20. Miyajima H, Takahashi Y, Kamata T, Shimizu H, Sakai N, Gitlin JD. Use of desferrioxamine in the treatment of aceruloplasminemia. Ann Neurol. 1997;41:404–7.
21. Pan PL, Tang HH, Chen Q, Song W, Shang HF. Desferrioxamine treatment of aceruloplasminemia: long-term follow-up. Mov Disord. 2011;26:2142–4.
22. Loreal O, Turlin B, Pigeon C, Moisan A, Ropert M, Morice P, Gandon Y, Jouanolle AM, Verin M, Hider RC, Yoshida K, Brissot P. Aceruloplasminemia: new clinical, pathophysiological and therapeutic insights. J Hepatol. 2002;36:851–6.
23. Mariani R, Arosio C, Pelucchi S, Grisoli M, Piga A, Trombini P, Piperno A. Iron chelation therapy in aceruloplasminaemia: study of a patient with a novel missense mutation. Gut. 2004;53:756–8.

24. Yonekawa M, Okabe T, Asamoto Y, Ohta M. A case of hereditary ceruloplasmin deficiency with iron deposition in the brain associated with chorea, dementia, diabetes mellitus and retinal pigmentation: administration of fresh-frozen human plasma. Eur Neurol. 1999;42: 157–62.
25. Skidmore FM, Drago V, Foster P, Schmalfuss IM, Heilman KM, Streiff RR. Aceruloplasminaemia with progressive atrophy without brain iron overload: treatment with oral chelation. J Neurol Neurosurg Psychiatry. 2008;79:467–70.
26. Kuhn J, Bewermeyer H, Miyajima H, Takahashi Y, Kuhn KF, Hoogenraad TU. Treatment of symptomatic heterozygous aceruloplasminemia with oral zinc sulphate. Brain Dev. 2007; 29:450–3.

Chapter 12
Chorea in Inherited Ataxias

Hélio A. Ghizoni Teive and Renato Puppi Munhoz

Abstract Inherited ataxias (IAs) can present with non-cerebellar signs and symptoms, especially those with movement disorders. This chapter will provide a review analyzing the presence of chorea in patients with inherited ataxias (IAs). Chorea can be found among different types of IA: autosomal recessive cerebellar ataxias (ARCAs), such as ataxia telangiectasia (more commonly), and Friedreich ataxia (more rarely). In the group of spinocerebellar ataxias (SCAs), chorea can be found more commonly in patients with SCA type 17, and in dentatorubral-pallidoluysian atrophy (DRPLA).

Keywords Chorea • Inherited ataxias • Ataxia telangiectasia • Friedreich's ataxia • Spinocerebellar ataxias

Introduction

Ataxia is a broad term whose meaning is literally "without order." The term locomotor ataxia was used by Duchenne, and has been since the nineteenth century, referring most commonly to motor incoordination [1, 2]. Ataxias may be classified as (1)

Financial Disclosure/Conflict of Interest/Funding Sources
The authors have nothing to disclose

H.A.G. Teive (✉)
Movement Disorders Unit, Neurology Service, Internal Medicine Department,
Hospital de Clínicas, Federal University of Paraná,
Rua General Carneiro 1103/102, Centro, 80060-150 Curitiba, PR, Brazil
e-mail: hagteive@mps.com.br, teiveads@mps.com.br

R.P. Munhoz
Movement Disorders Unit, Neurology Service, Internal Medicine Department,
Hospital de Clínicas, Federal University of Paraná,
Rua General Carneiro 1103/102, Centro, 80060-150 Curitiba, PR, Brazil

Department of Neurology, University of Toronto, Toronto, ON, Canada

F.E. Micheli, P.A. LeWitt (eds.), *Chorea*,
DOI 10.1007/978-1-4471-6455-5_12, © Springer-Verlag London 2014

cerebellar, if the cerebellum and its afferent or efferent projections are affected; (2) sensory (with the presence of Romberg's sign), if the proprioceptive pathways are affected; (3) frontal, a rare form related to cerebello-frontal injury; (4) thalamic, due to a damaged cerebello-thalamo-cortical loop; and (5) vestibular, a controversial entity that has been attributed to labyrinthine dysfunction [3, 4].

The most commonly recognized form, cerebellar ataxia (CA), represents a syndrome that includes several signs and symptoms such as gait ataxia (especially tandem gait), dystasia, astasia, dysmetria, dysdiadochokinesia, dyssynergy, movement decomposition, dysarthria (scanning or staccato speech), titubation, hypotonia, pendular reflexes, rebound (Holmes phenomenon), eye movement abnormalities (gaze-evoked nystagmus, saccades, and smooth pursuit disorders), kinetic or intentional tremor, and cognitive dysfunction (cerebellar cognitive affective syndrome) [4, 5].

CAs can be classified as primary, idiopathic, acquired or secondary, or sporadic. Primary CAs are further subdivided into congenital and hereditary categories, including autosomal recessive (ARCA), autosomal dominant (ADCA), currently designated spinocerebellar ataxias (SCAs), X-linked CAs, and mitochondrial ataxias.

Congenital ataxias (CoA), which represent the outcomes of cerebellar malformations, comprise a heterogenous group of CAs embracing a variety of conditions that can be separated according to the specific type of malformation (i.e., midline or unilateral cerebellar malformations, and pontocerebellar hypoplasia). The most widely known forms are the Dandy-Walker (extensive cystic dilatation of the posterior fossa that is contiguous with the forth ventricle), the Dandy-Walker variants, Chiari malformations, vermis dysgenesis (Cogan congenital oculomotor apraxia, Lhermitte-Duclos disease, and so on), verminal agenesis (Joubert, Debakan-Arina, Walker-Warburg, and Gillespie syndromes), and also several other disorders that cause ponto-cerebellar hypoplasia. Joubert's syndrome is a rare, autosomal recessive disease characterized by a congenital malformation of the hindbrain, which can be identified on head MRI as the "molar tooth" sign [6].

Autosomal recessive cerebellar ataxias (ARCAs) are part of a heterogeneous group of IAs [1, 5, 7, 8]. ARCAs are typically characterized by cerebellar and spinal cord degeneration, showing a relatively early age of onset. Clinically, balance and gait abnormalities, incoordination, action tremor, and dysarthria are almost always present. Additional neurological and non-neurological signs and symptoms also may be found. The pathogenesis of ARCA most commonly involves "loss of function" of cellular proteins related to cerebellar or brainstem development, structural maintenance, mitochondrial function, cell cycle and homeostasis, or DNA repair [7–9].

Spinocerebellar ataxias (SCAs) are a large and complex heterogeneous group of autosomal dominant degenerative disorders characterized by progressive degeneration of the cerebellum and its afferent and efferent connections. Other nervous system structures are typically affected, including the basal ganglia, brainstem nuclei, pyramidal tracts, posterior column and anterior horn of the spinal cord, and peripheral nerves [2–5, 10, 11].

The wide range of clinical manifestations in SCAs include cerebellar gait and limb ataxia, with dysmetria, dysdiadochokinesia, intention tremor, dysarthria, and nystagmus. In addition, patients may have extra-cerebellar signs, such as dementia, epilepsy, visual disorders, peripheral neuropathy, ophthalmoplegia, pyramidal signs, and movement disorders. Among the latter are parkinsonism, dystonia, myoclonus, and chorea [2–5, 10, 11].

X-linked spinocerebellar ataxias are very rare forms of ataxia caused by X-linked recessive gene mutations. Currently, the most clinically relevant and common form is the fragile X permutation tremor ataxia syndrome (FXTAS) described by Hargerman and colleagues in 2001 [12]. The syndrome occurs predominantly in males, over 50 years of age, and is characterized by the presence of action tremor with prominent kinetic component, CA, cognitive dysfunction, and occasionally with parkinsonism and autonomic dysfunction. Brain magnetic resonance shows increased T2 signal intensity in the middle cerebellar peduncles in the majority of patients. FXTAS is caused by intermediate expansions (between 50 and 200 repeats) of a CGG trinucleotide in the fragile X mental retardation 1 (FMR1) gene. FMR1 is the same gene that causes fragile X syndrome, the most common inherited form of mental retardation (full mutation range >200 CGG repeats) [12].

Mitochondrial ataxias usually combine cerebellar and sensory ataxia (along with other features) and are due to the abnormalities of mitochondrial DNA. These forms of ataxia include maternally inherited heredoataxias arising from point mutations in genes coding for RNAs, respiratory chain subunits, or deletions/duplications of the mitochondrial DNA. This group includes myoclonic epilepsy associated with ragged-red fibers (MERRF); neuropathy, ataxia and retinitis pigmentosa (NARP); Kearns-Sayre Syndrome (KSS); mitochondrial myopathy, encephalopathy, lactic acidosis, and stroke-like episodes (MELAS); infantile onset spinocerebellar ataxia (IOSCA); and mitochondrial recessive ataxia syndrome (MIRAS). MIRAS is caused by the mutation in the mitochondrial DNA polymerase gamma (POLG) [13].

Movement disorders other than those related to ataxia may be prominent clinical manifestations in some forms of IAs and are often extremely challenging for clinicians to arrive at an accurate diagnosis [14, 15].

This is especially true for chorea, which is derived from the Greek word "choreia" (to dance). Chorea is movement disorder characterized by involuntary brief jerky movements, which appear at random in the affected body part or parts [16]. The aim of this study was to evaluate the presence of chorea in inherited ataxias, including congenital ataxias, X-linked cerebellar ataxias and mitochondrial ataxias and, particularly, in ARCA and SCA.

Chorea in ARCA

ARCA comprises a large group of rare diseases, the most frequent being Friedreich's ataxia (FA). Other forms of ARCA include ataxia telangiectasia and ataxia with oculomotor apraxia (types 1 and 2).

Ataxia Telangiectasia

Ataxia telangiectasia (AT) represents the second most common form of ARCA, with a prevalence of 1.5–2.5/100.000. AT is caused by a mutation of the ATM gene, on chromosome 11q22-23. In clinical practice, an elevated serum level of

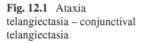

Fig. 12.1 Ataxia
telangiectasia – conjunctival
telangiectasia

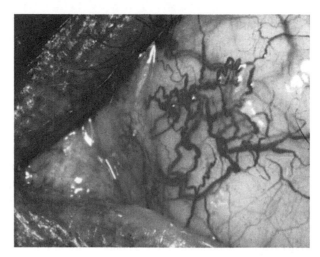

α-fetoprotein is a typical laboratory finding and cerebellar atrophy is very commonly found on neuroimaging, especially MRI [4, 7–9].

Generally, the disease onset ranges from 2 to 4 years of age, with a classical clinical picture characterized by the presence of progressive cerebellar ataxia, oculomotor apraxia, and oculocutaneous telangiectasia (Fig. 12.1). Other problems include recurrent sinopulmonary infections, immunodeficiency (posing a high risk of malignancy such as leukemia and lymphoma), and hypersensitivity to ionizing radioactivity [4, 7–9]. Different types of movement disorders have been described in patients with AT, including dystonia, myoclonus, parkinsonism, and chorea [17–19]. In 1996, Klein et al. described case reports of AT presenting with chorea in early childhood and lacking telangiectasia. On follow-up, these patients developed cerebellar ataxia and oculomotor apraxia [18]. Teive et al. 2010 studied a Brazilian series of ten patients from six different families with genetically confirmed AT. They demonstrated cerebellar ataxia and an elevated level of serum α-fetoprotein in all patients. Chorea was found in 4 (40 %) of patients and ocular telangiectasia in 3 (30 %) of cases. More recently, Nissenkorn et al. studied 17 children with genetically confirmed AT, who manifested various movement disorders such as ataxia, parkinsonism, myoclonus, and chorea. The authors demonstrated that treatment using amantadine, an N-methyl-d-aspartate receptor blocker, was partially effective for symptomatic benefit and had only mild side effects [19].

Friedreich's Ataxia

The prevalence of FA ranges from 1 in 30,000 to 1 in 50,000 in most populations, and it is considered the most common hereditary ataxia overall. In about 98 % of patients, FA is caused by a triplet GAA expansion within the first intron of the

Fig. 12.2 Friedreich's ataxia – spinal cord MRI, T1-weighted, sagittal view, with cervical spinal cord atrophy

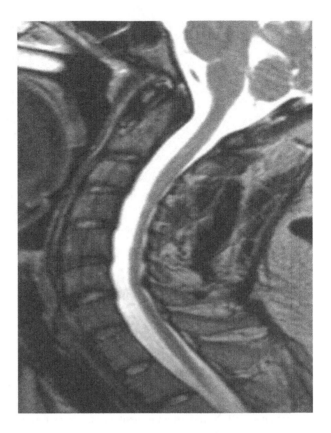

frataxin gene on chromosome 9q13. The repeat expansions range from 90 to 1,300 and there is an inverse correlation of age of onset, severity of disease, and associated systemic symptoms with the size of the repeat expansions. The age of onset ranges from 5 to 15 years, and the disease is typically characterized by early-onset progressive gait ataxia, dysarthria, loss of vibration and proprioceptive senses, profound areflexia, abnormal eye movements, and the presence of Babinski's sign [4, 7–9]. Systemic manifestations include cardiomyopathy, diabetes, scoliosis, and pes cavus. Additional FA phenotypes include late onset FA (LOFA), very late onset FA (VLOFA), ataxia with FA retained reflexes (FARR), and the Acadian form of FA (with slow progression). Other rare manifestations of FA are pure sensory ataxia, ataxia with spasticity and spastic paraplegia, and chorea [4, 7–9]. Magnetic resonance imaging commonly shows cervical spinal cord atrophy, without cerebellar atrophy (Fig. 12.2).

Generalized chorea was described for the first time in two patients with FA in 1998, by Hanna et al., from the Institute of Neurology of Queen Square (David Marsden's group). Both patients had genetically confirmed FA and exhibited generalized chorea in the absence of cerebellar signs [20]. In 2002, Zhu et al. reported the

Table 12.1 Autosomal recessive cerebellar ataxias – frequency of chorea

ARCA	Frequency of chorea
Ataxia telangiectasia	Common
Friedreich's ataxia	Rare
AOA1	Common, but transient
AOA2	Rare
Others: XP	Rare

ARCA autosomal recessive cerebellar ataxia, *AOA1* ataxia with oculomotor apraxia 1, *AOA2* ataxia with oculomotor apraxia 2, *XP* xeroderma pigmentosum

case of a patient with genetically confirmed FA, who presented with chorea associated with myoclonus, and later developed classic FA signs [21]. Spacey et al., in 2004, reported a ten-year-old patient from Malaysia with genetically confirmed FA associated with generalized chorea [22].

Others ARCA

Chorea has been described very rarely in other forms of ARCA. Cerebellar ataxia with oculomotor apraxia type 1 (AOA1) is an ARCA associated with hypoalbuminemia and hypercolesterolemia. AOA1 represents the most frequent ARCA in Japan and the second most common in Portugal. Chorea is observed in 79 % of cases at onset of the disease, but frequently disappeared during its course [23]. On the other hand, in patients with ataxia with oculomotor apraxia type 2 (AOA2), chorea is found in only 9.5 % of the patients [24]. Xeroderma pigmentosum, a very rare inherited neurodermatological disorder, also represents a very rare cause of ataxia and chorea [25] (Table 12.1).

Spinocerebellar Ataxias

Several movement disorders, including myoclonus, dystonia, parkinsonism, tremor, and chorea, have been described in different types of SCAs, particularly SCAs types 1, 2, 3, 6, 7, 12, 14, 17, 19, 20, 21, 27, and DRPLA. Chorea has been reported in cases of SCAs types 1, 2, 3, 14, 17, and DRPLA [14, 15].

SCA Type 17

SCA type 17 is a rare, autosomal dominant cerebellar ataxia caused by the expansion of a triplet of CAG repeats coding for polyglutamine stretches. The phenotype is characterized by gait ataxia, pyramidal signs, cognitive dysfunction, and movement disorders, including parkinsonism, dystonia, and chorea. Chorea is particularly common in patients with SCA type 17, with a prevalence between 20 and 66 %. The clinical picture of SCA type 17 is very similar to that of Huntington's disease (HD), and, currently, SCA type 17 represents the HD-like syndrome type 4 [15, 26].

Fig. 12.3 Spinocerebellar ataxia – brain MRI, T1-weighted, sagittal view, with cerebellar atrophy

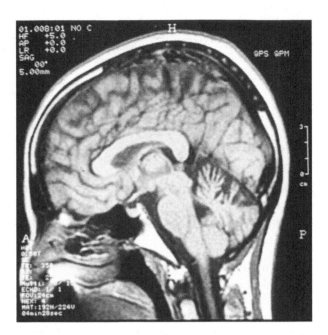

DRPLA

Dentatorubral-pallidoluysian atrophy (DRPLA) represents a rare form of autosomal dominant neurodegenerative disorder, related to a triplet CAG repeat expansion of the ATN1 gene on chromosome 12p13.31. Patients with juvenile onset DRPLA present with cerebellar ataxia, seizures (progressive myoclonus epilepsy), and cognitive disorder. Patients with DRPLA with onset after 20 years of age commonly present with cerebellar ataxia, dementia, and chorea. DRPLA is also known as HD-like syndrome type 3 [27].

Others SCAs

More rarely, chorea can be found in patients with SCAs type 1, 2, 3, and 14. In a Brazilian series of 104 families with SCAs, with genotype–phenotype correlations, chorea was not found [28] (Fig. 12.3). More recently, Moro et al. studied 169 Brazilian families with SCAs (total 378 patients), looking specifically for movement disorders, and again chorea was not described in this study [29] (Table 12.2).

Table 12.2 Spinocerebellar ataxias – frequency of chorea

SCA	Frequency of chorea
SCA type 17	Very common
DRPLA	Very common
SCA type 3	Rare
Other SCAs	Rare

SCA spinocerebellar ataxia, *DRPLA* dentatorubral-pallidoluysian atrophy

Table 12.3 Other inherited ataxias – frequency of chorea

Inherited ataxia	Frequency of chorea
Mitochondrial ataxia (MIRAS/POLG)	Rare
X-linked ataxia	Rare
FXTAS	Rare

MIRAS mitochondrial recessive ataxia syndrome, *POLG* polymerase gamma, *FXTAS* fragile X-associated tremor/ataxia syndrome

Chorea in Other IA

Mitochondrial ataxias in general combine cerebellar and sensory ataxia, due to mitochondrial DNA abnormalities. Mutations in the mitochondrial DNA polymerase gamma (POLG) can cause a pleomorphic spectrum of neurological abnormalities, including movement disorders [4, 13]. In 2012, Synofzik et al. studied 13 patients with POLG-associated ataxia. The authors demonstrated that chorea was present in 31 % of cases in this series [30].

X-linked ataxias, particularly fragile X tremor-ataxia syndrome (FXTAS) represents a new disorder, manifesting predominantly in males patients with cerebellar ataxia, associated with postural tremor and other movement disorders, including parkinsonism, and more rarely chorea [4, 12] (Table 12.3).

Conclusions

Chorea, a very peculiar movement disorder, can be presented in different types of IA, including ARCA, such as ataxia telangiectasia, more commonly, and FA, more rarely. Chorea is also present in patients with AOA1 and AOA2. In the group of SCAs, chorea can be found in patients with SCA type 17 and DRPLA. More rarely, chorea can be found in patients with mitochondrial ataxias and X-linked ataxias (FTXAS).

References

1. Harding AE. Classification of the hereditary ataxias and paraplegias. Lancet. 1983;1:1151–5.
2. Klockgether T. Sporadic ataxia with adult onset: classification and diagnostic criteria. Lancet Neurol. 2010;9:94–104.
3. Soong BW, Paulson HL. Spinocerebellar ataxias: an update. Curr Opin Neurol. 2007;20:438–46.
4. Perlman SL. Spinocerebellar degenerations. Handb Clin Neurol. 2011;100:113–40.
5. Teive HAG, Munhoz RP, Ashizawa T. Inherited and sporadic ataxias. In: Albanese A, Jankovic J, editors. Hyperkinetic movement disorders. Oxford: Wiley-Blackwell; 2012. p. 279–95.
6. Parisi MA, Doherty D, Chance PF, Glass IA. Joubert syndrome (and related disorders) (OMIM 213300). Eur J Hum Genet. 2007;15:511–21.
7. Palau F, Espinós C. Autosomal recessive cerebellar ataxias. Orphanet J Rare Dis. 2006;1:47.

8. Fogel BL, Perlman S. Clinical features and molecular genetics of autosomal recessive cerebellar ataxias. Lancet Neurol. 2007;6:245–57.
9. Di Donato S, Gellera C, Mariotti C. The complex clinical and genetic classification of inherited ataxias. II. Autosomal recessive ataxias. Neurol Sci. 2001;22:219–28.
10. Durr A. Autosomal dominant cerebellar ataxias: polyglutamine expansions and beyond. Lancet Neurol. 2010;9:885–94.
11. Teive HAG. Spinocerebellar ataxias. Arq Neuropsiquiatr. 2009;67:1133–42.
12. Leehey MA, Hagerman PJ. Fragile X-associated tremor/ataxia syndrome. Handb Clin Neurol. 2012;103:373–86.
13. Finsterer J. Mitochondrial ataxias. Can J Neurol Sci. 2009;36:543–53.
14. Schöls L, Peters S, Szymanski S, et al. Extrapyramidal motor signs in degenerative ataxias. Arch Neurol. 2000;57:1495–500.
15. van Gaalen J, Giunti P, van de Warremburg BP. Movement disorders in spinocerebellar ataxias. Mov Disord. 2011;26:792–800.
16. Donaldson IM, Marsden CD, Schneider SA, Bhatia KP. Chorea. In: Donaldson IM, Marsden CD, Schneider SA, Bhatia KP, editors. Marsden's book of movement disorders. Oxford: University Press; 2012. p. 730.
17. Worth PF, Srinivasan V, Smith A, et al. Very mild presentation in adult with classical cellular phenotype of ataxia telangiectasia. Mov Disord. 2013;28:524–8.
18. Klein C, Wenning GK, Quinn NP, Marsden CD. Ataxia without telangiectasia masquerading as benign hereditary chorea. Mov Disord. 1996;11:217–20.
19. Nissenkon A, Hassin-Baer S, Lerman SF, Levi YB, Tzadok M, Ben-Zeev B. Movement disorder in ataxia-telangiectasia: treatment with amantadine sulfate. J Child Neurol. 2013;28:155–60.
20. Hanna MG, Davis MB, Sweeney MG, et al. Generalized chorea in two patients harboring the Friedreich's ataxia gene trinucleotide repeat expansion. Mov Disord. 1998;13:339–40.
21. Zhu D, Burke C, Leslie A, Nicholson GA. Friedreich's ataxia with chorea and myoclonus caused by a compound heterozygosity for a novel deletion and the trinucleotide GAA expansion. Mov Disord. 2002;17:585–9.
22. Spacey SD, Szczygielski BI, Young SP, Hukin J, Selby K, Snutch TP. Malaysian siblings with Friedreich ataxia and chorea: a novel deletion in the frataxin gene. Can J Neurol Sci. 2004;31:383–6.
23. Le Ber I, Moreira MC, Rivaud-Péchoux S, et al. Cerebellar ataxia with oculomotor apraxia type 1: clinical and genetic studies. Brain. 2003;126:2761–72.
24. Anheim M, Monga B, Fleury M, et al. Ataxia with oculomotor apraxia type 2: clinical, biological and genotype/phenotype correlation study of a cohort of 90 patients. Brain. 2009;132:2688–98.
25. Adamec D, Xie J, Poisson A, Broussolle E, Thobois S. Xeroderma pigmentosum: a rare cause of chorea. Rev Neurol. 2011;167:837–40.
26. Tsuji S. Spinocerebellar ataxia 17 (SCA17). In: Pulst S-M, editor. Genetics of movement disorders. San Diego: Academic/Elsevier; 2003. p. 139–41.
27. Tsuji S. Dentatorubral-Pallidoluysian Atrophy (DRPLA). In: Pulst S-M, editor. Genetics of movement disorders. San Diego: Academic/Elsevier; 2003. p. 143–50.
28. Teive HA, Munhoz RP, Arruda WO, et al. Spinocerebellar ataxias: genotype-phenotype correlations in 104 Brazilian families. Clinics. 2012;67:443–9.
29. Moro A, Munhoz RP, Arruda WO, Raskin S, Teive HAG. Movement disorders in spinocerebellar ataxias in a cohort of Brazilian patients. Eur Neurol. 2014.
30. Synofzik M, Srulijes K, Godau J, Berg D, Schöls L. Characterizing POLG ataxia: clinics, electrophysiology and imaging. Cerebellum. 2012;11:1002–11.

Chapter 13
Chorea in Prion Diseases

Marie-Claire Porter and Simon Mead

Abstract Prion diseases are neurodegenerative disorders associated with misfolding of prion protein. They are transmissible, often rapidly progressive, fatal conditions that occur in both humans and animals. Sporadic, acquired, and inherited forms of prion disease exist. One of the distinct characteristics of all human prion diseases is their clinical and pathological heterogeneity; however, there are clinical features that are common to all forms and these include progressive cognitive impairment and movement disorders, including chorea. Chorea occurs most frequently in variant Creutzfeldt-Jakob disease (vCJD), an acquired form of prion disease, but has been reported to occur in both the sporadic and inherited forms of the disease too. Inherited prion diseases are caused by autosomal dominant mutations in the prion protein gene (*PRNP*) and can be mistaken for Huntington's disease (HD). It is important that a diagnosis of prion disease is considered if HD gene testing is negative. This chapter aims to give a general overview of the clinical features, investigations, and pathophysiology of the human prion diseases, accompanied by representative case reports of patients seen at the NHS National Prion Clinic, UK, in whom chorea was a prominent feature.

Keywords Prion • Creutzfeldt-Jakob • PRNP • Chorea • Myoclonus

M-C. Porter
The National Prion Clinic, The National Hospital for Neurology
and Neurosurgery, University College London Hospitals NHS Trust, London, UK

S. Mead (✉)
MRC Prion Unit, Department of Neurodegenerative Disease, UCL Institute
of Neurology, University College London, Queen Square, London WC1N 3BG, UK
e-mail: s.mead@prion.ucl.ac.uk

F.E. Micheli, P.A. LeWitt (eds.), *Chorea*,
DOI 10.1007/978-1-4471-6455-5_13, © Springer-Verlag London 2014

Introduction

Prion diseases, also known as transmissible spongiform encephalopathies, are a group of neurodegenerative conditions that exist in both humans and animals. They are self-propagating, progressive, and uniformly fatal neurodegenerative disorders. Prions, "proteinaceous infectious particles," were christened by Prusiner in 1982; it is now well established that they are the causative agents of prion diseases [26]. The naturally occurring prion protein, PrP^C, "C" denoting the cellular form, is a cell surface protein expressed in a wide range of cell types, particularly in neuronal and immune cells. PrP^C is encoded by the prion protein gene (*PRNP*). Prion diseases are associated with the build up in the brain and other organs of an abnormal form of the prion protein (PrP^{Sc}), "Sc" denoting the scrapie isoform. PrP^{Sc} is known to arise via introduction from the external environment in acquired forms of prion disease and through autosomal dominantly inherited mutations of *PRNP*. Once PrP^{Sc} is present an autocatalytic process is thought to occur whereby normal PrP^C is converted to PrP^{Sc} resulting in accumulation of the misfolded pathogenic protein in the brain [6]. Unlike PrP^C, PrP^{Sc} is resistant to protease action and has distinct biochemical properties including a propensity to aggregate in tissue, which in some circumstances manifests as PrP-amyloid deposition. The underlying mechanism of how the neurodegenerative process occurs in prion disease is still unknown.

Human prion diseases show marked phenotypic variability. In inherited prion disease this can be explained in part by the mutation itself; however, a major factor in accounting for this diversity is the existence of distinct prion "strains." By analogy with strains of infectious organisms such as influenza or tuberculosis, prion strains are isolates which can be serially propagated and which account for distinct pathological and clinical characteristics [4, 8]. The identification in 1996 of a new human prion strain and the demonstration that this was the same strain that was present in bovine spongiform encephalopathy (BSE) led to the confirmation of a link between the cattle prion disease and a novel human disease variant Creutzfeldt-Jakob disease (vCJD) and a public health crisis [9, 5, 15]. In addition to this a common *PRNP* polymorphism, encoding either methionine or valine at codon 129 is a strong susceptibility factor for human prion disease [24]. In vCJD all pathologically confirmed cases have been homozygous at codon 129 [21]. In some forms of inherited prion diseases, the codon 129 genotype has been found to influence both disease onset and duration [19, 28].

Clinical Features

Human prion diseases can be divided into 3 subtypes: sporadic Creutzfeldt-Jakob disease (sCJD); the inherited clinical syndromes, Gerstmann-Straussler-Scheinker (GSS), familial CJD, and fatal familial insomnia (FFI); and acquired forms such as vCJD, kuru, and iatrogenic CJD.

Sporadic Prion Disease

Sporadic Creutzfeldt-Jakob disease accounts for around 85 % of all cases of prion disease. sCJD occurs in every country, with a surprisingly similar annual incidence of approximately 1–2 cases per million. The disease usually affects people between the ages of 45 and 75 years with an average age of onset of around 65 years. The clinical course of the disease is typically rapidly progressive, with a median survival time of 4–5 months [25]. The majority of cases are fatal within 6 months; however, the clinical course can vary and in some individuals the disease duration can be longer, in a minority over 3 years. Patients classically present with symptoms of cognitive decline accompanied by a movement disorder. In many cases of sCJD, the onset of disease is insidious; prodromal symptoms such as mood or sleep disturbance are often reported several months prior to the onset of more definite symptoms and signs. A number of clinical subtypes have been described. The Heidenhain variant of sCJD is associated with visual symptoms that typically progress rapidly to complete blindness. The Brownell-Oppenheimer variant presents with prominent cerebellar disturbance, in addition to these are documented classical and thalamic disease phenotypes.

At present the etiology of sCJD is unclear. It may arise from a spontaneous conversion of PrP^C to PrP^{Sc}, from a somatic mutation occurring in *PRNP*, or it is possible that some sCJD may also result from an unknown environmental source. The common *PRNP* polymorphism present at codon 129 is known to affect both susceptibility to developing sCJD and the disease phenotype. The majority of patients who develop sCJD are homozygous for either methionine or valine at codon 129 [24]. In sCJD the duration of clinical disease is heavily influenced by both the codon 129 genotype and partially by the accompanying PrP^{Sc} strain.

Acquired Prion Disease

Kuru

Human to human transmissibility was first recognized in the 1950s during the investigation of an epidemic of spongiform encephalopathy in the Fore linguistic group and surrounding populations located in the Eastern Highlands Province of Papua New Guinea.

Kuru resulted from the practice of ritualistic cannibalism. In the Fore population women and children were primarily affected, as they were the ones to consume the brain and other internal organs of deceased relatives during cannibalistic feasts. It is thought that the trigger for the kuru epidemic may have come about through the recycling of prions from a member of the population who had sCJD. The age at disease onset is wide and ranges between 4 and over 60 years. The mean incubation period for kuru is approximately 12 years, but there are cases that have occurred after a minimum exposure of 41 years. Between 1957 and 2004 over 2,700 cases of kuru were reported [7]. The main features of the disease are cerebellar ataxia, tremor, and dementia.

Iatrogenic CJD

Iatrogenic transmission of prions has also been identified as a means of acquiring prion disease, with several routes of transmission having been identified.

Cadaveric-derived pituitary growth hormone was used as treatment for growth problems in children, it is thought that a number of the pituitary glands harvested were from patients with sCJD albeit undiagnosed at the time [3]. The practice of using cadaveric-sourced growth hormone ceased in 1985 when artificially synthesized growth hormone became available. At present in the UK there are approximately 2–6 new cases of growth hormone induced iCJD diagnosed each year. The incubation period is long, with some cases occurring 30 years after exposure. The duration of illness is variable but usually between 8 and 18 months. Cerebellar symptoms are a prominent feature of the disease followed by late onset dementia.

There have been a small number of cases of iCJD occurring in women treated for fertility problems secondary to human-derived pituitary gonadotrophin.

Human-derived dura was used for dural repair surgery until 1992 when a synthetic form was introduced. Although cases have been reported worldwide, the majority of reported patients with iCJD secondary to infected dural grafts have been in Japan, a minimal number of cases have been reported in the UK. The incubation period is highly variable, between 14 months and 25 years. The presenting clinical symptoms are usually memory and cognitive problems; the majority of cases follow a clinical course similar to that seen in sCJD. Other routes of transmission include neurosurgical procedures, inadequately sterilized intracerebral electrodes, and corneal grafting.

Variant Creutzfeldt-Jakob Disease

The first cases of vCJD were recognized as such in 1995, 12 years after the identification of bovine spongiform encephalopathy (BSE) [30, 9]. The disease was initially thought to be an unusual form of sCJD; however, as the number of cases grew and both atypical clinical and pathological features were described, the new disease was initially termed new variant CJD, subsequently simply vCJD. It is well established that vCJD is caused by dietary exposure to the same prion strain which is responsible for BSE [9, 5]. The number of clinical cases of vCJD is around 220 to date with the majority of cases having occurred in the UK, a small number in comparison to the large number of people who have been exposed to BSE prions. A recent study suggested that approximately 1 in 2,000 individuals exposed has abnormal prion protein present in appendix lymphoreticular tissue [13]. There is therefore considerable uncertainty as to whether the epidemic is over given the potential for extremely long incubation periods in prion diseases.

In contrast to sCJD, vCJD typically affects a younger population, with a median age of onset of 28 years, and a range of 12–74 years, the incidence occurs equally in males and females [14]. In addition to the younger age of onset, the clinical presentation of vCJD differs substantially from the sporadic form; the vCJD phenotype

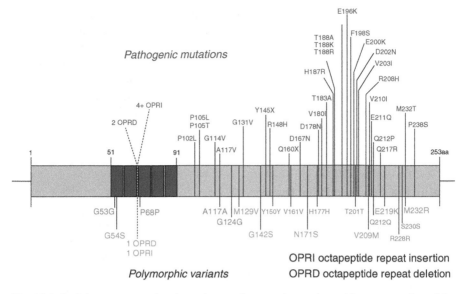

Fig. 13.1 Definite or suspected pathogenic mutations are shown above this representation of the prion protein gene in *red*. Neutral (*green*) and known or possible prion disease susceptibility/modifying polymorphisms (*blue*) are shown below

is a less rapidly progressive disease, with an average duration of illness of 14 months. Early reported clinical features are often psychiatric, with mood disturbance, irritability, and anxiety being commonly reported symptoms; in addition to this many patients are symptomatic with peripheral limb pain and/or paraesthesia. As the disease progresses the commonest neurological features to develop are cerebellar ataxia, dysarthria, cognitive impairment, and movement abnormalities such as chorea, dystonia, and myoclonus. Chorea in particular is one of the clinical features that aids in distinguishing the disease from sCJD and is included in the WHO diagnostic criteria.

Inherited Prion Disease

Ten to fifteen percent of cases of human prion disease are inherited. In the majority of cases, there is a family history of neurodegenerative disease. All mutations occur in *PRNP*, and the pattern of inheritance is autosomal dominant. To date over 30 pathogenic mutations have been identified (see Fig. 13.1); both point mutations and alterations in an octapeptide repeat motif expansion are known to be causative.

Worldwide the most prevalent point mutations are E200K, P102L, and D178N. Of the repeat expansions the 6-octapeptide repeat insertional (OPRI) mutation is by far the most common [20].

There is a very wide range in age of clinical onset of inherited prion disease, from adolescence to the extremes of old age, although most cases manifest in middle age (35–65 years). The phenotype of the disease is characterized by the specific mutation; however, even within families there can be significant variability both in the clinical characteristics of the disease and in the age of onset. Newly diagnosed patients will often have no family history of prion disease; this is due in part to the fact that some mutations, particularly those with later onset display reduced penetrance. Prion disease is often misdiagnosed as some other form of neurodegenerative disease.

Features of Inherited Prion Disease

Similar to sCJD and the acquired forms of prion disease, a prodrome of subtle symptoms and signs is often present before the onset of more definite features of the disease. These symptoms include fatigue, insomnia, behavioral change, and sensory symptoms. Prion diseases are now more commonly named after the mutation that causes them; they were previously classified under specific clinical syndromes, Gerstmann-Straussler-Scheinker (GSS), familial Creutzfeldt-Jakob disease, and fatal familial insomnia (FFI).

The GSS phenotype is most commonly caused by the P102L (proline to leucine) point mutation. Age of onset is between 20 and 68 years; the average disease duration is 3 years, ranging from 1 to 10 years. Typical features of the disease are early onset peripheral sensory disturbance accompanied by lower limb weakness and limb and truncal ataxia. Cognitive decline tends to manifest later on in the disease process. A third of patients present with behavioral or psychiatric features [28]. Other *PRNP* mutations that have been reported as having a GSS phenotype include the following: P105L, A117V, and a number of the octapeptide repeat mutations (5, 6, 7, 8, and 9 OPRI).

FFI is associated with the D178N missense prion protein mutation particularly with a codon 129 methionine allele on the same haplotype [22]. Disease onset is typically between 30 and 65 years and the disease duration between 6 months to 2 years. Clinical features of FFI include sleep disturbance: insomnia and accompanying excessive daytime somnolence, as well as motor signs and dysautomnia. On polysomnography patients have been found to have markedly disrupted sleep. Autonomic symptoms include temperature, blood pressure, and pulse rate dysregulation; sweating; impotence; increased salivation; and urinary retention.

The E200K *PRNP* mutation is most commonly associated with the familial CJD syndrome. The clinical phenotype is almost indistinguishable from sCJD [16]. Most patients follow a rapidly progressive course of illness; cognitive decline, cerebellar signs, and myoclonus are all features of the clinical presentation. Some patients have been found to have a peripheral neuropathy present; apart from this the clinical presentation is remarkably similar to that seen in sCJD. Although uncommon, at least in the UK, the D178N PRNP mutation, on the V allele, and the 4-OPRI mutation also present with rapidly progressive illnesses that are both pathologically and phenotypically similar to that seen in sCJD.

Chorea in Prion Disease

Chorea is one of a number of movement disorders that occur in prion disease. The basal ganglia and the thalamus are commonly affected in prion disease neuropathologically, and these areas show key diagnostic MRI signal changes in the early stages of disease. Chorea occurs in both sCJD and inherited prion disease; however, it has more commonly been reported to present in patients with vCJD. Chorea usually manifests itself in the later stages of the illness; despite this observation there have been case reports where chorea occurring has either been a presenting feature or occurred at an early stage of the disease process [2, 23]. Other movement disorders that occur in prion disease include ataxia, myoclonus, tremor, dystonia, hemiballismus, and alien limb.

Sporadic CJD

Movement disorders are common in patients with sCJD. Myoclonus has been reported as occurring in between 70 and 80 % of patients and ataxia in 80 % [12]. Although criteria exist for diagnosing sCJD, clinical features are highly variable and it may be more helpful to characterize the nature of movement disorders in sCJD in order to improve accuracy of diagnosis [29]. Chorea is not the most frequent clinical finding in sCJD but it has been reported as occurring in up to 11 % of patients [12]. In patients recruited to the National Prion Monitoring Cohort, a National Prion Clinic UK study, chorea has been identified as occurring in 7 % of those patients diagnosed with sCJD. Although chorea is an unusual early or presenting sign of sCJD, this is recognised [10].

Variant CJD

Movement disorders are a key clinical feature of vCJD and commonly include myoclonus, ataxia, and chorea. Chorea is a more common feature in vCJD than other forms of prion disease and forms part of the WHO diagnostic criteria. Chorea has been reported as occurring in 25–50 % of patients symptomatic with vCJD, typically presenting between 6 and 11 months after the onset of disease [27]. Although early onset symptoms in vCJD are predominately psychiatric accompanied by peripheral sensory disturbance, there have been case reports where chorea has been described as occurring either in the early stages of the disease or as a presenting symptom.

Mckee and Talbot published a case report of chorea as a presenting feature in a 27-year-old male [18]. The patient's initial symptoms were slurred speech, gait unsteadiness, and fidgeting from the onset of the illness. The fidgeting movements progressed to striking florid continuous choreiform movements in all limbs as well as the patient's neck. As the illness progressed the patient developed cognitive problems and myoclonus in addition to persistent chorea; he died 8 months after the disease onset; the neuropathology at postmortem confirmed the diagnosis of vCJD.

Another case, reported by Bowen et al, described a young woman who presented with a 4-month history of behavioral disturbance and cognitive decline; these symptoms were followed by early onset florid generalized choreiform movements and imbalance. The chorea was so pronounced that it interfered with activities of daily living. The diagnosis of vCJD was confirmed by tonsil biopsy [2]. These case reports described are beneficial not only in illustrating the variable onset time for chorea but also in detailing the clinical nature of the symptom itself.

Inherited Prion Diseases

Inherited prion disease is one of the differential diagnoses to consider in individuals who are suspected of being symptomatic with HD without genetic confirmation. The Huntington's disease-like 1 (HDL1) disorder is a HD phenocopy, first described in 1998 by Xiang F et al. A family of 6 young adults (mean age 29.7 years) presented with personality change, cognitive decline, and motor disturbance; chorea was described as being particularly prominent in 4 individuals [23]. The 8 octapeptide repeat expansion (8-OPRI) was found in all family members who underwent *PRNP* genotyping. Other case reports that have described clinical symptoms in symptomatic patients with the 8-OPRI mutation have not described chorea as a prominent feature; instead the clinical phenotype has been dominated by psychiatric symptoms [17]. This contrast of observed clinical findings in a specific genotype illustrates how heterogeneous the clinical features of prion disease can be even when the same mutation is causative.

Chorea although less common has been observed in a number of other *PRNP* mutations; these include E200K, D178N, 6OPRI, and P102L genotypes.

Investigations in Prion Disease

Sporadic CJD

Signal change on magnetic resonance imaging (MRI) is a key diagnostic finding. The signal change is most evident on axial diffusion-weighted imaging (DWI), although high signal is also apparent on both fluid-attenuated imaging (FLAIR) and T2-weighted sequences. The classical MRI appearances in sCJD are T2-weighted or FLAIR hyperintensity of the caudate and putamen as well as diffuse cortical hyperintensity (Fig. 13.2). This combination is reported to be 90 % sensitive for sCJD [31]. In sCJD early electroencephalogram (EEG) findings are often nonspecific. Diffuse slowing or frontal rhythmic delta activity is commonly detected. In approximately two thirds of patients with sCJD, the most characteristic finding on the EEG is repetitive triphasic periodic sharp wave complexes (PSWCs). The sensitivity for making this finding increases with test repetition and occurs more commonly when myoclonus is also present. PSWCs are more frequently observed in the middle and later stages of the disease.

On cerebrospinal fluid examination (CSF) the cell count and protein are usually normal in patients with prion disease. 14-3-3 is present in all eukaryotic cells and is found in most patients with rapidly progressive sCJD. The presence of 14-3-3 in a suspect case supports the diagnosis of sCJD; however, it is not specific to prion disease and may be elevated in other conditions including stroke, encephalitis, and seizure disorders. S100b proteins are found in the central nervous system glia; similarly to the 14-3-3 proteins they are raised in rapidly progressive CJD but again the specificity to the disease is low. Other CSF markers such as tau and neuron specific enolase (NSE) may also be elevated in sCJD.

The real-time quaking-induced conversion (RT-QUIC) is an amplification assay that was developed for the detection of PrP^{Sc} in CSF; the reported sensitivity of detecting abnormal PrP is around 80–90 % [1].

As a number of inherited prion disease mutations can mimic the sCJD phenotype, *PRNP* analysis should be carried out in all patients. A definitive diagnosis of sCJD requires examination of brain tissue; however, brain biopsy is usually only performed if there is a strong suspicion of an alternative and potentially treatable diagnosis.

Variant CJD

The MRI is important in the diagnosis of vCJD. Approximately 70 % of patients with vCJD present with the classical pulvinar sign, which is the presence of symmetrical high signal in the pulvinar and dorsomedial areas of the thalamus. The signal intensity in the pulvinar is classically higher in relation to that seen in the basal ganglia and cortex. The presence of the pulvinar sign is part of the WHO diagnostic criteria for vCJD.

In individuals with vCJD, EEG findings if present are usually nonspecific. PSWCs are not typically observed, and therefore if prion disease is suspected, the absence of PSWCs would point more towards a diagnosis of sCJD. Routine CSF analysis is usually normal. The presence of 14-3-3 is less commonly found than in sCJD; however, its absence does not help differentiate between the two.

The direct detection assay is a blood test that was developed to identify prion protein in the blood of patients with vCJD. It has high specificity to vCJD and a sensitivity of 70 % [11]. Tonsil biopsy has proved useful in the diagnosis of vCJD,

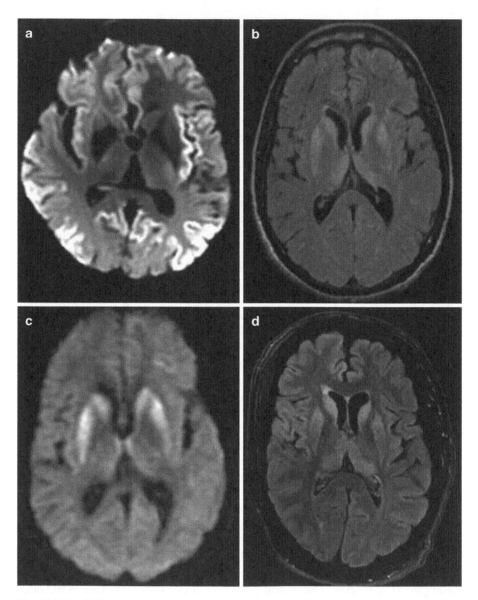

Fig. 13.2 (**a**) Diffusion-weighted imaging (*DWI*) showing extensive and patchy cortical ribbon in sCJD. (**b**) FLAIR images and (**c**) DWI showing high signal in the basal ganglia and thalamus in sCJD. (**d**) FLAIR images in iCJD showing cortical, caudate, putamen, and thalamic high signal

prion protein can be identified on examination of lymphoreticular tissue, and PrP[Sc] can be detected by immunohistochemistry.

It is especially important that *PRNP* analysis is carried out in young patients presenting with clinical features of vCJD as there are several forms of inherited prion disease that have a young age of onset.

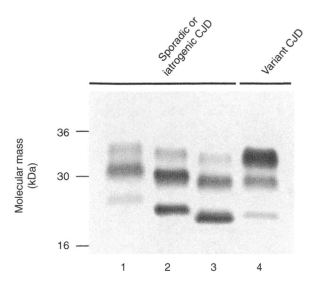

Fig. 13.3 Western blot demonstrating the presence of PrP^Sc in brain and tonsil from patients with sCJD and vCJD. The presence of PrP^Sc is revealed after proteinase K, (*PK*) treatment, which digests the normal form of PrP (PrP^C) but not the pathological form (PrP^Sc). Three common PrP^Sc types (1–3) in sporadic CJD can be distinguished by differing fragment sizes of the three PrP glycoforms

Inherited Prion Disease

To make a definitive diagnosis of inherited prion disease, *PRNP* analysis is required. As 10–15 % of cases of prion disease are inherited and a family history of inherited prion disease is absent in up to 50 % of cases, diagnostic genetic testing should be considered in all suspected cases of human prion disease. There are 3 types of mutations: point mutations that lead to an amino acid substitution, premature stop codon, and insertional octapeptide repeat mutations. When *PRNP* sequencing is performed, the whole gene is sequenced and therefore diagnosis is absolute.

Findings on MRI are variable and depend upon the underlying mutation present. The E200K mutation will often show the classical signal change that presents in sCJD. Atrophic changes are the most common finding in other forms of inherited prion disease, such as the 6-OPRI or P102L mutations.

Pathology

Macroscopically the cerebral hemispheres in prion disease are often of normal appearance although atrophy occurs in long standing cases. On microscopic examination prion disease is characterized by spongiform change, neuronal loss, astrocytosis, and abnormal PrP deposition, sometimes as amyloid. The pattern of disease observed accompanied by the PrP^Sc analysis by Western blot is diagnostically important in differentiating between specific forms of prion disease (Figs. 13.3 and 13.4).

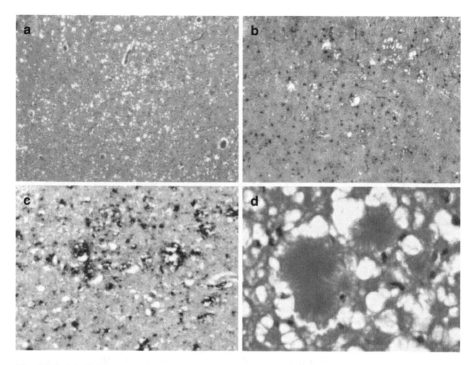

Fig. 13.4 (**a**) Cortical spongiform change in sCJD. (**b**) Gliosis and spongiform change in sCJD (GFAP). (**c**) Perivacuolar PrP staining in sCJD (ICSM35). (**d**) Florid plaques in vCJD (H&E, ICSM35) (Images were provided by Professor Sebastian Brandner)

Treatment

There is no disease modifying therapy currently available for prion disease. Several experimental agents have been assessed for therapeutic efficacy in clinical trials; these include intraventricular pentosan, quinacrine, and tetracyclic compounds such as doxycycline. So far none of these treatments have been found to halt the progression of prion disease. In regard to symptomatic treatment of chorea, no specific medication has been found to be particularly effective. Standard anti-choreic medications can be used but should be done so with care to minimize unwanted side effects such as worsening coexisting parkinsonian symptoms. There is one case report that found haloperidol effective in reducing both dystonia and chorea in a patient with prion disease [10]

Case Histories

Described below are 3 case histories in patients diagnosed with prion disease in which chorea was a prominent clinical feature.

Case 1

A 36-year-old right-hand-dominant man presented with a 5-month history of unusual behavior. He used to work as a security guard before the onset of his symptoms. He had no past medical history. He left school at the age of 16 years with few exams. He was previously fit and well and drank a moderate amount of alcohol. There was no family history of neurodegenerative disease.

Seven months prior to presentation, the patient lost his job; unusually for him he did not seek further employment and appeared low in mood. Two months later he began to repeat the same phrase a number of times and carry out repetitive behavior. He became disorientated in time and place. And shortly after, his gait became abnormal and he became unsteady on his feet and was noticed to have choreiform movements. He had evidence of dyspraxia; his parents commented on his inability to use the TV remote control.

On examination his MMSE score was 10/30; there were obvious problems with episodic memory, orientation, comprehension, and frontal executive function. On neurological examination he was mildly dysarthric; he had a supranuclear gaze paresis and abnormal saccades. He had choreiform movements present in his upper limbs and was dyspraxic. Coordination was poor with gross finger–nose–finger ataxia.

An MRI showed symmetrical high signal present in the basal ganglia and thalamus on DWI and FLAIR; the signal intensity was more pronounced in the pulvinar region of the thalamus. The EEG showed nonspecific slowing, on CSF examination the cell count and protein were normal, and 14-3-3 was found to be present. A tonsil biopsy was performed and PRP^{Sc} was seen in the tonsil follicles; on western blot PrP^{Sc} strain type 4 was identified leading to a diagnosis of vCJD.

Case 2

A 59-year-old left-hand-dominant female presented with an 8-month history of memory impairment, low mood, and anxiety; she was initially diagnosed with depression, but despite treatment with antidepressants symptoms persisted. Approximately 2 months later the patient became unsteady on her feet and developed problems with navigation. Six months after the onset of her initial symptoms, the patient began to display fidgeting movements in her upper limbs and had developed speech difficulties.

On examination the patient had global cognitive impairment present. The MMSE score was 11/30. On inspection the patient displayed constant fidgeting movements in her upper limbs to the extent that she put her hands under her thighs to try and control the involuntary movements; dystonic posturing was also present in both the upper and lower limbs. Speech was dysarthric and dysphasic. Both limb and gait ataxia were present.

There was bilateral symmetrical signal change present in the caudate on the MRI scan. Neither an EEG nor CSF examination was carried out. *PRNP* sequencing identified a mutation present at codon 178 linked to a valine allele at codon 129, confirming a diagnosis of inherited prion disease.

Case 3

A 58-year-old lady born and educated in Hong Kong but living in the UK for the last 40 years worked as a chef until she retired 18 months before presentation. There was no past medical history and a family history only positive for Parkinson's disease in her mother. She was a nonsmoker and had minimal alcohol intake.

The patient presented with a 6-month history of dizziness followed by problems with balance and a progressive gait disturbance. 4 months after onset the patient developed word finding difficulties. Her husband was frequently woken up at night by jerking movements that had developed in her upper and lower limbs.

An accurate cognitive examination was not possible secondary to the patient's severe dysphasia and apraxia. On neurological examination there was a supranuclear gaze paresis and a pout reflex present. Neck tone was increased. Tone was increased in the right upper and lower limbs; the deep tendon reflexes were brisk. The patient was reviewed 6 weeks later. At this point she had become bedbound and mute. On examination there were involuntary choreiform movements present in the head and neck. Tone was increased throughout but again was more pronounced on the right side in comparison to the left. Reflexes were brisk and the plantar response was flexor bilaterally.

The MRI head scan showed striking signal change on DWI and FLAIR sequences in the basal ganglia, thalamus bilaterally, hippocampus, and frontoparietal cortex. The EEG was reported as being nonspecifically slow. CSF examination was acellular, 14-3-3 was positive, S100-b was raised at 2.3 mg/ml (ref range 0–0.6), and tau >6,000. A diagnosis of probable sporadic CJD was made based on the clinical history and accompanying investigation findings.

Conclusion

Human prion diseases are phenotypically heterogeneous conditions, and chorea is a feature that has been described in the inherited, sporadic, and variant forms of the disease. Although chorea is a less prominent feature than other movement disorders such as myoclonus and ataxia, it is a symptom commonly seen in vCJD and forms part of the WHO diagnostic criteria. Prion disease and *PRNP* analysis should be included in the differential diagnosis of chorea especially in individuals where a diagnosis of HD, DRPLA, and SCA17 is being considered but genetic testing has proven negative.

References

1. Atarashi R, Wilham JM, Christensen L, Hughson AG, Moore RA, Johnson LM, et al. Simplified ultrasensitive prion detection by recombinant PrP conversion with shaking. Nat Methods. 2008;5:211–2.
2. Bowen J, Mitchell T, Pearce R, Quinn N. Chorea in new variant Creutzfeldt-Jakob disease. Mov Disord. 2000;15:1284–5.

3. Brown P, Preece M, Brandel JP, Sato T, McShane L, Zerr I, et al. Iatrogenic Creutzfeldt-Jakob disease at the millennium. Neurology. 2000;55:1075–81.
4. Bruce ME. Scrapie strain variation and mutation. Br Med Bull. 1993;49:822–38.
5. Bruce ME, Will RG, Ironside JW, McConnell I, Drummond D, Suttie A, et al. Transmissions to mice indicate that 'new variant' CJD is caused by the BSE agent. Nature. 1997;389: 498–501.
6. Collinge J. Prion diseases of humans and animals: their causes and molecular basis. Annu Rev Neurosci. 2001;24:519–50.
7. Collinge J, Whitfield J, McKintosh E, Beck J, Mead S, Thomas DJ, Alpers MP. Kuru in the 21st century–an acquired human prion disease with very long incubation periods. Lancet. 2006;367(9528):2068–74.
8. Collinge J, Clarke A. A general model of prion strains and their pathogenicity. Science. 2007;318:930–6.
9. Collinge J, Sidle KC, Meads J, Ironside J, Hill AF. Molecular analysis of prion strain variation and the aetiology of 'new variant' CJD. Nature. 1996;383:685–90.
10. Donmez B, Cakmur R, Men S, Oztura I, Kitis A. Coexistence of movement disorders and epilepsia partialis continua as the initial signs in probable Creutzfeldt-Jakob disease. Mov Disord. 2005;20(9):1220–3.
11. Edgeworth JA, Farmer M, Sicilia A, Tavares P, Beck J, Campbell T, et al. Detection of prion infection in variant Creutzfeldt-Jakob disease: a blood-based assay. Lancet. 2011;377(9764): 487–93.
12. Edler J, Mollenhauer B, Heinemann U, Varges D, Werner C, Zerr I, et al. Movement disturbances in the differential diagnosis of Creutzfeldt-Jakob disease. Mov Disord. 2009;24: 350–6.
13. Gill ON, Spencer Y, Richard-Loendt A, Kelly C, Dabaghian R, Boyes L, et al. Prevalent abnormal prion protein in human appendixes after bovine spongiform encephalopathy epizootic: large scale survey. BMJ. 2013;347:f5675.
14. Heath CA, Cooper SA, Murray K, Lowman A, Henry C, Macleod MA, et al. Diagnosing variant Creutzfeldt-Jakob disease: a retrospective analysis of the first 150 cases in the UK. J Neurol Neurosurg Psychiatry. 2011;82(6):646–51.
15. Hill AF, Desbruslais M, Joiner S, Sidle KC, Gowland I, Collinge J, et al. The same prion strain causes vCJD and BSE. Nature. 1997;389:448–50. 526.
16. Kahana E, Zilber N. Do Creutzfeldt-Jakob disease patients of Jewish Libyan origin have unique clinical features? Neurology. 1991;41:1390–2.
17. Laplanche JL, El Hachimi KH, Durieux I, Thuillet P, Defebvre L, Delasnerie-Lauprêtre N, et al. Prominent psychiatric features and early onset in an inherited prion disease with a new insertional mutation in the prion protein gene. Brain. 1999;122:2375–86.
18. McKee D, Talbot P. Chorea as a presenting feature of variant Creutzfeldt-Jakob disease. Mov Disord. 2003;18:837–8.
19. Mead S, Poulter M, Beck J, Webb T, Campbell T, Linehan J, et al. Inherited prion disease with six octapeptide repeat insertional mutation–molecular analysis of phenotypic heterogeneity. Brain. 2006;129:2297–317.
20. Mead S. Prion disease genetics. Eur J Hum Genet. 2006;14:273–81.
21. Mead S, Poulter M, Uphill J, Beck J, Whitfield J, Webb TE, et al. Genetic risk factors for variant Creutzfeldt-Jakob disease: a genome-wide association study. Lancet Neurol. 2009;8:57–66.
22. Montagna P, Gambetti P, Cortelli P, Lugaresi E. Familial and sporadic fatal insomnia. Lancet Neurol. 2003;2:167–76.
23. Moore RC, Xiang FQ, Monaghan J, Han D, Zhang ZP, Edström L, et al. Huntington disease phenocopy is a familial prion disease. Am J Hum Genet. 2001;69:1385–8.
24. Palmer MS, Dryden AJ, Hughes JT, Collinge J. Homozygous prion protein genotype predisposes to sporadic Creutzfeldt-Jakob disease. Nature. 1991;352:340–2.
25. Pocchiari M, Puopolo M, Croes EA, Budka H, Gelpi E, Collins S, et al. Predictors of survival in sporadic Creutzfeldt-Jakob disease and other human transmissible spongiform encephalopathies. Brain. 2004;127:2348–59.
26. Prusiner SB. Novel proteinaceous infectious particles cause scrapie. Science. 1982;216: 136–44.

27. Spencer MD, Knight RS, Will RG. First hundred cases of variant Creutzfeldt-Jakob disease: retrospective case note review of early psychiatric and neurological features. BMJ. 2002; 324:1479–82.

28. Webb TE, Poulter M, Beck J, Uphill J, Adamson G, Campbell T, et al. Phenotypic heterogeneity and genetic modification of P102L inherited prion disease in an international series. Brain. 2008;131:2632–46.

29. Weller M, Aguzzi A. PRION DISEASES Movement disorders reveal Creutzfeldt-Jakob disease. Nat Rev Neurol. 2009;5:185–6.

30. Will RG, Ironside JW, Zeidler M, Cousens SN, Estibeiro K, Alperovitch A, et al. A new variant of Creutzfeldt-Jakob disease in the UK. Lancet. 1996;347:921–5.

31. Zerr I, Kallenberg K, Summers DM, Romero C, Taratuto A, Heinemann U, et al. Updated clinical diagnostic criteria for sporadic Creutzfeldt-Jakob disease. Brain. 2009;132:2659–68.

Chapter 14
Inherited Metabolic Disorders Causing Chorea

Mônica Santoro Haddad

Abstract This chapter will review clinical features and diagnostic aspects of patients with inherited metabolic diseases (IMDs) presenting with chorea, among various and complex movement disorders, that can be seen in IMDs. Most of these conditions are rare and inherited and many of them manifest in childhood and adolescence and, less frequently, in adults .The term IMD can be used broadly to refer to almost any cellular process, including purine or pyrimidine metabolism, steroid metabolism, lysosomal or peroxisomal dysfunction or storage deficits, porphyrin metabolism, neurotransmitter diseases, mineral accumulation diseases and others. Disorders affecting amino acids, carbohydrates or lipids will be emphasized. The main aspects regarding clinical approach to identify the underlying IMDs will be revised. The available genetic data will be shown, besides pathophysiological and treatment aspects of the more important IMDs.

Keywords Inherited metabolic disorder • Chorea • Amino acid • Carbohydrate • Lipid

Introduction

There are a number of metabolic diseases that can result in movement disorders. Most of these conditions are rare and inherited and many of them begin in childhood and adolescence although some occur in adults [1].

This chapter will review the clinical features and diagnostic aspects for patients with inherited metabolic diseases (IMDs) presenting with chorea. Chorea can also

M.S. Haddad
Department of Neurology, Hospital das Clinicas da Faculdade
de Medicina da Universidade São Paulo, São Paulo, SP, Brazil
e-mail: mshaddad@uol.com.br

F.E. Micheli, P.A. LeWitt (eds.), *Chorea*,
DOI 10.1007/978-1-4471-6455-5_14, © Springer-Verlag London 2014

be associated with other movement disorders and terms such as choreoathetosis, athetosis, and ballismus are occasionally used. From a practical point of view, patients with IMDs frequently present chorea together with other symptoms like slower movements (athetosis) or more proximal and abrupt movements (ballismus). In addition, choreatic movements may be combined with myoclonus, ataxia, or parkinsonism in the same individual [2].

The term IMD can be used broadly to refer to almost any cellular process, including purine or pyrimidine metabolism, steroid metabolism, lysosomal or peroxisomal dysfunction or storage deficits, porphyrin metabolism, neurotransmitter diseases, and mineral accumulation diseases, among others [1, 3]. This chapter will describe disorders affecting amino acids, carbohydrates, or lipids.

General Clinical Aspects

Various complex movement disorders may be present in IMDs. The appropriate identification and classification is crucial for the diagnosis of the underlying IMD. The movement disorder may be the only apparent symptom, but much more frequently other neurological and systemic symptoms are present [4]. The same IMDs can appear with different movement disorders and the same movement disorder can be present in different IMDs. Dystonia seems to be more frequent than chorea in these patients and it is usually associated with parkinsonian and/or pyramidal signs. These symptoms may vary according to the age of onset and the duration of the disease [4]. We could consider Lesch-Nyhan disease, GM1 gangliosidosis type 3, and glutaricaciduria type 1 as examples. Young patients tend to suffer from axial hypotonia and hyperkinetic limbs, whereas older patients predominantly have rigid-akinetic parkinsonism and dystonia [1, 4]. The importance of age of onset will be discussed below.

Although chorea is unusual chorea is unusual in IMDs, it is crucial to make the correct diagnosis of the underlying IMD because some of them are treatable [3]. A rational approach is recommended. Instead of searching for all IMDs in each patient, it is more practical to search for diagnostic clues through clinical and neurological examinations. These few findings are more relevant to the etiological diagnosis and thay can be classified by the manner of onset (abrupt or insidious), heritability, associated clinical features, and abnormalities seen in brain magnetic resonance images (MRI). These suggest the need to consider metabolic disorders in the differential diagnosis of chorea etiology [3, 5].

Clinical Onset

The form of presentation is important for chorea and other movement disorders. They can be defined as acute, paroxysmal, and/or in chronic progressive forms. Usually, IMDs do not only cause chorea. The vast majority of patients are children that suffer from several movement disorders with some degree of encephalopathy among other non-neurological signs [1, 6]. Some of these patients start with a state

Table 14.1 Systemic manifestation of IMDs

Short stature
Weight loss
Delayed puberty
Multiendocrine problems
Immunological deficiency
Bone abnormalities (face/trunk/limbs)
Anemia/thrombocytopenia
Abnormal hair growth/pigmentation
Hyperventilation
Hepato/splenomegaly
Recurrent vomiting

of emergency showing prominent hyperkinesias with abrupt onset and encephalopathy in a life-threatening situation. IMDs should be suspected in critical and acute situations, for example, if lactic acidosis is identified. Also, a genetic problem may appear after a significant stress like febrile status, trauma, or an invasive medical procedure in children with IMDs.

The acute onset of a movement disorder can be seen in other non-genetic etiologies, but we should always consider inherited diseases like Wilson's disease, biotin-responsive basal ganglia disease, pyruvate dehydrogenase deficiency, and glutaric aciduria type 1 [3, 7, 8]. When the movement disorder is paroxystic, monoamine metabolism disorders, GLUT-1 deficiency, and pyruvate dehydrogenase deficiency may be the cause. Some of these will be described in more detail below. Chorea may also present insidiously in a patient with developmental delays who demonstrates a loss of skills or perhaps has some systemic problems or stigmas (see below). The appearance of a movement disorder in these circumstances is strongly reminiscent of an IMD [3].

Heritability

Although there are some autosomal dominant and X-linked disorders in this group, IMDs are usually autosomal recessive disorders with crucial enzyme mutations. Therefore, the absence of other affected members in the family does not exclude an IMD. The childhood death of an older sibling, parent, or relative should be looked for in a thorough investigation of the family tree [1]. However, it should be stressed that inheritance does not necessarily mean the IMD is congenital. Many IMDs manifest in childhood, in adolescence and into adulthood.

Associated Clinical Features

Table 14.1 summarizes some important systemic features that can help diagnosing IMDs. There are no major systemic problems that could be caused by IMDs. Usually, these manifestations are considered stigmas (like short stature) and they may be present in other family members [1, 3].

Table 14.2 IMDs causing chorea by age of onset

Infancy, childhood, and adolescence
Lesh-Nyhan syndrome
Wilson's disease
Hemoglobinopathies
Leigh disease
Amino acid disorders:
Glutaricacidemia/aciduriatype I, cystinuria, homocystinuria, phenylketonuria, argininosuccinicacidemia, propionicacidemia, etc.
Carbohydrate disorders:
Mucopolysaccharidoses,mucolipidoses,galactosemia,pyruvatedehydrogenase deficiency, GLUT-1 deficiency syndrome, etc.
Lipid disorders: sphingolipidosis, metachromatic leukodystrophy GM1 and GM2 gangliosidosis, ceroidlipofuscinosis, etc.
Niemann-Pick type C
Adolescence and adulthood
Wilson's disease
Familial calcification of basal ganglia
Late onset GM1 gangliosidose (type 3)
Glutaricaciduria
Niemann-Pick type C

Clinical and Therapeutic Aspects of IMDs Presenting with Chorea

As said above, chorea is a rare manifestation in IMDs. In this section, we will describe those IMDs that do, rarely, present with chorea. The most relevant diseases are dependent on the age of onset, shown in Table 14.2. Some diseases that present with chorea are inherited metabolic disorders like neurodegeneration with brain iron accumulation, neuroferritinopathies, Wilson's disease, and aceruloplasminemia, which are described in detail in other chapters. The main IMDs that we will describe below are found in Table 14.3 with inheritance and genetics factors.

Many IMDs appear in the first 24 months of life usually after the neonatal period [1]. General neurological symptoms include hypotonia, developmental delays, or regressions. Those that appear before the 9th month do not usually manifest chorea. When movement disorders are present, they are generally mixed including ataxia, dystonia, and choreoathetosis. Between 9 and 24 months, chorea is more frequently observed [2].

The clinical features of glutaric aciduria type 1 can be highly variable as in other IMDs. Phenotypic heterogeneity occurs within families. Frequently, it manifests with generalized dystonia but chorea is sometimes present. In addition, some features of encephalopathy may appear in critical moments. These patients may also have episodic ketoacidosis. In early infancy, the presentation is often catastrophic with macrocephaly at or shortly after birth, followed by a rapid deterioration of neurological conditions with hypotonia, irritability, and feeding difficulties combined with subdural bleeding after mild head trauma. Severe episodes of status

Table 14.3 Inheritance and genetic mutations of IMDs

Disease	Inheritance	Genetic mutation
Pyruvate-dehydrogenase deficiency	X-linked	E1-Alpha subunit of pyruvate dehydrogenase
Glutaricaciduria type 1	AR	Glutaryl-CoA dehydrogenase
3 Methylglutaconicaciduria	AR	AUH gene
17-beta-hydroxysteroid dehydrogenase X deficiency	X-linked dominant	17-beta-hydroxysteroid dehydrogenase gene
Propionicaciduria	AR	Propionic-CoA carboxylase deficiency
Homocystinuria	AR	Multiple mutations
Leigh syndrome	X-linked recessive	E1-alpha complex gene; others
Lesh-Nyhan syndrome	X-linked recessive	HPRT1 gene
Metachromatic leukodystrophy	AR	Arylsulfatase A gene
Sulfocysteinuria	AR	Sulfite oxidase gene
Niemann-Pick type C	AR	NPC1, NPC2
GM1 Gangliosidosis	AR	GLB1
Creatine synthesis deficiency (Guanidinoacetatemethyl transferase)	AR	GAMT
Methemoglobinemia type II	AR	CYB5R3

AR autossomic recessive

dystonicus can be accompanied by rabdomyolysis and hyperthermia. Occasionally, the manifestation may be more serious in later childhood or even in adults with an appearance of leukoencephalopathy [9, 10, 11]. MRI shows frontotemporal cortex atrophy, dilatation of Sylvian fissures and striking striatal degeneration with hypodensities in the lenticular nuclei. The physiopathology of the disease is unclear, but it is probably a mutation in the gene that encodes for the glutaryl-coenzyme. A dehydrogenase may cause glutamate-mediated excitotoxicity in the striatum. Treatment is only supportive. A reduction of lysine, tryptophan, and hydroxylysine (glutarigenic amino acids) have been used along with riboflavin and l-carnitine supplementation, but this has to be done before the occurrence of the striatal injury [3, 4].

Chorea is also rarely seen in other aminoacidopathies including 3-methylglutaconicaciduria, D-2-hydroxyglutaric academia, succinate-semialdehyde dehydrogenase deficiency, homocystinuria, cystinuria, arginosuccinic acidemia, Hartnup disease, and others. These children usually have developmental delays, hypotonia, epileptic manifestations, and different stigmas. Diagnosing the correct underlying disease is very important because some of them are treatable and it is important to make the correct genetic counseling [2, 3].

Creatine metabolism disorders, or brain creatine deficiency syndromes, include guanidine acetate methyltranferase deficiency and arginine glycine amidinotransferase deficiency (both autosomal recessive diseases) and an X-linked creatine transporter defect. Affected children have developmental delays or regression, severe language and behavioral disturbances, and dyskinetic movements. Seizures are prominent in some. For children with genetic recessive forms, treatment with creatine supplementation is helpful [12, 13, 14].

Among the purine metabolism disorders, we would like to call attention to Lesch-Nyhan disease (LND) which is an X-linked recessive disorder associated with multiple heterogeneous mutations in the gene for the enzyme hypoxanthine-guanine phosphoribosyltransferase (HPRT), that is deficient. [15]. The same mutation can lead to different phenotypes and the underlying disorder appears to be on dopaminergic pathways. The most important symptoms are the presence of hyperuricemia, which if not treated leads to renal stones and gout, self-injurious behaviors, and a constellation of other neurological problems. These symptoms can occur together or alone, which makes diagnosis difficult. Self-injurious behaviors are a hallmark of LND, but they can occur in other disorders like chorea acanthosytosis. Typically, these behaviors begin at the age of 2 or 3 years, but some individuals can be asymptomatic until late in their teenage years. There are signs of choreoathetosis, mental retardation, and spasticity in a minority of cases but the most striking symptom is severe dystonia present in all patients with axial hypotonia, which occurs in the first year of life. Chorea can be seen in half the patients while 25 % also have ballismus. Ocular motility is abnormal. Diagnosis is made upon clinical suspicion and elevated levels of uric acid in urine, but this feature may be absent in some patients requiring an enzymatic activity assay and molecular genetic testing for confirmation. Treatment is aimed at reducing hyperuricemia with hydration and allopurinol, and preventing the self-injurious behavior. Movement disorders are very difficult to treat in these patients although some of them improve on levodopa [16, 17, 18, 19]. Bilateral high-frequency stimulation of the internal pallidum with two targets (sensorimotor and limbic) was shown to be effective in reducing dystonia and self-injury [20].

Mitochondrial diseases are the most common metabolic disorders but are also very difficult to diagnose [3]. Multi-systemic involvement and increased lactate levels in serum and CSF are good clues since the identification of a precise genetic cause is rare. Movement disorders are often seen because striatal neurons are highly sensitive to energy failure due to respiratory chain defects [21, 22, 23]. The Leigh syndrome (subacute necrotizing encephalomyelopathy) is a complex of genetic disorders related to mitochondrial function, which is also usually autosomal recessive. Clinically, it appears as a progressive encephalopathy before the age of 2 (in 80 % of cases) with hypotonia and psychomotor delay. These patients have unexplained hyperventilation and other respiratory problems like oculomotor palsies, nystagmus, optic atrophy, and ataxia. Movement disorders including chorea, athetosis, tremor, dystonia (more common as generalized or multifocal), and myoclonus can be prominent in some cases even in the initial manifestation. Some patients show the movement disorder only during lactic acidosis episodes. Pathology findings include symmetric necrotic lesions with demyelination and gliosis affecting the basal ganglia, diencephalon, cerebellum, and brainstem. MRI images show alterations in symmetric areas with an increased signal in the putamen and, occasionally, in other structures. Lactate levels are elevated in arterial blood and cerebrospinal fluid. Unfortunately, diverse therapies, including vitamins, coenzyme Q10, and dichloroacetate have failed to produce clinical improvement. Treatment is symptomatic and supportive, and patients usually die within 5 years of disease onset [24, 25, 26, 27].

A mitochondrial disorder should also be suspected in a young patient with hyperglycemic chorea-ballism. A diabetic teenager with chorea and a mutation of polymerase gamma I (POLG1) has been described [28].

The pyruvate dehydrogenase complex deficiency is an X-linked recessive disease that results in an early-onset encephalopathy. The newborns have severe lactic acidosis and, subsequently, develop significant neurologic deficits such as those seen in Leigh syndrome. Later onset forms usually appears as paroxysmal neurologic episodes commonly after exercise or fever. The children develop mental retardation, pyramidal signs, and intermittent ataxia or dystonia had also been described. These may happen very frequently, from minutes to 1 h, affecting the tongue, limbs, and face. Chorea has also been described in adult onset affecting the face, tongue, and limbs [29, 30]. These patients can also present stereotypical movement disorders (tapping fingers or feet). Images of the brain can show a bilateral lesion of the putamen. Diagnosis is confirmed by pyruvate dehydrogenase activity assay, and a disproportionate increase in CSF lactate is another clue. Some cases of episodic ataxia respond to thiamine supplementation and a ketogenic diet [31, 32].

Another rare energy metabolism disorder that appears with choreic movements is glucose transport defect: GLUT-1 deficiency syndrome, which mediates glucose transportation across the blood-brain barrier. This is an autosomal dominant disorder (gene SLC2A1), and three main phenotypes have been described. The classic form features difficult to control seizures, microcephaly, mental retardation, ataxia and spasticity (which improves with carbohydrate intake), an ataxic and dystonic syndrome with or without seizures, and paroxysmal exercise induced dyskinesia (PED) with seizures. The PED can begin between 3 and 30 years of age, and the movements are choreoathetotic or dystonic and always induced by exercise [33, 34, 35, 36]. These movements most commonly affect the legs, which leads to walking difficulties and/or falls, which usually resolves in up to 15 min, but it may take several hours. Food and rest improve the symptoms that usually tend to become less severe with aging. Diagnosis is made from a low CSF glucose level after 12 h of fasting with a low CSF/serum ratio. The seizures and the paroxysmal movement disorders are treated with a ketogenic diet [37, 38, 39].

Although extremely rare, biotin-responsive basal ganglia disease is important because it is treatable. All cases begin in infancy or adolescence. After a febrile illness, or episodes of vomiting or diarrhea, there is a sub-acute onset of encephalitic manifestations of unknown origin [3]. If not treated, the patient persists with encephalopathy, loss of developmental milestones and speech, inability to swallow, and tetraparesis. Severe extrapyramidal and pyramidal signs appear, with non-fluctuating dystonia and choreoathetotic movements as well as severe rigidity at rest, which leads to an opisthotonic posture. Some patients only develop intermittent limb dystonia. MRI shows hallmark features in all cases with central destruction of the caudate heads and partial or complete loss of the putamen. The laboratory investigation is normal. A biotin transporter deficiency is responsible and the genetic defect is in the SLC19A3 gene [40].

Niemann-Pick disease encompasses a variety of lysosomal lipid storage diseases. The type C (NPC) usually begins between 2 and 10 years of age and has normal levels of sphingomyelinase. The lysosomal accumulation is of unesterified cholesterol and glycosphingolipids. There is abnormal lipid storage in the brain causing a ballooned appearance. The vast majority of the mutations of the NPC gene are on chromosome 18q11-12 [41, 42]. Movement disorders are present in almost 60 % of the cases, typically with progressive generalized dystonia. Chorea and parkinsonism are usually mild and can appear alone or along with dystonic movements. Rarely, movement disorders are the core or predominant symptoms and are typically associated with cognitive, behavioral, and cerebellar dysfunctions. Juvenile onset is from between the ages of 10 and 30 years. Supranuclear vertical gaze palsy and hepatosplenomegaly are also key diagnostic features. Epilepsy, cataplexy, and deafness can also be found while peripheral neuropathy is rare. Death occurs usually after a decade from onset because of dysphagia and its complications. Brain MRI shows normal or mild periventricular white matter hyperintensity and cortical, cerebellar, and brainstem atrophy. Cultured fibroblasts can be tested for abnormal cholesterol esterification. Unfortunately, there is no specific treatment available [43, 44, 45, 46].

Other neuronal storage disorders include GM1 and GM2 gangliosidosis and Gaucher disease, among others. These diseases usually appear with ataxia, gait disorders, dystonia, and parkinsonism. In adults, GM1 gangliosidosis is also associated without mental deterioration and a prolonged survival. Eventually, athetoid movements as well as other clinical and neurological exuberant signs have also been described [47, 48, 49, 50].

Calcification of the basal ganglia can occur in many sporadic and familial diseases and is often associated with several kinds of movement disorders, including parkinsonism, dystonia, and chorea in 20 % of all cases. Patients can also have seizures, cognitive and behavioral disturbances. The term Fahr disease is non-specific and describes numerous heterogeneous diseases including those with calcium and mitochondrial metabolic alterations [4]. Familial cases of idiopathic basal ganglia calcification are described and inheritance is autosomal dominant. Recently, various mutations in the SLC20A2 have been described as a major cause of familial idiopathic basal ganglia calcification. This gene is implied in phosphate homeostasis and this can open therapeutic possibilities in the next few years. At the present time, the therapy of basal ganglia calcification is merely symptomatic but when the metabolic disturbance is known (e.g., hypoparathyroidism) it can be treated, and better results should be expected [51].

Conclusions

Movement disorders, including chorea, can be caused by IMDs. Although rare, this group of heterogeneous diseases can cause chorea especially in the pediatric population. When a child has chorea and symptoms are not immune-mediated, IMDs should be considered even if family history is negative. The vast majority has symptoms other than movement disorders and the metabolic cause or defect has to be treated specifically when feasible. The symptomatic treatment of involuntary movement in patients with IMDs is the same as those in other clinical settings and it can helps improve the quality of life of these patients.

References

1. Gilbert DL. Inherited metabolic diseases causing chorea in childhood. In: Walker RH, editor. The differential diagnosis of chorea. Oxford: Oxford University Press; 2011. p. 231–61.
2. Singer HS, Mink JW, Gilbert DL, Jankovik J. Chapter 15: Inherited metabolic disorders associated with extrapyramidal symptoms. In: Singer HS, Mink JW, Gilbert DL, Jankovik J, editors. Movement disorders in childhood. 1st ed. Philadelphia: Saunders; 2010. p. 164–204.
3. Grabli D, Auré K, Vidailhet M, Roze E. Chapter 23: Movement disorders in neurometabolic diseases. In: Gálvez-Jiménez N, Tuite PJ, editors. Uncommon causes of movement disorders. 1st ed. Cambridge: Cambridge University Press; 2011. p. 245–57.
4. Walker RH. Other choreas. In: Schapira AHV, Lang AET, Fahn S, editors. Movement disorders, vol. 4. Philadelphia: Saunders; 2010. p. 558–86.
5. Engbers HM, Berger R, van Hasselt P, de Koning T, de Sain-van der Velden Mg, Kroes HY, et al. Yield of additional metabolic studies in neurodevelopmental disorders. Ann Neurol. 2008;64(2):212–7.
6. McDonald L, Rennie A, Tolmie J, Galloway P, Mc WR. Investigation of global developmental delay. Arch Dis Child. 2006;91(8):701–5.
7. Jinnah HA, Visser JE, Harris JC, Verdu A, Larovere L, Ceballos-Picot I, et al. Delineation of the motor disorder of Lesch-Nyhan disease. Brain. 2006;129(Pt5):1201–17.
8. Roze E, Paschke E, Lopez N, Eck T, Yoshida K, Maurel-Ollivier A, et al. Dystonia and parkinsonism in GM1 type 3 gangliosidosis. Mov Disord. 2005;20(10):1366–9.
9. Gitiaux C, Roze E, Kinugawa K, Flamand-Rouviere C, Boddaert N, Apartis E, et al. Spectrum of movement disorders associated with glutaric aciduria type 1: a study of 16 patients. Mov Disord. 2008;23(16):2392–7.
10. Hedlund GL, Longo N, Pasquali M. Glutaric acidemia type 1. Am J Med Genet C Semin Med Genet. 2006;142C(2):86–94.
11. Strauss K, Puffenberger E, Robinson DL, Morton DH. Type I glutaric aciduria, part 1: natural history of 77 patients. AM J Med Genet C Semin Med Genet. 2003;121(1):38–52.
12. Stöckler S, Holzbach U, Hanefeld F, Marquardt I, Helms G, Requart M, et al. Creatine deficiency in the brain: a new treatable inborn error of metabolism. Pediatr Res. 1994;36:409–13.
13. Stockler S, Schutz PW, Salomons GS. Cerebral creatine deficiency syndromes: clinical aspects, treatment and pathophysiology. Subcell Biochem. 2007;46:149–66.
14. DeGrauw TJ, Cecil KM, Byars AW, Salomons GS, Ball WS, Jakobs C. The clinical syndrome of creatine transporter deficiency. Mol Cell Biochem. 2003;244:45–8.
15. Wl N. Disorders of purine and pyrimidine metabolism. Mol Genet Metab. 2005;86:25–33.
16. Jinnah HA, Harris JC, Reich SG, et al. The motor disorder of Lesch –Nyhan disease. Mov Disord. 1998;13:98.
17. Lloyd KG, Hornykiewicz O, Davidson L, Shannak K, Farley I, Goldstein M, et al. Biochemical evidence of dysfunction of brain neurotransmitters in the Lesch-Nyhan syndrome. N Engl J Med. 1981;305:1106–11.
18. Serrano M, Perez-Duenas B, Ormazabal A, Artuch R, Campistol J, Torres RJ, et al. Levodopa therapy in Lesch-Nyhan disease patient: pathological, biochemical, neuroimaging, and therapeutic remarks. Mov Disord. 2008;23:1297–300.
19. Watts RW, Spellacy E, Gibbs DA, Allsop J, McKeran RO, Slavin GE. Clinical, postmortem, biochemical and therapeutic observation on the Lesch-Nyhan syndrome with particular reference to the neurological manifestations. Q J Med. 1982;51:43–78.
20. Cif L, Biolsi B, Gavarini S, Saux A, Robles SG, Tancu C, et al. Antero-ventral internal pallidum stimulation improves behavioral disorders in Lesch-Nyhan disease. Mov Disord. 2007;22(14):2126–9.
21. Lyon G, Kolodny EH, Pastores GM. Neurology of hereditary metabolic diseases of children. 3rd ed. New York: McGraw-Hill; 2006.
22. Debray FG, Lambert M, Chevalier I, Robitaille Y, Decarie JC, Shoubridge EA, et al. Long-term outcome and clinical spectrum of 73 pediatric patients with mitochondrial diseases. Pediatrics. 2007;119:722–33.

23. Wallace DC, Murdock DG. Mitochondria and dystonia: the movement disorder connection. Proc Natl Acad Sci U S A. 1999;96:1817–9.
24. Macaya A, Munell F, Burke RE, De Vivo DC. Disorders of movement in Leigh syndrome. Neuropediatrics. 1993;24(2):60–7.
25. DiMauro S, De Vivo DC. Genetic heterogeneity in Leigh syndrome. Ann Neurol. 1996;40:5–7.
26. De Vivo DC. Leigh syndrome: historical perspective and clinical variations. Biofactors. 1998;7:269–71.
27. Finstere J. Leigh and Leigh-like syndrome in children and adults. Pediatr Neurol. 2008;39:223–35.
28. Hopkins SE, Sonmoza A, Gilbert DL. Rare autosomal dominant POLG1 mutation in a family with metabolic strokes, posterior column spinal degeneration, and multiendrocrine disease. J Child Neurol. 2010;25(6):752–6.
29. Dahl HH, Maragos C, Brown RM, Hansen LL, Brown GK. Pyruvate dehydrogenase deficiency caused by deletion of a 7-bp repeat sequence in the El alpha gene. Am J Hum Genet. 1990;47(2):286–93.
30. Mellick G, Price L, Boyle R. Late-onset presentation of Pyruvate dehydrogenase deficiency. Mov Disord. 2004;19:727–9.
31. Debray FG, Lambert M, GAgne R, Maranda B, Laframboise R, MacKay N, et al. Pyruvate dehydrogenase deficiency presenting as intermittent isolated acute ataxia. Neuropediatrics. 2008;39:20–3.
32. Bindoff LA, Birch-Machin MA, Farnsworth L, Gardner-Medwin D, Lindsay JG, Turnbull DM. Familial intermittent ataxia due to a defect of the E1 component of pyruvate dehydrogenase complex. J Neurol Sci. 1989;93(2–3):311–8.
33. Brockmann K. The expanding phenotype of GLUT1- deficiency syndrome. Brain Dev. 2009;31(7):545–52.
34. Schneider SA, Paisan-Ruiz C, Garcia-Gorostiaga I, Quinn NP, Weber YG, Lerche H, et al. GLUT1 gene mutations cause sporadic paroxysmal exercise-induced dyskinesias. Mov Disord. 2009;24(11):1684–8.
35. Suls A, Dedeken P, Goffin K, Van Esch H, Dupont P, Cassiman D, et al. Paroxysmal exercise-induced dyskinesia and epilepsy is due to mutations in SLC2A1, encoding the glucose transporter GLUT1. Brain. 2008;131(Pt7):1831–44.
36. Zorzi G, Castellotti B, Zibordi F, Gellera C, Nardocci N. Paroxysmal movement disorders in GLUT1 deficiency syndrome. Neurology. 2008;71:146–8.
37. Wang D, Pascual JM, Yang H, Engelstad K, Jhung S, Sun RP, et al. Glut-1 deficiency syndrome: clinical, genetic, and therapeutic aspects. Ann Neurol. 2005;57:111–8.
38. Klepper J, Leiendecker B. Glut1 deficiency syndrome-2007 update. Dev Med Child Neurol. 2007;49:707–16.
39. Frieldman JR, Thiele EA, Wang D, Levine KB, Cloherty EK, Pfeifer HH, et al. Atypical GLUT1 deficiency with prominent movement disorder responsive to ketogenic diet. Mov Disord. 2006;21:241–5.
40. Ozand PT, Gascom GG, Al Essa M, Joshi S, Al Jishi E, Bakheet S, et al. Biotin-responsive basal ganglia disease: a novel entity. Brain. 1998;121(Pt7):1267–79.
41. Sevin M, Lesca G, Baumann N, Millat G, Lyon-Caen O, Vanier MT, et al. The adult form of Niemann-Pick/diseases type C. Brain. 2007;130(Pt1):120–33.
42. Scheretlen DJ, War J, Meyer SM, Yun J, Puig JG, Nyhan WL, et al. Behavioral aspects of Lesch-Nyhan disease and its variants. Dev Med Child Neurol. 2005;47(10):673–7.
43. Ory DS. Niemann-Pick type C. A disorder of cellular cholesterol trafficking. Biochim Biophys Acta. 2000;1529:331–9.
44. Fink JK, Filling-Karz MR, Sokol J, Cogan DG, Pikus A, Sonies B, et al. Clinical spectrum of Niemann-Pick disease type C. Neurology. 1989;39:1040–9.
45. Uc EY, Wenger DA, Jankovic J. Niemann-Pick disease type C. Two cases and an update. Mov Disord. 2000;15:1199–203.
46. Carstea ED, Morris Jam Coleman KG, Loftus SK, Zhang D, Cummings C, et al. Niemann-Pick C1 disease gene: homology to mediators of cholesterol homeostasis. Science. 1997;277(5323): 228–31.

47. Brunetti-Pierri N, Scaglia F. GM1 gangliosidosis: review of clinical, molecular, and therapeutic aspects. Mol Genet Metab. 2008;94:391–6.
48. Yoshida K, Oshima A, Sakuraba H, Nakano T, Yanagisawa N, Inui K, et al. GM1 gangliosidosis in adults: clinical and molecular analysis of 16 Japanese patients. Ann Neurol. 1992;31:328–32.
49. Campdelacreu J, Munoz E, Gomez B, Pujol T, Chabás A, Tolosa E. Generalised dystonia with na abnormal magnetic resonance imaging signal in the basal ganglia: a case of adult-onset GM1 gangliosidosis. Mov Disord. 2002;17:1095–7.
50. Oates CE, Bosch EP, Hart MN. Movement disorders associated with chronic GM2 gangliosidosis. Case report and review of the literature. Eur Neurol. 1986;25(2):154–9.
51. Hsu SC, Sears RL, Lemos RR, Quintáns B, Huang A, Spiteri E, et al. Mutations in SLC20A2 are a major cause of familial idiopathic basal ganglia calcification. Neurogenetics. 2013;14(1):11–22.

Chapter 15
Drug-Induced Chorea

Federico E. Micheli

Abstract A large number of drugs have been involved in the induction of a wide variety of movement disorders including parkinsonism, akathisia, tics, myoclonus, tremor, dystonia, and choreic movements. Onset can be acute, subacute, or insidious. Involuntary movements may present in isolation or as a part of a more generalized neurological or systemic condition.

Drug-induced choreas occur mainly in two circumstances, mostly in psychotic patients treated with typical neuroleptics and in parkinsonian patients in the long-term treatment with levodopa. In both cases, the diagnosis is obvious and the current therapeutic strategies are outlined in this chapter. The relationship between the drug intake and the appearance of the movement disorder is not so clear in cases of tardive dyskinesia, especially when the patient is not psychotic and he or she receives dopamine blockers for other circumstances. In such instances, the diagnosis is more troublesome and a careful search for drug intake should be done.

In this chapter, the main causes of drug-induced choreas are analyzed as well as their possible therapeutic approaches.

Keywords Drug-induced chorea • Tardive dyskinesia • Levodopa-induced dyskinesias • Dopamine • Antipsychotic drugs • Neuroleptics

F.E. Micheli, MD, PhD
Parkinson's Disease and Movement Disorders Program,
Neurology Department, University of Buenos Aires,
Ciudad Autónoma de Buenos Aires, Buenos Aires, Argentina

Neurology Department, Hospital de Clínicas José de San Martín,
University of Buenos Aires, Ciudad Autónoma de Buenos Aires,
Buenos Aires, Argentina
e-mail: fmicheli@fibertel.com.ar

F.E. Micheli, P.A. LeWitt (eds.), *Chorea*,
DOI 10.1007/978-1-4471-6455-5_15, © Springer-Verlag London 2014

Introduction

Drug-induced movement disorders (DIMD) comprise a heterogeneous group of movements that have been widely acknowledged as derived from diverse drugs. Curiously, in most reviews on the subject, dyskinesias as a side effect of levodopa (LID) have been neglected as a DIMD. As we find no reason to do so and considering its frequency, they will be discussed in this chapter.

A large number of drugs have been involved in the induction of a wide variety of movement disorders including parkinsonism, akathisia, tics, myoclonus, tremor, dystonia, and choreic movements. In the wide majority of cases, there is a temporal relationship between the drug intake and the appearance of the movements, but in some, a chronic exposure to the drug is the only sign that leads to the diagnosis. In such cases, even discontinuing the drug might not be enough to stop the movement disorder, as some of them are persistent or even permanent. In these cases, it is the phenomenology of the movement and/or the associated symptoms that could be the key to the diagnosis.

In this context, diagnosis also relies on a thorough medical history, as movements are mostly, but not exclusively, caused by central dopamine-receptor-blocking agents. Other drugs liable to induce movement disorders are anticonvulsants, antidepressants, anticholinergics, drugs of abuse, and those also liable to induce an imbalance in serotonin, noradrenalin, and cholinergic neurotransmission.

The time frame in which these movement disorders develop is also unpredictable as some of them can appear acutely within hours after the drug intake (i.e., dystonic reactions), while others do so after months of exposure to the offending drug.

Chorea is an involuntary movement disorder that can be derived from diverse etiologies including infectious, degenerative, metabolic, immunological, paraneoplastic, autoimmune [1], vascular [2], and drug induced (DIC) as well as idiopathic in some cases.

Despite increasing awareness of the prevalence estimates of Huntington's disease (HD) in different populations [2, 3], DIC is at present one of the most common causes of sporadic chorea [4]. While sporadic unilateral or focal manifestations are secondary to basal ganglia structural lesions, drug-induced cases are by and large bilateral.

Diverse drugs can trigger choreic movements, particularly in vulnerable patients, including levodopa in those with Parkinson's disease (PD) or dopamine-receptor antagonists in both psychotic and nonpsychotic patients.

In the long term, most PD patients develop motor fluctuations and dyskinesias mostly featuring dystonia and choreic movements. In these patients, chorea is a consequence of both the striatal denervation caused by the disease and oral pulsatile dopaminergic replacement therapy.

Psychotic patients can also develop a variety of movement disorders as an acute reaction to neuroleptic intake or after long-term use of these agents and represent the most common cause of choreic movements in non-parkinsonian cases (Table 15.1). Choreic movements in such cases are known as tardive dyskinesias (TD) and usually involve the oromandibular area.

Table 15.1 Drugs that can cause chorea

Drug categories	Drug descriptions
Typical antipsychotics	Haloperidol, droperidol
	Chlorpromazine, fluphenazine, trifluoperazine promethazine
	Chlorprothixene, flupenthixol, zuclopenthixol, clotiapine
Atypical antipsychotics	Olanzapine, risperidone, clozapine, aripiprazole, quetiapine, ziprasidone, paliperidone
Other dopamine-receptor blockers	Metoclopramide, sulpiride, tiapride
CNS stimulants	Amphetamine, cocaine, methylphenidate, pemoline
Catecholamine-depleting agents	Tetrabenazine
Antiparkinsonian agents	Levodopa, dopamine agonists, amantadine, anticholinergics
Calcium antagonists	Cinnarizine, flunarizine, nifedipine, nimodipine, verapamil, diltiazem, nitrendipine
Benzodiazepines and related drugs	Diazepam, clorazepate, flunitrazepam, triazolam, zopiclone
Antihypertensive medications	Reserpine and alpha-methyldopa
Anti-inflammatories	Sulfasalazine
Antidepressants	Amitriptyline, imipramine, clomipramine, desipramine, paroxetine, sertraline, venlafaxine
Antiemetic drugs	Cyclizine, metoclopramide, thiethylperazine
Antibiotics	Ciprofloxacin, levofloxacin
Antihistamine	Cyproheptadine
Hormones	Levothyroxine, estrogens
Others	Manganese, ethanol, lithium, digoxin
	Baclofen, methadone, theophylline
	Aminophyline

In addition, drugs other than antipsychotics, many of which have central antido-paminergic activity (i.e., metoclopramide), have been held responsible for causing movement disorders.

In this review, we will independently analyze these two groups of disorders.

Classification

DIC are classified according to different features:

Onset

Acute

Chorea occurs shortly after exposure to a drug and reaches the highest intensity rapidly. It can have a rapid evolution and get to such degrees of severity as to require hospitalization.

Acute hemichorea is usually due to an ischemic structural lesion (i.e., vascular) or metabolic disorder (i.e., diabetes). Acute drug reactions are mainly dystonic but can be choreic or have a mixed pattern. They are mostly triggered by antipsychotic drugs (APDs), but both adults and children seem to be particularly vulnerable to amphetamines. In addition, pemoline [5], cocaine, and crack [6] can induce acute chorea. Although a rare complication, cases with laryngospasm and compromise of the airway have also been reported [7].

Protracted/Tardive

The gradual onset of chorea is rarely caused by structural lesions, and HD as well as DIC should be considered in the differential diagnosis. In the latter, involuntary movements appear after months or even years of exposure to a drug and may be persistent or even permanent after drug discontinuance.

Pattern of Body Part Involvement

The body area involved is an important clue to the origin of the choreic movements and even their etiology:

(a) *Generalized chorea* in peak-of-dose dyskinesia/s versus *lower limb* involvement in diphasic dyskinesias, both in PD and caused by levodopa.
(b) *Oromandibular chorea* in TD due to chronic exposure to neuroleptics.

Early Versus Late Onset

Early onset is characteristic of acute reactions derived from neuroleptic drugs. Conversely, late onset is typical of TD.

Relationship with Possible Causative Drugs: Dopaminergic, Antidopaminergic, Others

APDs and other dopamine-blocking agents are the cause of TD, and levodopa/dopamine agonists in PD are the cause of LID. However, several other drugs can induce choreic movements.

Patient's Background

It is also important to take into account the patient's background when considering differential diagnoses as patients with psychotic disorders are liable to receive neuroleptics while patients with PD will receive levodopa or dopamine agonists. If none of these is the case, a thorough search for every possible drug exposure should be carried out.

There is little information about epidemiology on DIC, but a study showed that out of 7,829 patients admitted in two general hospitals in Italy between 1993 and 1996, 23 cases had sporadic choreas, 5 of them had DIC, 6 had vascular chorea, 1 had Sydenham's chorea, and in 4 patients, the etiology could not be determined [8]. However, patients with PD and dyskinesias were not included in this report. It was concluded that chorea is not a rare disorder among neurological department admissions and DIC is one of the most frequent causes.

In this review, we will focus in two different conditions, LID in PD and TD mainly in psychotic patients. Curiously, TD in schizophrenic patients treated with neuroleptics and LID in PD patients share similar clinical features. Both conditions are induced by chronic exposure to drugs that target dopaminergic receptors (antagonists in TD and agonists in LID) and cause pulsatile, nonphysiological stimulation of these receptors.

It has been hypothesized that both conditions might share genetic risk factors because certain genetic variants exert a pleiotropic effect, influencing susceptibility to TD as well as to LID, but this hypothesis has not yet been proven [9].

Levodopa-Induced Dyskinesias

Levodopa remains the most effective agent for improving motor symptoms in PD, but its chronic use is associated with the emergence of motor fluctuations and dyskinesias. Despite a broad range of dosages and administration schedules and after a variable initial period of stable response with levodopa and dopamine agonists (the "levodopa honeymoon period"), most patients with PD develop motor fluctuations and LID (Fig. 15.1a–c). It was initially estimated that after beginning treatment with levodopa, 10 % of patients per year developed these complications and that after 5 years more than 50 % suffer motor fluctuations and several types of LID [10]. The spectrum of the phenomenology of dyskinesias is broad and includes chorea, choreoathetosis, ballism, and dystonia [11].

The clinical pattern of LID varies and has been classified by phenomenology and the relation to the timing of levodopa intake and includes peak-of-dose (also known as high-dose dyskinesia as it will improve with a reduction in the dose of levodopa) and diphasic dyskinesias (known as "low-dose" dyskinesia, in which case, it will probably improve increasing the levodopa dose or adding a dopamine agonist). Dystonia usually takes place in the "OFF" periods and also improves by increasing the dopamine stimulation. The most common motor fluctuations are end-of-dose deterioration ("wearing OFF"), while abrupt, unpredictable shifts between "ON" and "OFF" states ("ON-OFF" phenomena) are also apt to occur (Fig. 15.2).

Peak-of-dose dyskinesias occur when the patient is experiencing a beneficial response of the parkinsonian symptoms during plasma levodopa peaks. They worsen with increases in dopaminergic (mainly levodopa) dose and improve with dose reduction. Peak-of-dose dyskinesias and "on-off" fluctuations are closely interconnected and often require opposite treatment approaches, as increasing dopaminergic

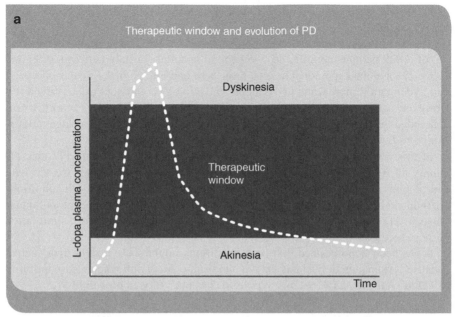

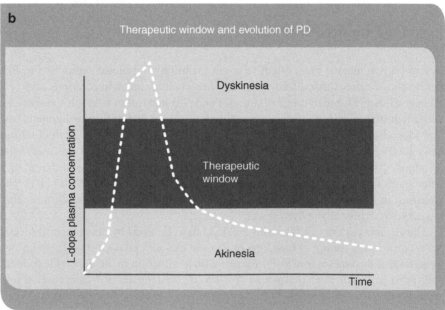

Fig. 15.1 (a) Sequential narrowing of the therapeutic window in patients with PD, showing that initially there is a wide range of levodopa concentrations which will benefit parkinsonian symptoms without causing dyskinesias, but in the long term (**b, c**), the risk of dyskinesias increases markedly

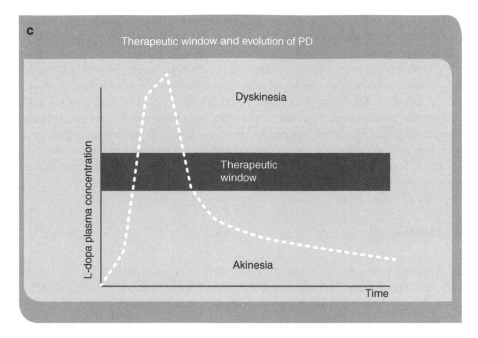

Fig. 15.1 (continued)

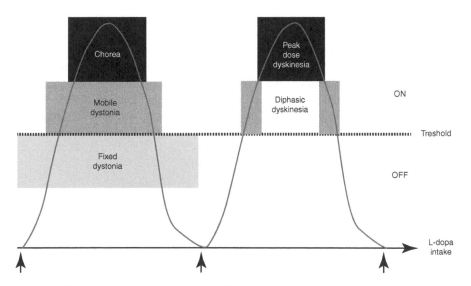

Fig. 15.2 Illustration of the various forms of levodopa-induced dyskinesia phenomenology and their relation to levodopa plasma levels. Chorea is usually described as peak-of-dose dyskinesia and presents when levodopa plasma levels are high

stimulation improves parkinsonian signs but worsens dyskinesias. This issue is one of the major complications in the treatment of PD and a major source of disability in advanced cases.

The phenomenology of peak-of-dose LID is mainly choreic, involving the upper limbs more which tends to be less disabling and less painful, while diphasic dyskinesias manifest as large-amplitude stereotypic, rhythmic, and repetitive movements, more often involving the legs that may coexist with tremor and other parkinsonian symptoms (Fig. 15.2). Both diphasic and peak-of-dose dyskinesias can be present in the same patient depending upon changes in plasma levodopa concentrations.

The symptomatology and therapeutic requirements are quite different between them, possibly due to different underlying physiological mechanisms.

Epidemiology

In the first 5 years of L-dopa therapy, up to 40 % of patients develop motor complications such as motor fluctuations and dyskinesia. In juvenile onset patients, 100 % have been shown to develop motor complications within 10 years of L-dopa therapy.

The overall risk of developing dyskinesia has been estimated to increase by 10 % per year in the first few years of treatment with levodopa.

Risk factors associated with increased occurrence of dyskinesias are a younger age at disease onset [12], particularly before the age of 50 [14, 15], longer disease duration with severe degeneration of nigrostriatal neurons, a longer duration of pulsatile dopaminergic treatment [13], and high levodopa dose [16].

The overall risk of developing dyskinesia has been estimated to increase by 10 % per year in the first few years of treatment with levodopa. However, the existence of profound interindividual heterogeneity suggests that there is a genetic predisposition and specific polymorphisms for dopamine receptors or dopamine transporters [17, 18] have been reported.

Pathophysiology

The pathophysiology underlying development of levodopa-induced dyskinesia is due to a combination of chronic levodopa therapy and PD progression. It does not develop in patients who do not have the disease and have mistakenly been given levodopa. It is believed that LID develops because of a combination of the following (Fig. 15.3):

(a) Progressive loss of nigrostriatal dopaminergic terminals results in changes in the central pharmacokinetics of levodopa.
(b) The changes in the peripheral pharmacokinetics of levodopa due to the irregular absorption associated to the limited passage across blood-brain barrier.

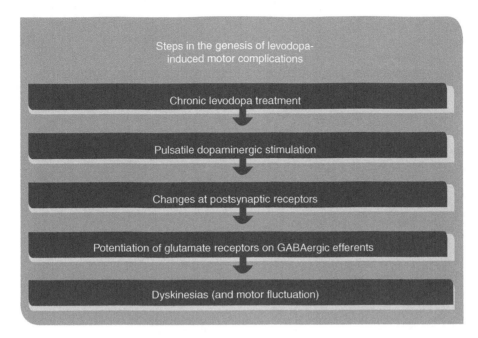

Fig. 15.3 Steps in the genesis of levodopa-induced motor complications

(c) The long-term use of levodopa causes changes in postsynaptic dopamine-receptor signaling and non-dopaminergic neurotransmitter systems.

The short plasma half-life of levodopa and its potential to induce pulsatile stimulation of striatal dopamine receptors are thought to induce dyskinesias. In support of this, parkinsonian patients on long-term treatment with levodopa markedly improve both motor fluctuations and dyskinesias with continuous dopaminergic stimulation [19]. Under these circumstances, reduced motor complications could be achieved by avoiding low plasma levodopa levels and would not be negatively exaggerated by high plasma levodopa concentrations.

Regrettably, the efforts to confirm this hypothesis have failed so far. Trials in early PD using Sinemet CR (CR-FIRST) [20] or the combination of levodopa with the COMT inhibitor entacapone (STRIDE-PD) [21] failed to reduce the risk of dyskinesia or have even increased it.

There has been no consensus regarding the underlying mechanisms of LID, and multiple pathways and neurochemical systems are probably involved, but major advances in the pharmacological and surgical management have recently been achieved. It is known that the firing pattern of the neurons of the internal globus pallidus (GPi) is greatly disturbed in PD [22]. After a unilateral injection of MPTP, in experimental parkinsonism, a decrease in the firing rate of the external globus pallidus (GPe) neurons, and a slight increase in their bursting activity, is observed. In the GPi, the main basal ganglia output structure, there is a considerable augmentation of both neuronal firing frequency and the number of bursting cells. After

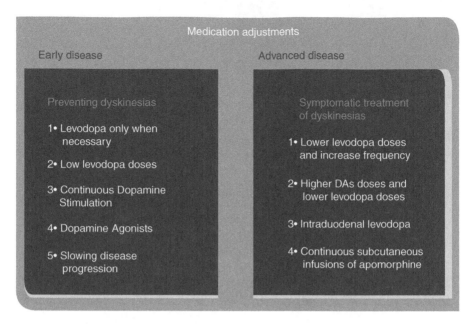

Fig. 15.4 Medication adjustments in different stages of the disease

L-dopa treatment, the firing frequency of GPi neurons decreases to levels even lower than the control intensity with a slight reduction in bursting activity [23]. Though the firing pattern modifications are also associated with the manifestation of dyskinesias, firing frequency seems to be decreased excessively during dyskinesias.

It looks as if the electrophysiological mechanism of dyskinesias involves both an excessive decrease in GPi firing frequency and a modification of the firing pattern in a state of dopamine deficiency [24].

Treatment

Management of LID remains one of the most difficult challenges in the treatment of advanced PD. Dyskinesias can only be controlled satisfactorily with standard oral therapy in some cases, while more complex strategies will be required in many others.

When considering the treatment options, the subtype and severity of dyskinesias as well as the patient's condition should be taken into account. As a rule, dopamine brain stimulation should be kept as stable as possible avoiding the dyskinesia threshold but above the therapeutic threshold (Fig. 15.4).

Those with refractory LIDs have to be considered for invasive approaches including continuous subcutaneous infusions of apomorphine [25, 26], intraduodenal L-dopa infusions [27], or deep brain stimulation (DBS) [28, 29].

Priming is the process by which the brain becomes sensitized such that each administration of dopaminergic therapy modifies the response to subsequent dopaminergic treatments. It is thought that oral levodopa primes for dyskinesias. The primary

Fig. 15.5 Risk factors for motor complications in PD

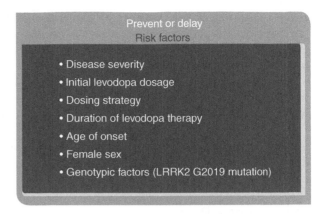

Prevent or delay
Risk factors

- Disease severity
- Initial levodopa dosage
- Dosing strategy
- Duration of levodopa therapy
- Age of onset
- Female sex
- Genotypic factors (LRRK2 G2019 mutation)

therapeutic strategy for managing LIDs in PD patients is to delay the introduction of levodopa therapy (Fig. 15.5). Once the brain has been primed to elicit LID, distinct but potentially overlapping processes are responsible for its expression.

Based on the belief that dyskinesias are induced by the chronic pulsatile stimulation of dopamine receptors in the presence of severe nigrostriatal dopaminergic depletion, drugs apt to generate a continuous stimulation of dopamine receptors can produce clear-cut benefits in motor fluctuations and dyskinesias.

Various drugs can be used as alternatives to levodopa as initial treatment of PD including rasagiline [30, 31], selegiline [32, 33], amantadine [34, 35], and especially dopamine agonists. They can be used both as monotherapy in early PD or as combination therapy in order to diminish levodopa requirements.

In this regard as an initial therapy, ropinirole with L-dopa supplementation compared with L-dopa alone was shown to delay the development of dyskinesia for up to 10 years [36, 37]. Comparing subjects randomized to initial treatment with ropinirole and levodopa, the incidence of dyskinesia was significantly lower in the ropinirole group with no significant differences in change in UPDRS activities of daily living or motor scores.

Likewise, a 4-year study showed that the initial treatment with pramipexole, for early PD compared with L-dopa, significantly reduced the incidence of dyskinesia (25 % versus 54 %, respectively) [38].

Long-term levodopa therapy coupled with disease progression leads to a narrowing of the therapeutic window. If the dose of levodopa is too low, the effect will be insufficient and will wear out quickly. On the contrary, if the dose is increased, dyskinesias will become noticeable (Fig. 15.5).

Oral Treatment

Adenosine A2A Antagonists

Brain adenosine A2A receptors have recently attracted considerable attention because of their interaction with the dopaminergic system and as potential targets for PD pharmacotherapy [39–41]. The pathogenesis of dyskinesias is linked to changes in the output of the striatum to the internal palladium via direct and indirect pathways. Indirect pathway neurons express adenosine A2A receptors and postmortem studies also show increased concentrations of striatal A2A mRNA expression in patients with PD and dyskinesias [42].

Istradefylline and preladenant, 2 selective A2A antagonists, are currently being tested as potential symptomatic agents in PD. The results of the first clinical trials with istradefylline [43] and preladenant [44] have recently become available.

A recent phase 2 study showed a reduction in OFF time compared with placebo with preladenant at doses of 5 and 10 mg twice daily, with no significant increase in dyskinesia [44].

Antiglutamatergic Drugs

Alterations of striatal N-methyl-D-aspartate-receptor (NMDAR) signaling are thought to be one of the main factors for the development of the adverse motor effects of long-term levodopa treatment, including dyskinesias [45]. It has been shown that N-methyl-D-aspartate (NMDA)-receptor antagonists reduce peak-of-dose dyskinesias in PD patients without worsening parkinsonian signs [46].

The antidyskinetic effects of drugs like *amantadine* and *dextromethorphan* are considered to be mediated at least in part through their antiglutamatergic properties.

Amantadine

Amantadine, an N-methyl-D-aspartate antagonist, is so far the only approved compound marketed for treating levodopa-induced dyskinesia, providing a sustained antidyskinetic effect without unacceptable side effects. Although it was initially thought that the effects of amantadine on LIDs wore off after an average of 5 months, long-term effects have been recently demonstrated in a randomized, double-blind, placebo-controlled parallel group trial [47].

Others

Levetiracetam

Levetiracetam (LEV), a drug with a pyrrolidone structure, inhibits neuronal hypersynchronization in experimental models of epilepsy. It acts through a novel mechanism [48] most likely by binding to a synaptic vesicle protein 2A (SV2A) located in presynaptic membranes presumably altering neurotransmitter exocytosis and synaptic vesicle turnover. There is some evidence that it can diminish the severity of dyskinesias without increasing the OFF time [49].

Sarizotan

Sarizotan is a drug with 5-HT(1A) agonist properties and high affinity for D(3) and D(4) receptors. It has antidyskinetic actions in rodent and primate models that have recently been translated into similar clinical effects in a double-blind placebo-controlled trial in PD patients [50].

Fluoxetine

Fluoxetine has also showed some antidyskinetic properties, but in some cases, it can aggravate parkinsonian symptoms.

Non-oral Drug Delivery

Apomorphine

Apomorphine is the most potent dopamine agonist, acting on both D1 and D2 receptors, but has a short-duration response. It can be delivered by means of continuous subcutaneous (SC) infusions leading to large reductions in the OFF time.

It also looks as if continuous SC apomorphine infusions can also desensitize PD patients to the hyperkinetic effects of pulsatile intermittent exogenous dopaminergic stimulation [51] (Fig. 15.6).

The time taken to reduce the severity of dyskinesias is probably related to the gradual reversal of plastic changes in the basal ganglia circuitry associated with dyskinesias. After 5 months, a reduction of over 50 % can be expected [52].

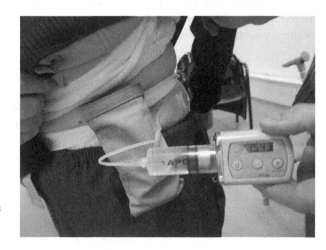

Fig. 15.6 Subcutaneous apomorphine infusion in Parkinson's disease patient via a portable battery-driven apomorphine pump using a prefilled syringe

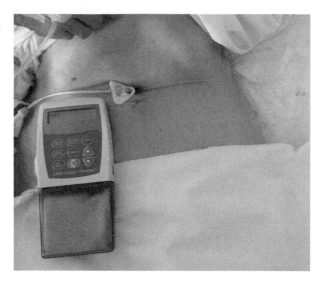

Fig. 15.7 Parkinson's drug pump: continuous delivery of levodopa/carbidopa to the intestine by means of an external pump

Unfortunately, apomorphine monotherapy can be attained only with doses up to 100 mg/day which are not tolerated by many patients who will require additional oral levodopa, not avoiding the pulsatile stimulation that is thought to cause dyskinesias.

IV Levodopa and Apomorphine

It has been shown that peak-of-dose dyskinesias and "ON-OFF" phenomena can be abolished if brain levels of levodopa are kept constant with continuous intravenous levodopa infusions, without impairing the relief of PD symptoms [53]. However, this is not a practical approach for the average patient.

IV apomorphine has also been used, and seems effective, but complication rates were high including intravascular thrombotic complications, secondary to apomorphine crystal accumulation, and require cardiothoracic surgery [54].

Intraduodenal Levodopa

Intraduodenal infusions have been investigated via a permanent percutaneous endoscopic gastrostomy, bypassing gastric emptying, which contributes to the erratic oscillations in levodopa. Reaching stable plasma levels reduces both the OFF time and severity of dyskinesias, depending on individual doses [55–57].

Evading peaks and troughs in plasma levels might broaden the therapeutic window, even in complicated patients (Fig. 15.7).

Transdermal Rotigotine

The transdermal patch is another non-oral alternative, a much more convenient option than the intravenous, subcutaneous, or intraduodenal formulations.

Although several drugs have been investigated, rotigotine is the only one available on the market. Rotigotine is a D2 and D3 receptor agonist and transdermal delivery produces stable dose-related plasma concentrations of the drug. It produces more continuous dopaminergic stimulation and may possibly have a role in preventing the incidence of dyskinesias if given as an early therapy in PD [58, 59].

IV Amantadine

Increased stimulation of the striatal glutamate receptors, particularly of the NMDA subtype, seems to be a major factor contributing to the expression of LID.

It has been shown that intravenous administration of amantadine significantly reduces LID without any loss of antiparkinsonian benefit from levodopa. This effect occurs acutely and does not require a period of time to develop. Brain levels of the drug achieved with therapeutic doses are sufficient to block NMDA receptors and explain its antidyskinetic efficacy.

In addition, it has been also shown that IV amantadine reduces dyskinesias in patients with HD [60, 61].

Botulinum Toxin

Functional chemodenervation with botulinum toxin type A, a first-line treatment for several forms of dystonia, may be considered a treatment strategy against LID when they affect cervical, oromandibular, or limb muscles; however, the benefit in choreic movements is debatable [62].

Surgery

Stereotactic surgery has become an established treatment for advanced Parkinson's disease with fluctuations and with dyskinesia poorly responsive to pharmacological treatment. It has been shown to have long-term efficacy.

It has also been shown to be superior to medical therapy alone even at early stages of PD, before the appearance of severe disabling motor complications [63], suggesting that it could be a therapeutic option for patients at an earlier stage than currently thought.

Most of the pharmacological therapies that work for peak-of-dose dyskinesias fail to do so in cases of diphasic dyskinesias, but both of them are abolished or greatly improved by surgery.

In the era of deep brain stimulation (DBS), ablative surgery is only considered in very rare circumstances in developed countries, but it is still an option in developing countries which have economic limitations.

Because of the high rate of side effects, bilateral lesions are not performed. Lesions are kept unilateral even when the surgical procedure is bilateral, performing a lesion only on one side and DBS on other. When comparing DBS with brain

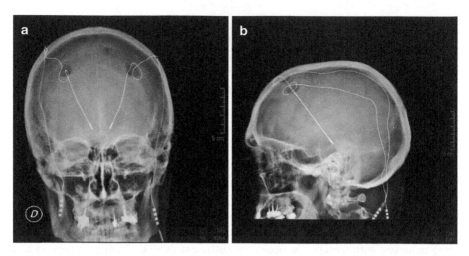

Fig. 15.8 (**a, b**) X-rays showing the electrodes implanted in the GPi in case of PD and deep brain stimulation

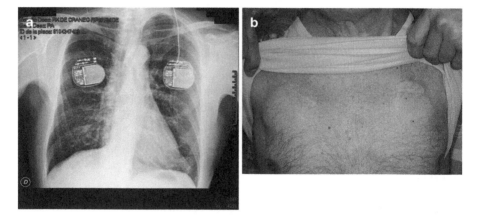

Fig. 15.9 (**a, b**) Localization of the implanted pacemaker in the thoracic subcutaneous tissue

lesioning (pallidotomy, thalamotomy, and subthalamotomy), the main advantages of DBS over lesioning include reversibility of the procedure, the possibility to program the stimulator, and the ability to perform bilateral procedures without risk of inducing dysarthria or dysphagia [64].

The target nuclei for surgery are either the internal segment of the globus pallidus (GPi) (Figs. 15.8a, b and 15.9a, b) or the subthalamic nucleus (STN). By different means, with both approaches, choreic movements can be greatly improved. In the former, movements lessen even when keeping the doses of levodopa high, while in the latter, they improve because the doses required to keep the patient in the ON state markedly diminish.

Fig. 15.10 Surgical
treatment for dyskinesias

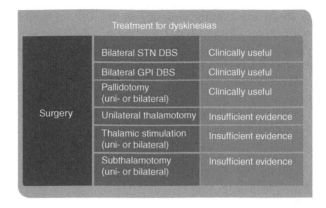

Patients with PD have similar improvement in motor function after either pallidal or subthalamic stimulation [63], but in the long term, those with subthalamic stimulation seem to have more psychiatric side effects.

Candidates for DBS should have PD and troublesome dyskinesias that cannot be adequately controlled with medical therapy. During the best ON periods, gait difficulties, instability, and speech problems should be minimal, reflecting a good response to levodopa with a normal cognitive and psychiatric condition.

Unilateral pallidotomy seems to have a long-lasting effect on contralateral dyskinesia lasting up to 13.5 years even when other PD symptoms had reappeared [65] (Fig. 15.10).

Tardive Dyskinesias

It is well known that neuroleptics are apt to induce acute-transient or chronic-persistent (tardive) movement disorders. In addition, other drugs, most of them with dopamine-receptor-blocking properties, have also been associated with similar unwanted effects.

Tardive syndromes comprise a wide variety of movement disorders that have been recently defined as having at least a history of 3-month total cumulative neuroleptic exposure, a presence of at least "moderate" abnormal involuntary movements in one or more body areas or at least "mild" movements in two or more body areas, and an absence of other conditions that might produce abnormal involuntary movements [66].

In clinical practice, this definition has been currently extended and includes exposure to drugs other than neuroleptics liable to cause similar movements (cinnarizine and flunarizine [67], metoclopramide [68, 69], etc.) (Table 15.1). Some of these movements including buccolingual movements are acknowledged as TD.

Several orodental alterations, such as temporomandibular joint dysfunction, static occlusal contacts, and denture condition, have been implicated in the genesis of orofacial TD. However, recent studies failed to find an association between these factors and TD [70].

In the course of the disease, patients with schizophrenia with or without the use of APDs can develop dyskinesias in up to 40 % of cases. In patients with psychiatric disorders other than schizophrenia, TD can also develop after treatment with APDs or other drugs with antidopaminergic properties [71].

TD rates with second-generation antipsychotics (SGAs) are considered to be low relative to first-generation antipsychotics (FGAs), even in the particularly vulnerable elderly population. However, risk estimates are unavailable for patients naive to FGA [72].

Epidemiology

Estimates of the prevalence rate of TD in patients receiving neuroleptics range from 0.5 to 70 %, with an average prevalence rate of 24 % [73]. The reported incidence of spontaneous dyskinesias in neuroleptic naïve first-episode schizophrenia is 4 %, but in such patients over the age of 60, it may be as high as 60 % [74]. However, prevalence rates in the range of 20–30 % in psychiatric patients chronically exposed to antipsychotic agents have been reported [75].

Unfortunately, despite the general knowledge that neuroleptics can cause movement disorders, these drugs remain the most effective means of treating psychotic symptoms and Tourette's syndrome, as well as for the management of behavioral disorders in developmentally disabled individuals.

A 5-year longitudinal study with a group of 169 schizophrenic outpatients treated with APDs with a prevalence of TD from 31 to 58 % was reported. In this group, parkinsonism was the best predictor of subsequent development of TD [76].

From a large number of the psychiatric and geriatric subjects, "spontaneous" dyskinesia was observed in 1.3 % of 400 healthy elderly people, 4.8 % among the geriatric inpatients and ranged from 0 to 2 % among psychiatric patients never exposed to neuroleptics. Prevalence ranged from 13.3 to 36.1 % in neuroleptic exposed patients. Age and gender were important factors concerning the prevalence of TD. In younger subjects, men had higher rates; among subjects over age 40, rates were higher for women. Edentulousness, neurological disorders, smoking habits, diabetes mellitus, and alcohol abuse were possible contributors for high rates in the elderly [77, 78].

Recent studies are directed at identifying factors that contribute to increased plasma concentrations of the offending drug like genetic polymorphisms of several enzymes that encode for enzymatic variants. The risk of TD is considerably lower than those using typical neuroleptics, and there is little risk of acute movement disorders traditionally associated with APDs.

Nevertheless, risks of TD with modern APDs have not met the expectations, particularly in the elderly population who are most vulnerable to this side effect [79].

Risperidone (and its active metabolite paliperidone), at high doses, may carry unusually high TD risk, whereas TD risk is low with clozapine, quetiapine, and probably aripiprazole. Risperidone is currently accepted to be an atypical neuroleptic devoid of extrapyramidal side effects and is widely prescribed in children and adolescents. However, this drug may have side effects that were previously unrecognized and cause severe movement disorders including chorea [80] and other forms of TD [81].

Diagnosis

TD is a complex hyperkinetic syndrome consisting of choreic, athetoid rhythmically abnormal involuntary movements derived from dopamine-blocking agents. Such involuntary movements are frequently associated with drug-induced parkinsonian symptoms [82]. It may be a disabling, persistent, or even permanent condition that results mainly from the use of dopamine-receptor D2-blocking agent therapy. Diagnosis can be challenging, as there is no single test for TD. The diagnostic process may require the thorough review of the medical history, a physical examination, and even a neuropsychological evaluation. The diagnostic process is complicated by the fact that symptoms can fluctuate, in some circumstances, in relation to emotional states which may be more apparent at some times than at others. An accurate diagnosis may require several subsequent examinations.

Choreic movements may appear from days to months after the introduction of the continuous use of APDs and persist for more than a month after the APDs are discontinued. Individual susceptibility certainly plays a major role in the development of this unwanted effect. They feature irregular stereotyped movements that are usually choreic and subject to temporary volitional control. Dystonia, akathisia, and tics are variants of the classic tardive syndrome. They occur in up to one-third of patients chronically exposed to APDs [70]. Symptoms include orobuccolingual movements and "piano-playing" movements of the limbs.

In some patients with chorea, including those with acanthocytosis, TD, and HD, psychiatric symptoms and movement disorders may coexist requiring a correct differential diagnosis. In addition, they may have received APDs making the diagnosis even more troublesome. Although it is not always possible to differentiate TD from other forms of choreic movements, there are some clues that can help to make an accurate diagnosis [83] as summarized below.

Stereotypy: It is typical of a drug-induced movement disorder and features repetition of the same movement over time [84].

Suppressibility: Involuntary movements can be suppressed or improved by an effort of will in patients with TD or LID whereas most patients with HD cannot. TD and L-dopa-induced chorea may be more pathophysiologically similar to each other than either is to HD [85].

Speech: It is usually affected in patients with HD but is not in cases of TD.

Pattern of body part involvement: Orobuccolingual movements are often first to appear in patients with TD and have a repetitive pattern, whereas in HD on occasion, they first appear in the face but they may resemble winking, smiling, grimacing, shrugging, or gesturing (Fig. 15.11a–c).

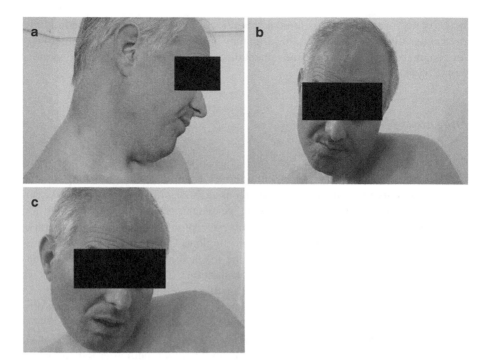

Fig. 15.11 (a–c) Choreic facial and neck movements in a patient with tardive dyskinesia, who markedly improved with botulinum toxin injections

Semi-purposiveness of abnormal movements: The association of involuntary limb movements in some voluntary activities is often observed in patients with HD, but they are rare in TD. This combination has been referred to as semi-purposeful movements because they may mimic a fragment of normal motor activity commonly of the automatic but voluntary kind [84].

Eye movements: No abnormalities are seen in patients with TD; however, they are an early sign of HD. They feature slow initiation of saccades, saccadic intrusions of pursuits, oculomotor impersistence, and impaired optokinetic nystagmus.

Motor impersistence: Oculomotor, tongue, and grip impersistence are highly suggestive of HD and are not seen in TD.

Associated movement disorders: Dystonia, tremor, or other associated involuntary movements are common in TD. Myoclonus can be present in juvenile or childhood HD patients [86]

Other neurological signs: Hung up reflexes and gait imbalance are frequently seen in HD [87, 88] but not in DIC.

Pathophysiology

The pathophysiology of TD remains poorly understood and treatment is often tricky. Diverse neurochemical hypotheses have been proposed for the development of TD.

These include dopaminergic hypersensitivity [89], disturbed balance between dopamine and cholinergic systems, involvement of serotonergic (5-HT) system [90], dysfunctions of striatonigral GABAergic neurons, and excitotoxicity.

After short-term treatment with APDs, D2 blockade results in the elimination of excess activity in the control pathway of motor behaviors, but the D1 select-on function is unchanged. Clinical manifestations in such cases are parkinsonian signs.

However, prolonged treatment with APD is believed to produce D2 upregulation and sensitization, resulting in hyperkinesis and TD.

Risk Factors

Oxidative stress has been one of the proposed mechanisms influencing TD risk and significant association between TD and the NADPH quinine oxidoreductase 1 (NQO1) gene Pro187Ser (C609T, rs1800566) polymorphism in Korean schizophrenia patients was reported [91–93]. Several associations between TD and quite a few genetic polymorphisms have been reported including dopamine D3 receptor [94], as mice show locomotor hyperactivation resembling the extrapyramidal side effects of neuroleptic treatment [95], cytochrome P450 (CYP) 2D6 [96], 5-HT2A receptor [97], and 5-HT2Creceptor genes [98].

However, recent evidence suggests that neither the NQO1 Pro187Ser nor the SOD2 Ala9Val appear to play a major role in TD risk although additional polymorphisms should be tested before the role of NQO1 and SOD2 in TD can be completely ruled out.

There is limited evidence of ethnic differences in the risk of TD. However, this could be due to confounding factors and further studies are needed to resolve this issue [75, 99, 100].

Elderly women and young men are at more risk of developing TD [101]. It has also been suggested that TD may be associated with positive symptoms in psychotic patients; but the severity of TD could be linked to the negative characteristics. The correlation between negative symptoms and TD severity may be influenced by the presence of parkinsonism [102]. Intermittent neuroleptic treatment increases the risk of TD which is three times greater for patients with more than two neuroleptic interruptions as for patients with no interruptions [103].

Treatment

The first issue to address is how to prevent this complication as the treatment is troublesome and the prognosis poor. Several studies have compared conventional and atypical APDs in order to decrease the risk of developing movement disorders and to prevent their presence and improve their treatment [104]. It is known that certain populations including females and the elderly seem to be more vulnerable to this side effect. TD rates with SGAs are considered to be low relative to FGAs.

However, this might not be the case with risperidone (and its active metabolite paliperidone), which at high doses may carry unusually high TD risk, whereas TD risk is low with clozapine, and perhaps quetiapine and aripiprazole [79].

Once TD has developed, the first seemingly rational step would be to discontinue the dopamine-receptor antagonist. However, it is well known that withdrawal may worsen dyskinesias in the short term (1 or 2 months), whereas adding APDs with stronger extrapyramidal side effects can reduce TD [105]. But little is known about what will happen in the long term after either continuing or discontinuing the APDs. In this context, the American Psychiatric Association Task Force recommends APD withdrawal only in patients who can tolerate it.

It is important to recall that many patients receive dopamine-blocking agents chronically without a formal indication, and in this case, they should obviously be discontinued.

Atypical neuroleptics are thought to cause less TD than typical APDs, but this has not been fully proven [106]. In cases where APDs are needed, most physicians would switch to atypical APDs though changing to atypical neuroleptics might not be necessarily useful.

Pharmacological Treatment of TD

Unfortunately, therapies that target pathogenesis are still lacking, but successful management of TD can be achieved in many cases with current pharmacological approaches.

Dopamine-Depleting Agents

Tetrabenazine

Tetrabenazine has proven useful in the treatment of hyperkinetic movement disorders, a term used to describe a wide array of movement disorders featuring unnecessary recurring or intermittent involuntary movements.

Tetrabenazine should be considered as a first choice in the treatment of TD as there is evidence that in the short term, it is useful [107]. However, no information is available on its long-term use and it can cause depression and parkinsonism [108, 109].

Amantadine

Amantadine, an N-methyl-D-aspartate antagonist, has been proven a useful agent in the treatment of peak-of-dose dyskinesias in PD patients. These two drug-induced dyskinesias are undoubtedly different as to the offending drugs and the underlying disease, but they share some characteristics in terms of clinical phenomenology, epidemiology, risk factors, and pathophysiological mechanisms.

In both instances, dysregulation occurring at the striatal level, the dopaminergic receptors, is involved in the pathogenesis, and similar therapeutic strategies to manage these disorders have been employed.

Amantadine has been also used in the treatment of HD [110–112].

A recent double-blind study in 22 TD cases showed that amantadine can be an effective and safe treatment for TD. The severity of TD movements improved significantly more than in those receiving placebo, as measured by the AIMS score [113]. Amantadine has been also reported to prevent or reduce the severity of TD when concomitantly used with neuroleptics [114].

Other Drugs

Acetazolamide and thiamine (A + T) have been reported to improve TD in one Class III study. The symptoms of TD and parkinsonism of 8 elderly and 25 younger chronic hospitalized mental patients were evaluated in a placebo-controlled, double-blind, crossover study. They were all kept on their psychoactive and antiparkinsonian medications. Both groups showed a significant decrease in scores on the Abnormal Involuntary Movement Scale (TD) and the Simpson-Angus Neurological Rating Scale (parkinsonism) while on A + T [115]. However, in clinical practice, they do not seem to be very useful.

Typical APDs

Although typical APDs possibly reduce TD symptoms, they are not recommended because of the risk of akinetic-rigid syndromes and possibly perpetuating TD. Safety data are unavailable concerning long-term use of typical APDs such as TD suppressive agents, and these drugs themselves can cause TD. These significant threats are more important than the potential benefits of any short-range use of typical APDs.

Atypical APDs

Atypical APDs are defined as compounds that have an antipsychotic effect with a low probability for inducing extrapyramidal side effects.

The term "atypical" was originally used to describe drugs that in animal models predict antipsychotic effects but do not produce catalepsy—most notably clozapine.

However, other APDs including quetiapine, olanzapine, and risperidone are also considered atypical and some of them may have a beneficial symptomatic effect on TD symptoms. While clozapine has hardly any effect, risperidone has shown to improve some patients. This is probably because this drug is not really a true atypical neuroleptic, is apt to cause TD and rigid akinetic symptomatology, and is certainly not tolerated by patients with PD.

Botulinum Toxin

Botulinum toxin has been successfully used to treat patients with tardive syndromes, especially those with a dystonic component [116].

Electroconvulsive Therapy

There is not enough information about electroconvulsive therapy in TD, but some case reports suggest that it might be useful [117, 118].

Surgery

As medical management is often insufficient and not devoid of side effects, there have been several case reports of deep brain stimulation (DBS) used to treat TD with favorable outcomes. A review of the literatures suggests that DBS is effective and safe for patients with treatment-resistant TD, giving rise to few psychiatric side effects. GPi seems to be the best target [119, 120].

Conclusions

In order to treat PD patients, the doses of levodopa should be kept as low as possible and introduced into therapy only when necessary.

A thorough search should be carried out in patients with sporadic, generalized chorea to rule out drugs as the potential cause of the movement disorder. The sooner the offending drug is withdrawn, the better the prognosis. It should be considered that after long-term treatment with APDs, choreic movements could be persistent or even permanent after discontinuation of the causative drug.

Drugs capable of inducing movement disorders should only be used when necessary and when there is no better alternative. Patients should be monitored regularly to diagnose choreic movements as soon as possible and to determine when to discontinue drug intake.

References

1. O'Toole O, Lennon VA, Ahlskog JE, Matsumoto JY, Pittock SJ, Bower J, et al. Autoimmune chorea in adults. Neurology. 2013;80(12):1133–44.
2. AhnE S, Scott RM, Robertson Jr RL, Smith ER. Chorea in the clinical presentation of moyamoya disease: results of surgical revascularization and a proposed clinicopathological correlation. J Neurosurg Pediatr. 2013;11(3):313–9.
3. Morrison PJ. Prevalence estimates of Huntington disease in Caucasian populations are gross underestimates. Mov Disord. 2012;27(13):1707–8.

4. Piccolo I, Sterzi R, Thiella G, Minazzi MS, Caraceni T. Sporadic choreas: analysis of a general hospital series. Eur Neurol. 1999;41(3):143–9.
5. Stork CM, Cantor R. Pemoline induced acute choreoathetosis: case report and review of the literature. J Toxicol Clin Toxicol. 1997;35(1):105–8.
6. Kamath S, Bajaj N. Crack dancing in the United Kingdom: apropos a video case presentation. Mov Disord. 2007;22(8):1190–1.
7. Rupniak NM, Jenner P, Marsden CD. Acute dystonia induced by neuroleptic drugs. Psychopharmacology (Berl). 1986;88(4):403–19.
8. Piccolo I, Thiella G, Sterzi R, Colombo N, Defanti CA. Chorea as a symptom of neuroborreliosis: a case study. Ital J Neurol Sci. 1998;19(4):235–9.
9. Greenbaum L, Goldwurm S, Zozulinsky P, Lifschytz T, Cohen OS, Yahalom G, et al. Do tardive dyskinesia and L-dopa induced dyskinesia share common genetic risk factors? An exploratory study. J Mol Neurosci. 2013;51(2):380–8.
10. Marsden CD, Parkes JD. Success and problems of long-term levodopa therapy in Parkinson's disease. Lancet. 1977;1(8007):345–9.
11. Bezard E, Brotchie JM, Gross CE. Pathophysiology of levodopa-induced dyskinesia: potential for new therapies. Nat Rev Neurosci. 2001;2(8):577–88.
12. Kostic V, Przedborski S, Flaster E, Sternic N. Early development of levodopa-induced dyskinesias and response fluctuations in young-onset Parkinson's disease. Neurology. 1991;41.2 (Pt 1):202–5.
13. Tambasco N, Simoni S, Marsili E, Sacchini E, Murasecco D, Cardaioli G, et al. Clinical aspects and management of levodopa-induced dyskinesia. Parkinsons Dis. 2012;2012:745947.
14. Lyons KE, Hubble JP, Tröster AI, Pahwa R, Koller WC. Gender differences in Parkinson's disease. Clin Neuropharmacol. 1998;21(2):118–21.
15. Zappia M, Annesi G, Nicoletti G, Arabia G, Annesi F, Messina D, et al. Sex differences in clinical and genetic determinants of levodopa peak-dose dyskinesias in Parkinson disease: an exploratory study. Arch Neurol. 2005;62(4):601–5.
16. Cervantes-Arriaga A, Rodríguez-Violante M, Salmerón-Mercado M, Calleja-Castillo J, Corona T, Yescas P, et al. Incidence and determinants of levodopa-induced dyskinesia in a retrospective cohort of Mexican patients with Parkinson's disease. Rev Invest Clin. 2012;64(3):220–6.
17. Gilgun-Sherki Y, Djaldetti R, Melamed E, Offen D. Polymorphism in candidate genes: implications for the risk and treatment of idiopathic Parkinson's disease. Pharmacogenomics J. 2004;4(5):291–306.
18. Kaiser R, Hofer A, Grapengiesser A, Gasser T, Kupsch A, Roots I, et al. L-dopa-induced adverse effects in PD and dopamine transporter gene polymorphism. Neurology. 2003;60(11):1750–5.
19. Stocchi F, Vacca L, Ruggieri S, Olanow CW. Intermittent vs continuous levodopa administration in patients with advanced Parkinson disease: a clinical and pharmacokinetic study. Arch Neurol. 2005;62(6):905–10.
20. Block G, Liss C, Reines S, Irr J, Nibbelink D. Comparison of immediate-release and controlled release carbidopa/levodopa in Parkinson's disease. A multicenter 5-year study. The CR First Study Group. Eur Neurol. 1997;37(1):23–7.
21. Stocchi F, Rascol O, Kieburtz K, Poewe W, Jankovic J, Tolosa E, et al. Initiating levodopa/carbidopa therapy with and without entacapone in early Parkinson disease: the STRIDE-PD study. Ann Neurol. 2010;68(1):18–27.
22. Boraud T, Bezard E, Bioulac B, Gross C. High frequency stimulation of the internal Globus Pallidus (GPi) simultaneously improves parkinsonian symptoms and reduces the firing frequency of GPi neurons in the MPTP-treated monkey. Neurosci Lett. 1996;215(1):17–20.
23. Boraud T, Bezard E, Guehl D, Bioulac B, Gross C. Effects of L-DOPA on neuronal activity of the globus pallidus externalis (GPe) and globus pallidus internalis (GPi) in the MPTP-treated monkey. Brain Res. 1998;787(1):157–60.
24. Boraud T, Bezard E, Bioulac B, Gross CE. Dopamine agonist-induced dyskinesias are correlated to both firing pattern and frequency alterations of pallidal neurones in the MPTP-treated monkey. Brain. 2001;124(Pt 3):546–57.

25. Deleu D, Hanssens Y, Northway MG. Subcutaneous apomorphine : an evidence-based review of its use in Parkinson's disease. Drugs Aging. 2004;21(11):687–709.
26. Poewe W, Wenning GK. Apomorphine: an underutilized therapy for Parkinson's disease. Mov Disord. 2000;15(5):789–94.
27. Meppelink AM, Nyman R, van Laar T, Drent M, Prins T, Leenders KL. Transcutaneous port for continuous duodenal levodopa/carbidopa administration in Parkinson's disease. Mov Disord. 2011;26(2):331–4.
28. Stern MB, Follett KA, Weaver FM. Randomized trial of deep brain stimulation for Parkinson disease: thirty-six-month outcomes; turning tables: should GPi become the preferred DBS target for Parkinson disease? Author response. Neurology. 2013;80(2):225.
29. Olanow CW, Rascol O, Hauser R, Feigin PD, Jankovic J, Lang A, et al. A double-blind, delayed-start trial of rasagiline in Parkinson's disease. N Engl J Med. 2009;361(13): 1268–78.
30. Hauser RA, Lew MF, Hurtig HI, Ondo WG, Wojcieszek J, Fitzer-Attas CJ, et al. Long-term outcome of early versus delayed rasagiline treatment in early Parkinson's disease. Mov Disord. 2009;24(4):564–73.
31. Linazasoro G. Rasagiline in Parkinson's disease. Neurologia. 2008;23(4):238–45.
32. Lohle M, Reichmann H. Controversies in neurology: why monoamine oxidase B inhibitors could be a good choice for the initial treatment of Parkinson's disease. BMC Neurol. 2011;11:112.
33. Koller WC. Initiating treatment of Parkinson's disease. Neurology. 1992;42.1 Suppl 1:33–8.
34. Lees A. Alternatives to levodopa in the initial treatment of early Parkinson's disease. Drugs Aging. 2005;22(9):731–40.
35. Kulisevsky J, Lopez-Villegas D. Initial treatment of Parkinson's disease. Rev Neurol. 1997;25 Suppl 2:S163–9.
36. Hauser RA, et al. Ten-year follow-up of Parkinson's disease patients randomized to initial therapy with ropinirole or levodopa. Mov Disord. 2007;22(16):2409–17.
37. Rascol O, Brooks DJ, Korczyn AD, De Deyn PP, Clarke CE, Lang AE, et al. Development of dyskinesias in a 5-year trial of ropinirole and L-dopa. Mov Disord. 2006;21(11):1844–50.
38. Holloway RG, Shoulson I, Fahn S, Kieburtz K, Lang A, Marek K, et al. Pramipexole vs levodopa as initial treatment for Parkinson disease: a 4-year randomized controlled trial. Arch Neurol. 2004;61(7):1044–53.
39. Hickey P, Stacy M. Adenosine A2A antagonists in Parkinson's disease: what's next? Curr Neurol Neurosci Rep. 2012;12(4):376–85.
40. Jenner P, Mori A, Hauser R, Morelli M, Fredholm BB, Chen JF. Adenosine, adenosine A 2A antagonists, and Parkinson's disease. Parkinsonism Relat Disord. 2009;15(6):406–13.
41. Cieslak M, Komoszynski M, Wojtczak A. Adenosine A(2A) receptors in Parkinson's disease treatment. Purinergic Signal. 2008;4(4):305–12.
42. Calon F, Dridi M, Hornykiewicz O, Bédard PJ, Rajput AH, Di Paolo T. Increased adenosine A2A receptors in the brain of Parkinson's disease patients with dyskinesias. Brain. 2004;127(Pt 5):1075–84.
43. Chen W, Wang H, Wei H, Gu S, Wei H. Istradefylline, an adenosine A(2)A receptor antagonist, for patients with Parkinson's Disease: a meta-analysis. J Neurol Sci. 2013;324(1-2):21–8.
44. Hauser RA, Cantillon M, Pourcher E, Micheli F, Mok V, Onofrj M, et al. Preladenant in patients with Parkinson's disease and motor fluctuations: a phase 2, double-blind, randomised trial. Lancet Neurol. 2011;10(3):221–9.
45. Errico F, Bonito-Oliva A, Bagetta V, Vitucci D, Romano R, Zianni E, et al. Higher free D-aspartate and N-methyl-D-aspartate levels prevent striatal depotentiation and anticipate L-DOPA-induced dyskinesia. Exp Neurol. 2011;232(2):240–50.
46. Elahi B, Phielipp N, Chen R. N-Methyl-D-Aspartate antagonists in levodopa induced dyskinesia: a meta-analysis. Can J Neurol Sci. 2012;39(4):465–72.
47. Wolf E, Seppi K, Katzenschlager R, Hochschorner G, Ransmayr G, Schwingenschuh P, et al. Long-term antidyskinetic efficacy of amantadine in Parkinson's disease. Mov Disord. 2010; 25(10):1357–63.

48. Klitgaard H. Levetiracetam: the preclinical profile of a new class of antiepileptic drugs? Epilepsia. 2001;42 Suppl 4:13–8.
49. Stathis P, Konitsiotis S, Tagaris G, Peterson D, VALID-PD Study Group. Levetiracetam for the management of levodopa-induced dyskinesias in Parkinson's disease. Mov Disord. 2011;26(2):264–70.
50. Goetz CG, Damier P, Hicking C, Laska E, Müller T, Olanow CW, et al. Sarizotan as a treatment for dyskinesias in Parkinson's disease: a double-blind placebo-controlled trial. Mov Disord. 2007;22(2):179–86.
51. Wenning GK, Bösch S, Luginger E, Wagner M, Poewe W. Effects of long-term, continuous subcutaneous apomorphine infusions on motor complications in advanced Parkinson's disease. Adv Neurol. 1999;80:545–8.
52. Manson AJ, Turner K, Lees AJ. Apomorphine monotherapy in the treatment of refractory motor complications of Parkinson's disease: long-term follow-up study of 64 patients. Mov Disord. 2002;17(6):1235–41.
53. Schuh LA, Bennett Jr JP. Suppression of dyskinesias in advanced Parkinson's disease. I. Continuous intravenous levodopa shifts dose response for production of dyskinesias but not for relief of parkinsonism in patients with advanced Parkinson's disease. Neurology. 1993;43(8):1545–50.
54. Manson AJ, Hanagasi H, Turner K, Patsalos PN, Carey P, Ratnaraj N, et al. Intravenous apomorphine therapy in Parkinson's disease: clinical and pharmacokinetic observations. Brain. 2001;124(Pt 2):331–40.
55. Nyholm D. Duodopa(R) treatment for advanced Parkinson's disease: a review of efficacy and safety. Parkinsonism Relat Disord. 2012;18(8):916–29.
56. Nyholm D, Johansson A, Aquilonius SM, Hellquist E, Lennernäs H, Askmark H. Complexity of motor response to different doses of duodenal levodopa infusion in Parkinson disease. Clin Neuropharmacol. 2012;35(1):6–14.
57. Nilsson D, Nyholm D, Aquilonius SM. Duodenal levodopa infusion in Parkinson's disease– long-term experience. Acta Neurol Scand. 2001;104(6):343–8.
58. Wright BA, Waters CH. Continuous dopaminergic delivery to minimize motor complications in Parkinson's disease. Expert Rev Neurothe. 2013;13(6):719–29.
59. Jenner P, McCreary AC, Scheller DK. Continuous drug delivery in early- and late-stage Parkinson's disease as a strategy for avoiding dyskinesia induction and expression. J Neural Transm. 2011;118(12):1691–702.
60. Heckmann JM, Legg P, Sklar D, Fine J, Bryer A, Kies B. IV amantadine improves chorea in Huntington's disease: an acute randomized, controlled study. Neurology. 2004;63(3):597–8.
61. Lucetti C, Del Dotto P, Gambaccini G, Dell' Agnello G, Bernardini S, Rossi G, et al. IV amantadine improves chorea in Huntington's disease: an acute randomized, controlled study. Neurology. 2003;60(12):1995–7.
62. Espay AJ, Vaughan JE, Shukla R, Gartner M, Sahay A, Revilla FJ, et al. Botulinum toxin type A for Levodopa-induced cervical dyskinesias in Parkinson's disease: unfavorable risk-benefit ratio. Mov Disord. 2011;26(5):913–4.
63. Follett KA, Weaver FM, Stern M, Hur K, Harris CL, Luo P, et al. Pallidal versus subthalamic deep-brain stimulation for Parkinson's disease. N Engl J Med. 2010;362(22):2077–91.
64. Okun MS, Vitek JL. Lesion therapy for Parkinson's disease and other movement disorders: update and controversies. Mov Disord. 2004;19(4):375–89.
65. Kleiner-Fisman G, Lozano A, Moro E, Poon YY, Lang AE. Long-term effect of unilateral pallidotomy on levodopa-induced dyskinesia. Mov Disord. 2010;25(10):1496–8.
66. Bhidayasiri R, Fahn S, Weiner WJ, Gronseth GS, Sullivan KL, Zesiewicz TA, et al. Evidence-based guideline: treatment of tardive syndromes: report of the Guideline Development Subcommittee of the American Academy of Neurology. Neurology. 2013;81(5):463–9.
67. Micheli F, Pardal MF, Gatto M, Torres M, Paradiso G, Parera IC, et al. Flunarizine- and cinnarizine-induced extrapyramidal reactions. Neurology. 1987;37(5):881–4.
68. Rao AS, Camilleri M. Review article: metoclopramide and tardive dyskinesia. Aliment Pharmacol Ther. 2010;31(1):11–9.

69. Kenney C, Hunter C, Davidson A, Jankovic J. Metoclopramide, an increasingly recognized cause of tardive dyskinesia. J Clin Pharmacol. 2008;48(3):379–84.
70. Girard P, Monette C, Normandeau L, Pampoulova T, Rompré PH, de Grandmont P, et al. Contribution of orodental status to the intensity of orofacial tardive dyskinesia: an interdisciplinary and video-based assessment. J Psychiatr Res. 2012;46(5):684–7.
71. Tenback DE, van Harten PN. Epidemiology and risk factors for (tardive) dyskinesia. Int Rev Neurobiol. 2011;98:211–30.
72. Woerner MG, Kane JM, Lieberman JA, Alvir J, Bergmann KJ, Borenstein M, et al. The prevalence of tardive dyskinesia. J Clin Psychopharmacol. 1991;11(1):34–42.
73. Kulkarni SK, Naidu PS. Pathophysiology and drug therapy of tardive dyskinesia: current concepts and future perspectives. Drugs Today (Barc). 2003;39(1):19–49.
74. Fenton WS, Blyler CR, Wyatt RJ, McGlashan TH. Prevalence of spontaneous dyskinesia in schizophrenic and non-schizophrenic psychiatric patients. Br J Psychiatry. 1997;171:265–8.
75. Al Hadithy AF, Ivanova SA, Pechlivanoglou P, Semke A, Fedorenko O, Kornetova E, et al. Tardive dyskinesia and DRD3, HTR2A and HTR2C gene polymorphisms in Russian psychiatric inpatients from Siberia. Prog Neuropsychopharmacol Biol Psychiatry. 2009;33(3): 475–81.
76. Chouinard G, Annable L, Ross-Chouinard A, Mercier P. A 5-year prospective longitudinal study of tardive dyskinesia: factors predicting appearance of new cases. J Clin Psychopharmacol. 1988;8(4 Suppl):21S–6.
77. Woerner MG, Correll CU, Alvir JM, Greenwald B, Delman H, Kane JM. Incidence of tardive dyskinesia with risperidone or olanzapine in the elderly: results from a 2-year, prospective study in antipsychotic-naive patients. Neuropsychopharmacology. 2011;36(8):1738–46.
78. Frascarelli M, Paolemili M, Gallo M, Parente F, Biondi M. Tardive dyskinesia: diagnosis, assessment and treatment. Riv Psichiatr. 2013;48(3):187–96.
79. Tarsy D, Lungu C, Baldessarini RJ. Epidemiology of tardive dyskinesia before and during the era of modern antipsychotic drugs. Handb Clin Neurol. 2011;100:601–16.
80. Carroll NB, Boehm KE, Strickland RT. Chorea and tardive dyskinesia in a patient taking risperidone. J Clin Psychiatry. 1999;60(7):485–7.
81. Bhimanil MM, Khan MM, Khan MF, Waris MS. Respiratory dyskinesi--n under-recognized side-effect of neuroleptic medications. J Pak Med Assoc. 2011;61(9):930–2.
82. Orti-Pareja M, Jiménez-Jiménez FJ, Vázquez A, Catalán MJ, Zurdo M, Burguera JA, et al. Drug-induced tardive syndromes. Parkinsonism Relat Disord. 1999;5(1-2):59–65.
83. Kumar H, Jog M. Missing Huntington's disease for tardive dyskinesia: a preventable error. Can J Neurol Sci. 2011;38(5):762–4.
84. Cummings JL, Wirshing WC. Recognition and differential diagnosis of tardive dyskinesia. Int J Psychiatry Med. 1989;19(2):133–44.
85. Walters AS, McHale D, Sage JI, Hening WA, Bergen M. A blinded study of the suppressibility of involuntary movements in Huntington's chorea, tardive dyskinesia, and L-dopa-induced chorea. Clin Neuropharmacol. 1990;13(3):236–40.
86. Rossi Sebastiano D, Soliveri P, Panzica F, Moroni I, Gellera C, Gilioli I, et al. Cortical myoclonus in childhood and juvenile onset Huntington's disease. Parkinsonism Relat Disord. 2012;18(6):794–7.
87. Valle-Lopez P, Canas-Canas MT, Camara-Barrio S. Psychiatric symptoms in a woman with chorea-acanthocytosis. Actas Esp Psiquiatr. 2013;41(2):133–6.
88. Delval A, Krystkowiak P. Locomotion disturbances in Huntington's disease. Rev Neurol (Paris). 2010;166(2):213–20.
89. Baldessarini RJ, Tarsy D. Pathophysiologic basis of tardive dyskinesia. Adv Biochem Psychopharmacol. 1980;24:451–5.
90. Meltzer HY. An overview of the mechanism of action of clozapine. J Clin Psychiatry. 1994;55(Suppl B):47–52.
91. Pae CU. Additive effect between quinine oxidoreductase gene (NQO1: Pro187Ser) and manganese superoxide dismutase gene (MnSOD: Ala-9Val) polymorphisms on tardive dyskinesia in patients with schizophrenia. Psychiatry Res. 2008;161(3):336–8.

92. Pae CU, Kim TS, Patkar AA, Kim JJ, Lee CU, Lee SJ, et al. Manganese superoxide dismutase (MnSOD: Ala-9Val) gene polymorphism may not be associated with schizophrenia and tardive dyskinesia. Psychiatry Res. 2007;153(1):77–81.
93. Pae CU, Yu HS, Kim JJ, Lee CU, Lee SJ, Jun TY, et al. Quinone oxidoreductase (NQO1) gene polymorphism (609C/T) may be associated with tardive dyskinesia, but not with the development of schizophrenia. Int J Neuropsychopharmacol. 2004;7(4):495–500.
94. Steen VM, Løvlie R, MacEwan T, McCreadie RG. Dopamine D3-receptor gene variant and susceptibility to tardive dyskinesia in schizophrenic patients. Mol Psychiatry. 1997;2(2): 139–45.
95. Rietschel M, Krauss H, Müller DJ, Schulze TG, Knapp M, Marwinski K, et al. Dopamine D3 receptor variant and tardive dyskinesia. Eur Arch Psychiatry Clin Neurosci. 2000;250(1): 31–5.
96. Kapitany T, Meszaros K, Lenzinger E, Schindler SD, Barnas C, Fuchs K, et al. Genetic polymorphisms for drug metabolism (CYP2D6) and tardive dyskinesia in schizophrenia. Schizophr Res. 1998;32(2):101–6.
97. Segman RH, Heresco-Levy U, Finkel B, Goltser T, Shalem R, Schlafman M, et al. Association between the serotonin 2A receptor gene and tardive dyskinesia in chronic schizophrenia. Mol Psychiatry. 2001;6(2):225–9.
98. Segman RH, Heresco-Levy U, Finkel B, Inbar R, Neeman T, Schlafman M, et al. Association between the serotonin 2C receptor gene and tardive dyskinesia in chronic schizophrenia: additive contribution of 5-HT2Cser and DRD3gly alleles to susceptibility. Psychopharmacology (Berl). 2000;152(4):408–13.
99. Xiang YT, Wang CY, Si TM, Lee EH, He YL, Ungvari GS, et al. Tardive dyskinesia in the treatment of schizophrenia: the findings of the Research on Asian Psychotropic Prescription Pattern (REAP) survey (2001–2009). Int J Clin Pharmacol Ther. 2011;49(6):382–7.
100. Lee J, Jiang J, Sim K, Chong SA. The prevalence of tardive dyskinesia in Chinese Singaporean patients with schizophrenia: revisited. J Clin Psychopharmacol. 2010;30(3):333–5.
101. van Os J, Walsh E, van Horn E, Tattan T, Bale R, Thompson SG. Tardive dyskinesia in psychosis: are women really more at risk? UK700 Group. Acta Psychiatr Scand. 1999;99(4): 288–93.
102. Yuen O, Caligiuri MP, Williams R, Dickson RA. Tardive dyskinesia and positive and negative symptoms of schizophrenia. A study using instrumental measures. Br J Psychiatry. 1996; 168(6):702–8.
103. van Harten PN, Hoek HW, Matroos GE, Koeter M, Kahn RS. Intermittent neuroleptic treatment and risk for tardive dyskinesia: Curacao Extrapyramidal Syndromes Study III. Am J Psychiatry. 1998;155(4):565–7.
104. Jesic MP, Jesić A, Filipović JB, Zivanović O. Extrapyramidal syndromes caused by antipsychotics. Med Pregl. 2012;65(11-12):521–6.
105. Shenoy RS, Sadler AG, Goldberg SC, Hamer RM, Ross B. Effects of a six-week drug holiday on symptom status, relapse, and tardive dyskinesia in chronic schizophrenics. J Clin Psychopharmacol. 1981;1(3):141–5.
106. Marshall DL, Hazlet TK, Gardner JS, Blough DK. Neuroleptic drug exposure and incidence of tardive dyskinesia: a records-based case-control study. J Manag Care Pharm. 2002;8(4):259–65.
107. Fahn S. Long-term treatment of tardive dyskinesia with presynaptically acting dopamine-depleting agents. Adv Neurol. 1983;37:267–76.
108. Paleacu D, Giladi N, Moore O, Stern A, Honigman S, Badarny S. Tetrabenazine treatment in movement disorders. Clin Neuropharmacol. 2004;27(5):230–3.
109. Tarsy D. Tardive dyskinesia. Curr Treat Options Neurol. 2000;2(3):205–14.
110. Bonelli RM, Hofmann P. A systematic review of the treatment studies in Huntington's disease since 1990. Expert Opin Pharmacother. 2007;8(2):141–53.
111. Wu J, Tang T, Bezprozvanny I. Evaluation of clinically relevant glutamate pathway inhibitors in in vitro model of Huntington's disease. Neurosci Lett. 2006;407(3):219–23.
112. Qin ZH, Wang J, Gu ZL. Development of novel therapies for Huntington's disease: hope and challenge. Acta Pharmacol Sin. 2005;26(2):129–42.

113. Pappa S, Tsouli S, Apostolou G, Mavreas V, Konitsiotis S. Effects of amantadine on tardive dyskinesia: a randomized, double-blind, placebo-controlled study. Clin Neuropharmacol. 2010;33(6):271–5.
114. Freudenreich O, McEvoy JP. Added amantadine may diminish tardive dyskinesia in patients requiring continued neuroleptics. J Clin Psychiatry. 1995;56(4):173.
115. Cowen MA, Green M, Bertollo DN, Abbott K. A treatment for tardive dyskinesia and some other extrapyramidal symptoms. J Clin Psychopharmacol. 1997;17(3):190–3.
116. Tarsy D, Kaufman D, Sethi KD, Rivner MH, Molho E, Factor S. An open-label study of botulinum toxin A for treatment of tardive dystonia. Clin Neuropharmacol. 1997;20(1):90–3.
117. Nobuhara K, Matsuda S, Okugawa G, Tamagaki C, Kinoshita T. Successful electroconvulsive treatment of depression associated with a marked reduction in the symptoms of tardive dyskinesia. J ECT. 2004;20(4):262–3.
118. Peng LY, Lee Y, Lin PY. Electroconvulsive therapy for a patient with persistent tardive dyskinesia: a case report and literature review. J ECT. 2013;29(3):e52–4.
119. Mentzel CL, Tenback DE, Tijssen MA, Visser-Vandewalle VE, van Harten PN. Efficacy and safety of deep brain stimulation in patients with medication-induced tardive dyskinesia and/or dystonia: a systematic review. J Clin Psychiatry. 2012;73(11):1434–8.
120. Spindler MA, Galifianakis NB, Wilkinson JR, Duda JE. Globus pallidus interna deep brain stimulation for tardive dyskinesia: case report and review of the literature. Parkinsonism Relat Disord. 2013;19(2):141–7.

Chapter 16
Structural Causes of Chorea

Jon Snider and Roger L. Albin

Abstract Chorea developing from structural causes is comparatively rare, though these cases have a variety of possible etiologies. Vascular lesions represent the most frequent cause, with the vast majority of clinical symptoms remitting spontaneously with time. Post-traumatic chorea can present acutely or with significant delay from the time of initial injury, with the most effective treatment being evacuation of local hematomas, if present. Chorea from cerebral palsy is often life-long, debilitating, and refractory to medical treatments, often necessitating Deep Brain Stimulation for symptom control. Infectious causes can vary widely in presentation based on the underlying disease state, with resolution usually dependent on effective treatment of the pathogen. Finally, chorea from brain tumors and other similar masses are quite uncommon, but can often be effectively managed with treatment of the underlying tumor. Regardless of the etiology, treatment with dopamine antagonists, benzodiazepines, or anti-spasticity medications can provide symptomatic benefit.

Keywords Structural • Chorea • Ischemic • Infectious • Post-traumatic • Cancer

J. Snider, MD (✉)
Department of Neurology, University of Michigan,
5023 BSRB, 109 Zina Pitcher Place, Ann Arbor, MI 48109-2200, USA
e-mail: josnider@umich.edu

R.L. Albin, MD
Department of Neurology, University of Michigan,
5023 BSRB, 109 Zina Pitcher Place, Ann Arbor, MI 48109-2200, USA

Neurology Service and GRECC, VAAAHS,
5023 BSRB, 109 Zina Pitcher Place, Ann Arbor, MI 48109-2200, USA
e-mail: ralbin@med.umich.edu

F.E. Micheli, P.A. LeWitt (eds.), *Chorea*,
DOI 10.1007/978-1-4471-6455-5_16, © Springer-Verlag London 2014

Introduction

Structural causes of chorea are comparatively rare and varied in etiology. Lesions eliciting choreiform movements can be vascular in origin, post-traumatic, infectious, neoplastic, or result from congenital lesions in cerebral palsy. Unifying these disparate etiologies for structural causes are common anatomic localizations, specifically the basal ganglia or connected structures such as the thalamus. Choreoathetosis is most strongly associated with pathology of the striatum and subthalamic nucleus. It is important to note that the majority of structural lesions in the basal ganglia and connected regions do not result in abnormal movements. The exception may be lesions of the subthalamic nucleus. The lack of strong association between structural lesions of the basal ganglia and choreoathetosis is clearly different from the typical clinical presentation of several neurodegenerative diseases involving the striatum, notably Huntington disease. This discrepancy between clinical manifestations of structural basal ganglia lesions and clinical manifestations of neurodegenerative disorders involving the striatum may be due to lack of structural lesion selectivity with structural lesions affecting several basal ganglia nuclei and other structures simultaneously. It is likely as well that chorea secondary to neurodegenerative processes results from selective loss of specific striatal neuron subpopulations and such selective striatal neuron loss is absent in typical structural pathology. Nonetheless, recognition and characterization of involuntary movements associated with structural brain pathology are valuable for localization, and sometimes treatment, of structural disease processes.

Vascular Causes of Chorea

Epidemiology

Though stroke is the most common structural cause of movement disorders, the frequency of poststroke movement disorders is relatively low. Incidences range from 0.1 to 1.0 % of poststroke patients, with prevalence around 1 %. Of patients developing postischemic movement disorders, hemichorea and hemiballismus are the most common (38 %), compared to dystonia (17–29 %) or some variety of tremors (25 %). Poststroke chorea occurs equally in men and women and tends to occur in older patients (mean 75 years of age) compared to dystonia, tremor, or parkinsonism (with a mean age of 48, 63, and 62, respectively) [1, 2].

Pathophysiology

The typical localizations of ischemic lesions causing chorea are the contralateral basal ganglia, though lesions in these structures do not consistently lead to subsequent movement disorders. The first series of vascular chorea was reported by Martin in 1927, who reviewed 13 cases of acute hemichorea; almost all of these patients died from pneumonia a short time after symptom onset. Autopsies demonstrated lacunar strokes involving the subthalamic nucleus or its efferents. In Ghika et al.'s review of 536 patients with stroke and significant basal ganglia involvement, only 29 manifested a movement disorder [2, 3]. The majority of patients, conversely, developing a movement disorder after stroke will have a basal ganglia lesion contralateral to the clinically affected body parts [1, 2, 4]. Lesions in different components of basal ganglia circuitry or linked structures are responsible for producing identical movement disorders. Based on case series and neuroimaging reports, the most common sites of lesions causing chorea are the contralateral lentiform nuclei, thalami, and subthalamic nuclei [1, 5, 6].

In Ghika et al.'s review, infarcts responsible for chorea are predominantly (66 %) small vessel and less than 1.5 mm in size. Lesions were primarily in the contralateral MCA distribution, though may involve PCA territory deep perforators, particularly involving the thalamus. Other less frequent etiologies include medium- or large-vessel atherosclerotic or cardioembolic strokes [1, 7]. There are primarily pediatric case reports of chorea as a complication of moyamoya in up to 6 % of cases of this disorder, a significantly higher percentage than in usual ischemic injury [8]. The greatest risk factors for lesions causing chorea are hypertension and heart disease, expected given the association of these risk factors with small-vessel lacunar infarcts that make up the majority of these cases [2].

The presumed mechanism of chorea in these cases is direct ischemic damage or local compression from vasogenic edema or hemorrhage in these regions. In some cases, chorea onset may be delayed in onset relative to the ictal stroke. The specific mechanism for delayed-onset chorea is not well understood, though some hypotheses include toxic-metabolic insults that evolve over time, slow apoptosis of basal ganglia cells adjacent to areas of infarct, or delayed rearrangement of basal ganglia neural circuitry after ischemic injury (Table 16.1) [4].

Clinical Manifestations and Diagnosis

Though there is wide variability in the time of onset of symptoms after stroke (ranging from days to years), choreiform movements tend to occur much earlier than other poststroke movement disorders. Choreoathetosis has a mean onset of 4.3 days after ictal stroke compared to more than 120 days in parkinsonism and anywhere from 1 day to 5 years with dystonia [1]. The clinical presentation is predominantly contralateral hemichorea, though in some instances, generalized chorea is reported. In Ghika et al.'s review of the Lausanne Stroke Registry, the majority of cases (56 %) involve both arm and leg, followed by face, arm, and leg; considerably rarer is isolated extremity involvement. The majority (82 %) of these infarcts were ischemic [2, 6]. Associated symptoms differ based on lesion location. In a review by Handley et al., patients presented with a variety of deficits including transient motor impairments, sensory loss (particularly in thalamic strokes), mental status changes, visual phenomenon, language difficulties, and bulbar symptoms [1, 5]. As noted previously, concomitant development of dystonias, tremor, or parkinsonism is also a possibility. Much rarer, though occasionally reported and difficult to differentiate from chorea, is associated focal myoclonus [2].

Table 16.1 Clinical and imaging characteristics in a series of patients with chorea secondary to ischemia

Patient no./ sex/age (years)	Clinical features	CT scan or MRI
1/M/85	Left hemichorea S, A	Bilateral temporal infarct
2/F/74	Right hemichorea S, A, BM	Right paramedian thalamic infarct, left accipital calcification
3/F/70	Left hemichorea S, A, BM	Left posterolateral thalamic infarct
4/M/77	Left hemichorea S, A, BM	Right paramedian putaminothalamic capsular infarct
5/F/67	Right hemichorea S, A, BM	Left posterolateral thalamic infarct
6/M/80	Generalized chorea S, A	Bilateral cortical temporal infarcts
7/F/77	Rights hemichorea S, A	Bilateral lenticular infarcts and bilateral frontal temporal infarcts
8/F/74	Left hemichorea S, A	Bilateral external capsular infarcts
9/M/75	Right hemichorea S, A	Left putaminocapsulo-thalamic hemorrhage
10/F/75	Right hemichorea S, A	Left frontotemporal hemorrhagic infarct and right striatal lacunar infarct
11/M/58	Right hemichorea S, A	Left posterolateral thalamocapsulo-lenticular hemorrhage
12/F/75	Left hemichorea S, A	Bilateral corona radiata infarcts
13/F/76	Left hemichorea S, A	Right thalamocapsular infarct
14/F/90	Right hemichorea S, A, BM	Left pallidal capsular infarct
15/M/78	Generalized chorea S, A, BM	Bilateral thalamic hematoma
16/M/71	Left hemichorea S, A	Bilateral lenticular and right thalamic infarcts
17/F/71	Right hemichorea S, A	Bilateral lenticular-striatal infarcts
18/F/83	Right hemichorea S, A, BM	Right cerebellar and left thalamic infarct
19/F/72	Left foot chorea S, A, dystonia	Right corona radiata and right lateral pons infarcts
20/F/53	Left hemichorea S, A, BM	Right posterolateral thalamic and subthalamic infarct

Reproduced with permission from Alarcon et al. [1], p 1570
Type of chorea: *S* spontaneous, *A* intensified by action, *BM* with ballistic movement, *CT* computed tomography, *MRI* magnetic resonance imaging

Treatment

The vast majority of cases of hemichorea secondary to vascular injury, fortunately, remit spontaneously or show at least partial improvement. Ghika et al. noted that 50 % of their case series had spontaneous remission in 2 weeks, with partial improvement in another 37 % [2]. Chung et al. found that 56 % of patients had total resolution of symptoms within 1–2 months, with the severity of symptoms decreasing over time in those who remained affected; patients with primarily cortical lesions had an even higher remission rate (86 %). Of those with persistent symptoms, most had isolated subthalamic nucleus, putamen, or caudate lesions [6]. In the modest experience of one of the authors (RLA), persistent poststroke choreoathetosis is most often associated with putaminal injury. In a small series reported by Alarcon et al., 10 % fully recovered and 75 % partially recovered [1]. Symptoms can cause significant distress in the short term, necessitating symptomatic management with antidopaminergic agents. Typical or atypical antipsychotics are the mainstay of treatment; however, there is risk of induction of parkinsonism and tardive dyskinesia [2]. Tetrabenazine affords similar protection without the risk of tardive dyskinesias. Anecdotal reports suggest that clonazepam, sodium valproate, and topiramate may be useful [5]. For medically refractory cases, lesional surgeries or deep brain stimulation to the thalamic VPN can be effective [5]. In the unusual setting of moyamoya disease, superficial temporal artery-middle cerebral artery bypass improves chorea [9].

Post-traumatic Chorea

Epidemiology

The prevalence of movement disorders in post-traumatic is quite variable in reported series with chorea around 4 % [10–12].

Pathophysiology

As with ischemic lesions, post-traumatic chorea results usually from injury to the basal ganglia. Both primary and secondary processes may cause relevant injury. Primary injury could consist of direct mechanical disruption including contusion, hemorrhagic compression, and diffuse axonal injury [10]. Sung et al. suggested that accumulation of gradually increasing pressure and ischemia from compression of the basal ganglia by traumatic injury in adjacent regions could explain case reports noting substantial time lags between onset of injury and clinical manifestation of symptoms; this theory is supported by a patient they studied who developed chorea months after onset of subdural hematoma and whose symptoms spontaneously resolved soon after evacuation of the hematoma [13]. Secondary injury may involve compromise of small vessels in the setting of edema and mass effect, hypoxia, hypotension, increased intracranial pressure, or neurotoxic effects of local cellular injury [10, 14]. Jankovic et al. suggested that post-traumatic chorea may arise from effects of brain plasticity; aberrant regenerative changes may eventually lead to delayed clinical manifestations [15]. There are a number of case reports of patients with subdural or epidural hematomas in the region of the basal ganglia resulting in chorea, with ipsilateral, contralateral, or bilateral chorea relative to the hematoma sites [13, 16–18]. In ipsilateral lesions, it is hypothesized that in some cases, there may actually be an occult contralateral ischemic lesion contributing to symptomatology [16].

Clinical Manifestations and Diagnosis

Similar to ischemic cases of chorea, onset of post-traumatic chorea can be acute, subacute, or even delayed by months to years [14]. Some studies note associations between the severity of head injury and manifestations of chorea; specifically, milder injuries were less likely to result in chorea but tend to have longer latencies to chorea onset [10–12]. Imaging that demonstrates hemorrhage involving or adjacent to basal ganglia structures may help confirm the diagnosis. However, in nonhemorrhagic cases, lesions may not be readily apparent. Diagnosis may be complicated when onset of symptoms occurs well after the inciting trauma. A careful history of preceding injuries is critical to identifying a potentially causal relationship. King et al., for example, reported a case of delayed post-traumatic hemiballismus occurring 2 years after initial injury and subsequently demonstrated an atrophied subthalamic nucleus at autopsy [14]. Kant et al. noted a patient who experienced closed head injury with subsequent hemiballismus without MRI correlate, and subsequent SPECT imaging reported consistent with a subthalamic lesion [19].

Treatment

For patients who are found to have acute or chronic hematomas potentially compromising the basal ganglia in the setting of chorea, evacuation of the hematoma often leads to marked or total resolution of symptoms, usually within days to weeks [13, 16, 17]. In patients without a clear hemorrhagic injury after trauma, the degree of resolution is variable. For medical management, case reports note the benefit of dopamine antagonists, as well as antiepileptics such as valproic acid and phenobarbital [20, 21]. For medically refractory post-traumatic chorea, a number of case reports document the benefit of stereotaxic thalamotomy. In the majority of these cases, there is significant improvement in chorea or hemiballismus soon after surgery, with benefits that persist over long-term follow-up. These benefits may come at the risk of worsening dysphagia or dysarthria [12, 22]. Though deep brain stimulation has been demonstrated to be useful in the treatment of choreiform dyskinesias, evidence of its efficacy in post-traumatic settings is limited. Hooper et al. demonstrated resolution of post-traumatic hemiballismus after placement of an electrode in a patient's subthalamic nucleus, though the fact that the benefit occurred and persisted prior to initiation of electrical stimulation of the implanted electrode suggests a lesional effect of surgery [23].

Cerebral Palsy

Epidemiology and Pathophysiology

Cerebral palsy is a term encompassing a wide range of nonprogressive motor abnormalities secondary to brain abnormalities incurred early in brain development. The incidence of cerebral palsy is 1–2.5 per 1,000 births, with varying etiologies including hemorrhagic-ischemic damage, other perinatal insults, and infection. Athetosis and chorea are seen in 25 % of cases of cerebral palsy. Basal ganglia injury is the root cause of athetosis, and the predominant structures involved vary with the etiology of cerebral palsy, with kernicterus affecting globus pallidus and STN, and hypoxic-ischemic injury tending to primarily impact the caudate and putamen. Pathologically, autopsies often demonstrate "status marmoratus," marbling of the basal ganglia by fibers that disrupt the normal architecture [24].

Clinical Manifestations and Diagnosis

Cerebral palsy can result in a number of overlapping motor phenomena, including chorea, dystonia, spasticity, ataxia, and weakness. Though the cerebral injury is static, choreiform movements in cerebral palsy are often delayed in onset from the time of insult. The evolution of motor phenomena in cerebral palsy reflects the intersection of further postnatal brain development and the underlying pathologies. In cases of athetosis, 50 % are apparent in the first year of life; however, symptom onset may be delayed as long as 5 years. Symptoms are usually bilateral but can be asymmetrical. Common associated features include dysarthria (66 %), drooling, strabismus (33 %), seizures (25 %), lower IQ (25 %), and sensorineural hearing loss. Cognitive impairment in children with athetotic cerebral palsy is often minimal or absent. MRI scans will detect abnormalities in affected structures in approximately two thirds of cases, but many remain occult on MRI. In addition, imaging in many cases of cerebral palsy shows various basal ganglia lesions without clinical manifestations of extrapyramidal symptomatology [24–26].

Treatment

Medical therapy for chorea in cerebral palsy is similar to treatments in other structural causes of chorea, specifically a focus on dopamine antagonists, benzodiazepines, and anti-spasticity medications as mainstays of treatment. Unfortunately, many cases of chorea in cerebral palsy are either resistant to these treatments or limited by intolerable side effects; these cases can often persist into adulthood, resulting in significant disability. For this reason, attention has shifted more recently to the use of deep brain stimulation in adults with residual refractory chorea from cerebral palsy. Vidailhet et al. performed a multicenter prospective study of 13 adults with largely isolated dystonia-chorea secondary to cerebral palsy without significant spasticity or intellectual deficits, who underwent bilateral DBS placement. At one year reevaluations, they noted that 8 out of 13 patients demonstrated statistically significant improvement in functional disability, pain, and mental health quality of life (overall 24 % improvement). No significant psychiatric or cognitive effects were noted. The remaining patients had smaller degrees of improvement or did not show change in symptoms. Patients who demonstrated improvement and were using oral pharmacotherapy for their chorea prior to the surgery did not require it at 1 year afterward. In this small series, the best results were obtained with electrode placement in the posterolateroventral region of the GPi. It should be noted that it was unclear how much of the benefit came from improvement of dystonia versus chorea, given that many participants had overlapping features [27].

Infectious Structural Causes of Chorea

While there are a myriad of possible infectious etiologies of chorea, the majority of the recent published literature describes chorea associated with HIV infection and concomitant toxoplasmosis infection affecting subcortical structures. HIV itself, however, or other features of HIV infections can cause structural pathology eliciting this chorea. A wide variety of viral, bacterial, fungal, and parasitic infections are also reported as occasional structural causes of chorea. Diagnosis in all cases can be based on serologic and CSF studies or cultures. Treatment and prognosis vary widely.

Much of the data on movement disorders in HIV was accumulated in the pre-HAART era and this is particularly true for infectious causes of movement disorders in HIV-infected individuals. For HIV-positive patients, movement disorders as a whole occur with a frequency of 2–3 %. Within this cohort, chorea is the most common or second most common movement abnormality [28, 29]. Case series have shown that the mean age of affected patients tends to be younger (mean age of onset 32.5 years) and largely male. The mean time from diagnosis of HIV to onset of symptoms was 28 months and in some cases was actually the presenting symptom of AIDS. HIV predominantly appears to be able to induce chorea via allowing opportunistic infections such as toxoplasmosis, cryptococcus, CMV, tuberculosis, and progressive multifocal leukoencephalopathy to affect basal ganglia structures. In the vast majority of these cases, the primary infectious agent is toxoplasmosis infection and abscesses; in fact, the development of hemichorea in an HIV-infected patient is almost pathognomonic of underlying cerebral toxoplasmosis. Though 50 % of patients with HIV and toxoplasmosis will have subcortical abscesses, of those patients, only 7.4 % will also develop chorea [4, 30].

While the majority of patients with chorea in the setting of toxoplasmosis infection and HIV have lesions in the subthalamic nucleus, localization is complicated by the fact that these patients usually have multiple lesions scattered throughout their subcortical regions. Other possible foci of infection can include caudate, putamen, thalamus, global pallidus, midbrain, and internal capsule (Fig. 16.1). Cerebral toxoplasmosis was well recognized prior to the AIDS epidemic; however, it was only after the spread of HIV that reports of movement disorders associated with this condition were reported. One hypothesis is that HIV itself causes concomitant damage to subcortical structures or enhances toxoplasmosis effects on the basal ganglia [30, 31].

There is also evidence that HIV alone, without coexisting opportunistic infections, may induce movement disorders through structural changes in the basal ganglia. PET studies demonstrated that basal ganglia are often one of the first brain

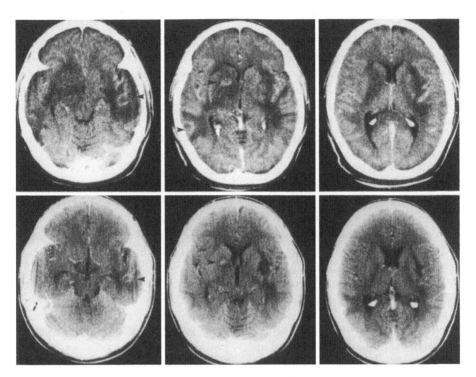

Fig. 16.1 Imaging demonstrates numerous ring-enhancing lesions from toxoplasmosis infection in an HIV-positive patient (*top*), with improvement following treatment (*bottom*) (Reproduced with permission from Nath et al. [31], p. 108)

structures affected in HIV, with autopsy studies revealing pathologic evidence of HIV infiltration into these structures. Gallo et al., Passarin et al., and Piccolo et al. reported patients with progressive HIV encephalopathy concurrent with generalized chorea. Though MRI performed by Piccolo et al. was negative for basal ganglia disease, Passarin et al. reported basal ganglia changes on MRI and hypometabolism of basal ganglia on PET scan; Gallo et al. reported autopsy data suggesting basal ganglia neuron loss and infiltration of microglia and astrogliosis [32–34].

Clinically, HIV infection often presents with focal neurologic findings as well as generalized cerebral dysfunction. Nath et al. reviewed cases of movement disorders in patients with cerebral toxoplasmosis. The most common reported movement disorder was unilateral chorea affecting proximal or distal extremity muscles; rare cases in which the chorea has been found to be generalized include bilateral toxoplasmosis and HIV encephalitis [31]. Movements may be acute or subacute in onset. Even more rare are situations in which facial movements are involved [30]. The mainstays of treatment in HIV-associated chorea include treatment of opportunistic infections, HAART therapy, and symptomatic treatment of the movements themselves. Case reports of patients with HIV encephalopathy show total resolution of chorea within months with appropriate antiretroviral therapy [33, 34]. In patients with HIV-/toxoplasmosis-induced chorea, focal neurologic symptoms usually resolved after adequate treatment of abscesses. However, in one series by Nath et al., symptoms remained refractory despite resolution of the infectious lesions; these patients required dopamine antagonists for symptomatic relief. Nath hypothesized that persistent gliosis or unrecognized effects of the HIV infection itself may be resulting in ongoing symptoms in these cases [31].

There are reported cases of neurosyphilis presenting with hemichorea, with a few modern-era patients reported. Binit et al. reviewed cases of neurosyphilitic chorea from 1917 to 2009, of which only half a dozen patients are reported; however, modern cases frequently involve coexisting HIV infections as well [35]. The clinical presentation of neurosyphilis chorea ranges from hemichorea to generalized chorea, often associated with more classic features of neurosyphilis such as general paresis and tabes dorsalis; rarely are structural lesions noted on MRI in these cases [35]. Patients with neurosyphilis and chorea generally see their movement disorder resolve with appropriate treatment of their infection with high-dose penicillin [35].

Progressive multifocal leukoencephalopathy occurs in 4–8 % of patients with AIDS, with multifocal demyelinating plaques in zones of high blood flow between gray and white matter, but can also involve the basal ganglia and be a rare cause of chorea. Case reports of patients with HIV-associated PML with chorea are almost entirely fatal [33].

There are case reports in which chorea developed in bacterial meningitis. Basal ganglia are generally spared in acute bacterial meningitis, explaining low frequency of movement disorders in this setting. Burstein et al. reported pediatric cases of bacterial meningitis presenting with generalized chorea, with causative organisms including *S. pneumoniae*, *N. meningitidis*, and *H. influenzae*. Initiation of movements began 1–2 weeks after disease onset. In all cases except the patient with meningococcal meningitis, there was preceding administration of anticonvulsants as prophylaxis within 1 week of onset of chorea, a possible alternative explanation for chorea [36]. Anticonvulsants are unlikely to be a significant confounder given that the specific antiepileptics used in these instances vary widely from case to case and symptoms persist even after anticonvulsants are discontinued [37]. The association between chorea and meningococcal meningitis was further supported by a pediatric case presenting with generalized upper extremity choreiform movements. MRI confirmed bilateral basal ganglia lesions secondary to cytotoxic edema from infection [38]. Clinically, chorea tends to be generalized in these instances [36]. Burstein et al.'s series of bacterial meningitis reported the time frame of resolution of generalized chorea ranging from days to months, while Scott et al. noted improvement but not total resolution after 8 weeks of treatment [36, 38].

Tuberculosis is one of the most common meningitides in the developing world and is associated with movement disorders in pediatric cases (18 % of patients), with the frequency of development of chorea in these situations being 4–11 % [39, 40]. In a significant number of patients, there are no identifiable MRI lesions, suggesting alternative mechanisms of injury. When MRI lesions are present, they tend to exist in the contralateral thalamus or basal ganglia (including ischemic damage to contralateral lenticular-striatal structures). Autopsies may reveal direct infiltration by basal exudates or tuberculomas into subthalamic and putaminal regions. Clinically, TB can present with either hemichorea or generalized chorea [39, 40]. Chorea from tubercular meningitis is self-limited in the majority of cases, usually within days to weeks; this may be due to destruction of other adjacent basal ganglia structures (particularly the globus pallidus) resulting in relative suppression of movement. Other case reports note remission of chorea after aggressive treatment of the tuberculosis with steroids and antibiotics [39–41].

Weeks et al. reported a case of cryptococcal meningitis associated with the onset of hemichorea; though CT head did not show definitive lesions in the basal ganglia, there were low attenuation lesions in the white matter that led them to theorize the presence of CT-occult disturbance of blood supply or local fungal invasion [42]. Namer et al. also documented a similar case of hemiballismus in cryptococcal meningitis, with contrast-enhanced CT showing hyperdense lesions in the head of the right caudate nucleus, the left internal capsule, and frontal and occipital lobes. Clinical presentation was contralateral hemichorea [43]. Patients with cryptococcal meningitis and chorea tend to either improve or show total resolution with appropriate treatment of the cryptococcal infection [42, 43].

There are reports of herpes simplex encephalitis (HSE) associated with movement disorders. Hargrave et al. describe a case series of 20 pediatric patients with HSE and subsequent delayed development of chorea weeks after initial onset of infection. MRIs in these cases did not demonstrate basal ganglia lesions at the time of presentation of the movement disorder. Kullnat et al. documented a case of HSE-associated chorea in which serial MRIs demonstrated the development of bilateral medial thalamus lesions over the course of weeks. The delayed onset of chorea weeks after HSE infection may be due to residual infection (though less likely, as repeat CSF HSV PCRs were most often negative at time of symptom onset). Alternatively, an immune-mediated process in response to the initial acute infection may occur, such as that seen in cases of Sydenham's chorea [37, 44]. Cases of HSE demonstrated chorea of all four extremities as well as facial chorea, with onset of symptoms tending to occur weeks after disease onset, usually in the setting of a clinical relapse [37, 44]. HSE-associated chorea also usually demonstrates resolution of the movement disorder over time. Hargrave et al. reported that at 8-month follow-up, there was self-limited resolution in 50 % of patients, persistence with improvement in 25 %, and persistence without improvement in 25 %; by 5-year follow-up, only 1 out of 20 patients had persistent symptoms. However, even with symptom resolution, there is still profound impairment from residual cortical damage as a result of the encephalitis; while the mortality rate of treated HSE is 10 %, in patients who presented with chorea it is increased to 20 % [37, 44].

Neurocysticercosis is a common CNS parasitic infection in developing nations. It is estimated that 25 % of cases have lesions affecting the basal ganglia, most commonly affecting the striatum and pallidum. The probability of basal ganglia involvement is proportional to the overall number of cysts. Despite significant basal ganglia disease burden, the vast majority is clinically silent. Possibilities for the lack of development of choreiform movements could include small cysts, slow growth, and adaptation to host tissue. Typical presenting symptoms of neurocysticercosis include seizures, headache, increased ICP, focal neurologic deficits, and dementia. When chorea is present, it is reported as hemichorea. Patients with neurocysticercosis often see significant improvement with steroids and albendazole, even though the lesion burden itself appears to remain stable [45].

Brain Tumors and Other Mass Lesions

Epidemiology and Pathophysiology

Tumors of the thalamus and basal ganglia are rare, estimated to make up 1–3 % of intracranial tumors. The majority of these tumors are astrocytomas. In patients with basal ganglia-thalamic tumors, roughly 1–7 % experience movement disorders. Krauss et al. examined a series of 225 patients with biopsy-proven astrocytomas (grades I–IV) involving the basal ganglia and thalamus. In this cohort, 9 % ($n=20$) had movement disorders, of whom 3 had chorea. More frequent movement disorders included tremor ($n=12$) and dystonia ($n=8$). Lesions were noted in a variety of subcortical structures including subthalamic nucleus, caudate, internal capsule, and thalamus. Though it is not known why so many patients with significant basal ganglia disease burden do not manifest chorea, it can be hypothesized that diffuse astrocytoma involvement diffusely affects striatal neurons, as opposed to the subpopulation-specific degeneration typical of neurodegenerative disorders with chorea [46].

Meningiomas may be associated with chorea. The most common sites for meningiomas causing movement disorders are the parasagittal region, the sphenoid ridge, and the cerebral convexity. Dieckmann et al. report a few cases of hemiballismus secondary to contralateral meningioma, most frequently presenting with significant edema and mass effect, likely causing compression around the basal ganglia. Various possible mechanisms for the production of movement disorders in these patients include temporary displacement of basal ganglia structures without permanent damage (supported by the fact that decompression leads to rapid resolution of symptoms and that autopsy data has not noted permanent pathologic changes to basal ganglia structures), compromise of dopaminergic pathways, neurotransmitter imbalance in the GABA system, or some unknown factor production by the tumor itself that exerts local effects in the basal ganglia [47]. Barriero et al. reported a rare case of a patient with a giant unruptured aneurysm with hemichorea secondary to direct impingement of the left thalamus, putamen, globus pallidus, and subthalamic nucleus from this mass lesion [48].

Metastatic disease causing chorea is much rarer, likely owing to its predilection for gray-white junctions distant from basal ganglia structures. Glass et al. review a number of cases of metastatic cancer presenting as hemichorea. In this series, cancer primaries were lung and breast (the primary cancers that most commonly metastasize to brain) with metastatic disease noted on autopsy in the contralateral subthalamic nucleus, caudate nucleus, putamen, or ventral thalamic nucleus [49]. Karampelas et al. reported a case of hemichorea with metastatic breast cancer seen on MRI to be invading the STN [50]. Ziania et al. noted a unique case of a patient with metastatic squamous cell cancer of the cervix involving the left basal ganglia on MRI, resulting in acute contralateral hemiballismus; this is notable also due to the unusual location of the metastasis. Though the incidence of brain metastases in cervical cancer has been reported as 0.8–5.0 % (with a mean time from diagnosis to brain metastasis of 18 months), this cancer tends to favor spread to the frontal or parietal lobes and not subthalamic structures [51]. Finally, Laing et al. have a case report of an 84-year-old female with acute chorea, with autopsy noting metastatic gastric adenocarcinoma with extensive intravascular dissemination with direct occlusion of vessels by neoplasm that caused numerous subcortical infarcts [52]. Hengstman et al. reported a case of a 68-year-old female with recently diagnosed mycosis fungoides who subsequently developed left hemichorea, with MRI demonstrating a single mass affecting the right caudate, internal capsule, and putamen; this was felt to be a metastatic lesion from her MF. This condition rarely presents with parenchymal involvement (0.9 % of cases), with only two reported cases of basal ganglia involvement and only one with clinical effects [53].

Clinical Manifestations and Diagnosis

The affected astrocytoma patients in Krauss et al.'s review all experienced hemichorea, including one case with facial involvement. Tissue analysis is often critical to confirm the diagnosis and guide proper treatment. Patients with meningiomas that provoke ballismus tended to present with hemiballismus, with lesions clearly seen on imaging and confirmed on subsequent resection and histopathology [47]. Case reports have generally shown that metastatic lung, breast, and cervical cancers manifest as contralateral hemichorea. These metastatic lesions often also present with typical symptoms of headache, nausea, vomiting, focal weakness, and mental status changes [49–52].

Treatment

Krauss et al. noted that the small group of astrocytoma-related choreas they reviewed had universal resolution of symptoms following aggressive treatment of astrocytoma with radiation therapy; this highlights that the optimal long-term treatment strategy for many malignancies provoking movement disorders is management of the underlying tumor [46]. Similarly, resection of meningiomas usually leads to rapid and total resolution of underlying chorea or hemiballismus [47]. While cases of metastatic cancer to the brain causing hemichorea or hemiballismus were universally fatal (usually within months to a year), some patients saw at least temporary improvement or resolution of abnormal movements with steroids and radiation therapy. Overall prognosis is limited by the prognosis of metastatic disease, which is often poor [49–52].

Conclusion

In summary, though generally rare etiologies of chorea, there are a wide variety of potential structural causes to consider when presented with a patient with this movement disorder. These conditions include vascular, traumatic, infectious, and mass lesions and can exhibit variable clinical presentations. Appropriate treatment varies markedly and depends on the correct diagnosis of the underlying etiology. Evaluation will be dictated by the setting in which chorea arises. Careful examination is important as it may disclose other useful neurologic signs. In many instances, prognosis for functional improvement and recovery is good with appropriate treatment. Furthermore, a number of symptomatic treatment options for refractory cases can also help to hasten return to baseline. Recent developments in deep brain stimulation also provide additional hope for patients with persistent chorea.

References

1. Alarcon F, et al. Post-stroke movement disorders: report of 56 patients. J Neurol Neurosurg Psychiatry. 2004;75(11):1568–74.
2. Ghika-Schmid F, et al. Hyperkinetic movement disorders during and after acute stroke: the Lausanne Stroke Registry. J Neurol Sci. 1997;146(2):109–16.
3. Martin JP. Hemichorea resulting from a local lesion of the brain (the syndrome of the Body of Luys). Brain. 1927;50(3–4):637–49.
4. Piccolo I, et al. Cause and course in a series of patients with sporadic chorea. J Neurol. 2003;250(4):429–35.
5. Handley A, et al. Movement disorders after stroke. Age Ageing. 2009;38(3):260–6.
6. Chung SJ, et al. Hemichorea after stroke: clinical-radiological correlation. J Neurol. 2004;251(6):725–9.
7. Sugiura A, Fujimoto M. Facial chorea and hemichorea due to cardiogenic cerebral embolism in the cortex and subcortical white matter. Rinsho Shinkeigaku. 2006;46(6):415–7.
8. Pavlakis SG, et al. Steroid-responsive chorea in moyamoya disease. Mov Disord. 1991;6(4):347–9.
9. Watanabe K, et al. Moyamoya disease presenting with chorea. Pediatr Neurol. 1990;6(1):40–2.
10. Krauss JK, Trankle R, Kopp K-H. Post-traumatic movement disorders in survivors of severe head injury. Neurology. 1996;47(6):1488–92.
11. O'Suilleabhain P, Dewey Jr RB. Movement disorders after head injury: diagnosis and management. J Head Trauma Rehabil. 2004;19(4):305–13.
12. Krauss JK, et al. The treatment of posttraumatic tremor by stereotactic surgery: symptomatic and functional outcome in a series of 35 patients. J Neurosurg. 1994;80(5):810–9.
13. Sung Y-F, Ma H-I, Hsu Y-D. Generalized chorea associated with bilateral chronic subdural hematoma. Eur Neurol. 2004;51(4):227–30.

14. King RB, Fuller C, Collins GH. Delayed onset of hemidystonia and hemiballismus following head injury: a clinicopathological correlation: case report. J Neurosurg. 2001;94(2):309–14.
15. Jankovic J. Post-traumatic movement disorders central and peripheral mechanisms. Neurology. 1994;44(11):2006.
16. Yoshikawa M, et al. Hemichorea associated with ipsilateral chronic subdural hematoma – case report. Neurol Med Chir. 1992;32(10):769–72.
17. Adler JR, Winston KR. Chorea as a manifestation of epidural hematoma: case report. J Neurosurg. 1984;60(4):856–7.
18. Kotagal S, Shuter E, Horenstein S. Chorea as a manifestation of bilateral subdural hematoma in an elderly man. Arch Neurol. 1981;38(3):195.
19. Kant R, Zailer D. Hemiballismus following closed head injury. Brain Inj. 1996;10(2):155–8.
20. Chandra V, Spunt AL, Rusinowitz MS. Treatment of post-traumatic choreo-athetosis with sodium valproate. J Neurol Neurosurg Psychiatry. 1983;46(10):963.
21. Drake Jr ME, Jackson RD, Miller CA. Paroxysmal choreoathetosis after head injury. J Neurol Neurosurg Psychiatry. 1986;49(7):837.
22. Bullard DE, Jr Nashold BS. Stereotaxic thalamotomy for treatment of posttraumatic movement disorders. J Neurosurg. 1984;61(2):316–21.
23. Hooper J, Simpson P, Whittle IR. Chronic posttraumatic movement disorder alleviated by insertion of meso-diencephalic deep brain stimulating electrode. Br J Neurosurg. 2001;15(5):435–8.
24. Albright AL. Spasticity and movement disorders in cerebral palsy. J Child Neurol. 1996;11(1 Suppl):S1–4.
25. Pranzatelli MR. Oral pharmacotherapy for the movement disorders of cerebral palsy. J Child Neurol. 1996;11(1 Suppl):S13–22.
26. Filloux FM. Neuropathophysiology of movement disorders in cerebral palsy. J Child Neurol. 1996;11(1 Suppl):S5–12.
27. Vidailhet M, et al. Bilateral pallidal deep brain stimulation for the treatment of patients with dystonia-choreoathetosis cerebral palsy: a prospective pilot study. Lancet Neurol. 2009;8(8):709–17.
28. Sporer B, et al. HIV–induced chorea. J Neurol. 2005;252(3):356–8.
29. de Mattos JP, et al. Movement disorders in 28 HIV-infected patients. Arq Neuropsiquiatr. 2002;60(3A):525–30.
30. Tse W, et al. Movement disorders and AIDS: a review. Parkinsonism Relat Disord. 2004;10(6):323–34.
31. Nath A, Hobson DE, Russel A. Movement disorders with cerebral toxoplasmosis and AIDS. Mov Disord. 1993;8(1):107–12.
32. Gallo BV, et al. HIV encephalitis presenting with severe generalized chorea. Neurology. 1996;46(4):1163–5.
33. Piccolo L, et al. Chorea in patients with AIDS. Acta Neurol Scand. 1999;100(5):332–6.
34. Passarin MG, et al. Reversible choreoathetosis as the early onset of HIV-encephalopathy. Neurol Sci. 2005;26(1):55–6.
35. Shah BB, Lang AE. Acquired neurosyphilis presenting as movement disorders. Mov Disord. 2012;27(6):690–5.
36. Burstein L, Breningstall GN. Movement disorders in bacterial meningitis. J Pediatr. 1986;109(2):260–4.
37. Kullnat MW, Morse RP. Choreoathetosis after herpes simplex encephalitis with basal ganglia involvement on MRI. Pediatrics. 2008;121(4):e1003–7.
38. Scott O, Hasal S, Goez HR. Basal ganglia injury with extrapyramidal presentation: a complication of meningococcal meningitis. J Child Neurol. 2012;28:1489–92.
39. Udani PM, Parekh UC, Dastur DK. Neurological and related syndromes in CNS tuberculosis clinical features and pathogenesis. J Neurol Sci. 1971;14(3):341–57.
40. Alarcón F, et al. Movement disorders in 30 patients with tuberculous meningitis. Mov Disord. 2000;15(3):561–9.
41. Kalita J, et al. Hemichorea: a rare presentation of tuberculoma. J Neurol Sci. 2003;208(1):109–11.

42. Weeks RA, Clough CG. Hemichorea due to cryptococcal meningitis. Mov Disord. 1995; 10(4):522.
43. Namer IJ, et al. A case of hemiballismus during cryptococcal meningitis. Rev Neurol. 1990;146(2):153.
44. Hargrave DR, Webb DW. Movement disorders in association with herpes simplex virus encephalitis in children: a review. Dev Med Child Neurol. 1998;40(9):640–2.
45. Karnik PS, et al. Neurocysticercosis-induced hemiballismus in a child. J Child Neurol. 2011;26(7):904–6.
46. Krauss JK, et al. Movement disorders in astrocytomas of the basal ganglia and the thalamus. J Neurol Neurosurg Psychiatry. 1992;55(12):1162–7.
47. Dieckmann M, et al. Hemiballismus due to contralateral sphenoid ridge meningioma. J Clin Neurosci. 1998;5(3):350–3.
48. Barreiro de Madariaga LM, et al. Arm chorea secondary to an unruptured giant aneurysm. Mov Disord. 2003;18(11):1397–9.
49. Glass JP, Jankovic J, Borit A. Hemiballism and metastatic brain tumor. Neurology. 1984;34(2):204.
50. Karampelas I, et al. Subthalamic nucleus metastasis causing hemichorea-hemiballism treated by gamma knife stereotactic radiosurgery. Acta Neurochir. 2008;150(4):395–7.
51. Ziainia T, Resnik E. Hemiballismus and brain metastases from squamous cell carcinoma of the cervix. Gynecol Oncol. 1999;75(2):289–92.
52. Laing RW, Howell SJL. Acute bilateral ballism in a patient with intravascular dissemination of gastric carcinoma. Neuropathol Appl Neurobiol. 1992;18(2):201–5.
53. Hengstman GJD, et al. Chorea due to mycosis fungoides metastasis. J Neurooncol. 2005; 73(1):87–8.

Chapter 17
Chorea: A Surgical Approach

Raul Martinez-Fernandez and Elena Moro

Abstract The term chorea refers to a hyperkinetic movement disorder that can present with a broad spectrum of clinical conditions. Choreic movements can be reversible (self-limited) or can be controlled by treating the underlying cause. However, especially in patients who have chorea secondary to neurodegenerative diseases, only symptomatic treatments can be effective (and not always successfully). In such instances, chorea can lead to severe disability with heavy impact on patients' quality of life. As a result, more aggressive treatment must be pursued. Functional neurosurgery has been applied for the treatment of movement disorders since the 1940s. The former ablative procedures have been replaced during the last 25 years by deep brain stimulation and additional functional neurosurgical techniques. When applied to chorea, some investigators report potential utility of these surgical approaches. However, in the absence of high level of scientific evidence, the surgical approach for treating chorea still remains empirical. Several other experimental approaches (such as cellular transplantation and delivery of neurotrophic factors) have showed promising results when applied to animal models of Huntington's disease, the most frequent neurodegenerative cause of chorea.

The aim of this chapter is to review from the literature the available data regarding neurosurgical approaches to chorea and to update the current status of experimental techniques with the promise for treating choreic conditions in the future.

Keywords Ablative neurosurgery • Cellular transplantation • Chorea • Deep brain stimulation • Huntington's disease • Neurotrophic factors

R. Martinez-Fernandez • E. Moro (✉)
Movement Disorder Unit, Department of Psychiatry and Neurology,
CHU de Grenoble, Joseph Fourier University, CS 10217, F-38000 Grenoble,
Cedex 9, 38043, France

INSERM, Unit 836, Institut des Neurosciences, F-38000 Grenoble, France
e-mail: emoro@chu-grenoble.fr

F.E. Micheli, P.A. LeWitt (eds.), *Chorea*,
DOI 10.1007/978-1-4471-6455-5_17, © Springer-Verlag London 2014

Introduction

Chorea is a hyperkinetic movement disorder characterized by involuntary, irregular, fluent, purposeless, rapid, non-stereotyped, and randomly distributed movements [1]. The intensity of these movements is variable and typically fluctuating, being influenced by several internal or external factors. Depending on the cause, chorea ranges from mild and almost imperceptible movements to severe movements of great amplitude and speed (e.g., ballism) that can confer major disability.

The term "chorea" derives from the Greek word *choreia* (χορεία in the ancient Greek spelling) that means "dance." It was formerly used to describe both organic and psychogenic motor disorders observed during the "dancing mania epidemic" coincident with and possibly related to the black plague that outbroke in central Europe in the Middle Ages. In the fourteenth century, Saint Vitus and the other 13 Holy Helpers were venerated and prayed to intercede for curing chorea (especially in France and Germany). This religious practice was the first known "treatment" for chorea [2]. More than 100 years later, Paracelsus (1493–1541) introduced the concept of *chorea naturalis* to distinguish between organic and psychogenic chorea (*chorea imaginativa* and *lasciva*) [3]. In the following century, Sydenham (1686) published his observations of postinfectious childhood chorea (and named it "chorea santi Viti") [2]. Another milestone in the history of studying chorea's phenomenology was made by George Huntington, an American physician. In his seminal essay "On chorea" [4] published in 1872, Huntington provided thorough descriptions of an adult-onset progressive chorea accompanied by neuropsychiatric symptoms and highlighted details of its hereditary nature. Huntington established the clinical features of the disease that would bear his name and started an era of intensive research on this disease. In the twentieth century, studies on chorea reached their peak with the identification in 1993 of the Huntington's disease (HD) gene as a result of an international collaborative effort [5] that further stimulated research on chorea pathophysiology.

The etiology of chorea is the determinant factor in the choice of treatment [6]. Indeed, treating the underlying cause can ameliorate the symptoms in conditions as in Wilson's disease or chorea induced by metabolic disturbances. Unfortunately, for other disorders in which etiologic therapy does not exist, only a few drugs are available to manage chorea [7]. Antidopaminergic agents are the most widely used medication for the symptomatic treatment of chorea. Subsequently, both classic and atypical neuroleptics have shown to lessen choreic movements induced by different causes [8–10]. The presynaptic dopamine depletor tetrabenazine is the only approved drug by the US Food and Drug Administration for the symptomatic treatment of HD [11], and it has also been demonstrated to be beneficial in other choreic disorders hyperglycemic-induced hemichorea-hemiballismus [12], benign hereditary chorea [13], tardive dyskinesia [6], poststroke dystonic choreoathetosis [14], and cerebral palsy chorea [15]. Valproic acid is usually used for post-streptococcal infection chorea [16] and it has also been tested with relative success in posttraumatic [17], postanoxic [18], and vascular hemichorea-hemiballism [19]. In chorea related to systemic immunological diseases (and Sydenham's chorea), corticosteroid therapy can also be effective [20–22]. However, in the cases that are resistant to treatment with medications, more aggressive therapy is required for the symptomatic control.

Brain surgery has been used for the treatment of various movement disorders since the 1930s [23]. The earliest large cerebral resections [24] were followed by the development of safer stereotaxic techniques [25] that were also applied to hyperkinetic disorders [26]. Since the late 1940s, several classic publications have reported variable outcomes of ablative neurosurgery in patients with chorea [27–32]. Spiegel and Wycis [33] achieved clinical improvement in three of nine choreic patients who were treated with pallidotomy and in one of three patients who underwent thalamotomy. Other techniques (such as pyramidotomy in the cerebral peduncle) were also tried [34]. Nevertheless, the majority of these studies are difficult to interpret because of the lack of standardized clinical assessments and unclear diagnosis as to the nature of the choreic syndrome. After a relative neglect of surgery for chorea for several decades, in the late 1980s, the surgical treatment of movement disorders was resurrected by the observations of Benabid and coworkers in 1987 about amelioration of parkinsonian tremor with high-frequency thalamic deep brain stimulation (DBS) [35]. Subsequently, DBS was also successfully applied to the subthalamic nucleus (STN) [36, 37] and to the globus pallidus internus (GPi) [38] for improving all levodopa-responsive signs in Parkinson's disease (PD). Shortly after, GPi DBS was also demonstrated to ameliorate dystonia, a condition that is pathologically more closely related to chorea than to PD [39]. DBS surgery has thus ushered in the modern era of functional neurosurgery.

In recent years, DBS has been extended to treat several psychiatric [40] and other neurological diseases, including those with disabling chorea. At the same time, some other experimental neurosurgical therapies such as cell transplantation [41] and gene therapy have arisen as potential treatments for movement disorders and, in a few instances, have been tested in choreic syndromes (mostly HD) [42, 43]. These new treatment approaches are paving the way for disease modification treatments and more effective focused therapy (see Table 17.1).

The aim of this chapter is to update the surgical treatments of chorea and outline the surgical experimental approaches with possible future therapeutic application.

Causes of Chorea

Chorea is usually identified with HD (previously and less accurately called Huntington's chorea because of the diversity of clinical symptomatology possible), but there are many other conditions, both genetic and sporadic, that can induce choreic movements. Nevertheless, the possible devastating implications and the peculiar clinical presentation of chorea have given this movement disorder a particular focus of interest.

Table 17.1 Surgical approaches in patients presenting with chorea

Type of surgery	Indication	Targets
Deep brain stimulation	Huntington's disease	Bilateral GPi
		Bilateral GPe
	Neuroacanthocytosis	Bilateral GPi
		Bilateral Vop
	Vascular HcHb	Unilateral Vim
		Unilateral Vo-c
		Unilateral GPi
	Child hyperkinesias of unknown cause	Bilateral GPi
	Cockayne syndrome	Unilateral Vim
	Cerebral palsy	Unilateral Vim
		Bilateral GPi
	Tardive dyskinesia	Unilateral GPi
		Bilateral GPi
	Senile chorea	Unilateral Vo-c/GPi
	PNKD	Unilateral Vim
		Bilateral GPi
Ablative neurosurgery	Huntington's disease (Westphal variant)	Bilateral pallidotomy
	Neuroacanthocytosis	Bilateral pallidotomy
	Vascular/lesional HcHb	Unilateral Vo-c thalamotomy
		Unilateral Vim thalamotomy
		Unilateral pallidotomy
		Subthalamotomy
		Zona incerta lesion
		Internal capsule lesion
		Ansa lenticularis lesion
	Tardive dyskinesia	Unilateral pallidotomy
		Bilateral pallidotomy
		Unilateral thalamotomy
Restorative cell transplantation	Huntington's disease	Striatal transplants of human fetal striatal tissue
Ex vivo NTF delivery	Huntington's disease	Intraventricular implantation of encapsulated NTFs

GPi globus pallidus internus, *GPe* globus pallidus externus, *Vop* posteroventral oralis thalamic nucleus, *Vim* ventralis intermedius nucleus of the thalamus, *Voc* Vo-complex of the thalamus, *HcHb* hemichorea-hemiballism, *PNKD* paroxysmal nonkinesigenic dyskinesias, *NTFs* neurotrophic factors

Choreiform syndromes can be classified in many ways, depending on etiology, clinical onset and evolution, distribution, age at onset, or presence of family history (i.e., genetic or nongenetic nature).

Beyond the "classical" form of HD, other inherited conditions can exhibit chorea as a predominant clinical feature. Some of these entities can be accompanied by neuropsychiatric signs and symptoms that resemble those of HD. These disorders have been classified as "HD-like syndromes" and represent about 1 % of cases with an apparent HD phenotype [44]. This group comprises relatively rare entities like dentatorubral-pallidoluysian atrophy [45], neuroacanthocytosis (NA) [46], benign hereditary chorea [47], neurodegeneration with brain iron accumulation (NBIA) syndromes [48, 49], spinocerebellar ataxia 1,2,3, and 17 [50], and HD-like syndromes 1–3 [44]. Chorea can also be present, but less frequently, in many other genetic conditions [51].

Between the acquired choreic syndromes, levodopa-induced dyskinesias (LID) in patients with PD are probably the most common cause of drug-induced chorea [52]. Many other drugs are known to induce chorea [51, 53], especially neuroleptic medications [54].

Acute-onset hemichorea is typically seen associated with cerebrovascular diseases (ischemic or hemorrhagic stroke in the subthalamic region) [55, 56]. Nonketotic hyperglycemia is the second most frequent cause [57]. Finally, a large number of conditions complete the list of acquired causes of chorea, including chorea gravidarum [58], metabolic disturbances [55], and systemic immune-mediated diseases [59, 60] and infections, such as AIDS [55]. Postinfectious Sydenham's chorea remains one of the most frequent etiologies especially in some populations [61]. Psychogenic chorea is one of the diagnostic possibilities for a patient without other explanation, and it has been described also in a patient with a known family history of HD [62].

Pathophysiology of Chorea

Pathophysiology and molecular mechanisms underlying chorea are described in other chapters of this book. Briefly, according to the classic model of basal ganglia pathophysiology [63, 64], chorea is the result of a disruption in the "indirect" striato-pallidal pathway (a connection that involves the globus pallidus externus, GPe, and the STN with the final net effect of increasing GPi activity). The outcome is the increase of GPe activity and the consequent inhibition of the STN and the GPi. This situation results in a disinhibition of the thalamocortical output from the basal ganglia to the premotor cortex areas (and thus cortical overactivation) [65]. In support of this hypothesis, the degeneration of the striatal medium spiny neurons in early stages of choreic HD has been suspected of producing the selective loss of striatal cells containing GABA/enkephalin that project to the GPe (i.e., part of the indirect pathway) rather than those containing substance P and projecting to GPi, a loss that seems to correlate with the development of chorea [66, 67]. On the other hand, in the HD akinetic-rigid type, the neuronal damage involves all striatal projections to both GPe and GPi [67]. However, these hypothetical scenarios remain incomplete since the improvement of LID and other hyperkinetic disorders following GPi lesions are paradoxical obstacles for the simple understanding of this model [68]. Moreover, some studies suggest that cortical hyperexcitability is not the proximate cause of choreic movements [69]. Mouse models of movement disorders have shed new insights by showing that the loss of associated firing and rare coincident spike bursting of the striatal medium spiny neurons are a pathological hallmark of

HD and correlate to disease severity [70, 71]. These findings are in keeping with the pathogenesis of LID [72]. As such, firing pattern abnormalities in striatal cells could be part of the pathological substrate of chorea, and the development of dyskinesia would not only be related to firing rate disturbance but also to changes in the firing pattern [72–74]. In the hyperkinetic condition, pallidotomy might lessen dyskinesia by interrupting the transmission of this abnormal choreogenic output from the GPi to the thalamus [7, 75, 76]. Nevertheless, more studies are needed to clarify the pathological phenomenon underlying choreic disorders.

Selected Diseases Associated with Chorea

Huntington's Disease

HD is a neurodegenerative autosomal dominant neuropsychiatric disease caused by an abnormal CAG trinucleotide repeat expansion in the *huntingtin* (*Htt*) protein gene, located in the short arm of chromosome 4 (4p 16.3) [5]. This disturbance results in the formation of an *Htt* mutant form, which accounts for the pathological substrate in the HD brain [77]. In the normal population, the triplet size ranges between 10 and 29 copies. Individuals with 27–35 repetitions do not develop the disease, but they can be unstable on replication, leading to a length increase throughout successive generations (the anticipation phenomenon), especially when transmitted through paternal line [78]. In subjects carrying between 36 and 40 repetitions, the penetrance can be incomplete and so, some carriers can remain unaffected. Beyond the presence of 41 CAG triplets, HD is inevitable [78, 79]. The instability of trinucleotide repeats is the base of the anticipation effect, since there is an inverse correlation between the number of repeats and the age of onset. Hence, the greater the CAG repeat size, the sooner the disease tends to manifest [80]. However, it has been shown that other genetic and environmental predictors can influence the age at onset [81, 82] and that the number of CAG repeats does not correlate well with clinical severity [80].

Although still considered a rare disease because of its relatively low incidence and prevalence (0.046–0.8/100,000/year and 0.4–5.7/100,000, respectively, depending on a population's variability) [83], HD remains the most common non-iatrogenic cause of chorea [51]. The onset of the clinical presentation is usually in the fourth decade. Juvenile onset of HD has been described (Westphal variant); it typically presents with parkinsonism rather than with chorea [84]. Behavioral and psychiatric symptoms often precede motor disturbances and progress during the course of the disease [85]. The suicide rate in this subtype of HD is five to ten times greater than that of the general population (5–10 % of HD individuals) [86, 87]. Suicide attempts have occurred for almost 25 % within the overall course of the disease [51, 88]. The subsequent development of dementia, particularly involving executive functions, is the other main clinical sign [85]. The clinical progression of HD leads to death about 15–20 years after the diagnosis, generally caused by pneumonia [89].

Despite the recent advances in the understanding of HD pathogenesis, there is currently no treatment to halt or slow the progression of the disease [90]. The promising results obtained with some molecules in preclinical experiments [91–94] have generally failed when tested in patients [95]. Among several pharmacological clinical trials with different drugs, amantadine [96], riluzole [97], and nabilone [98] have each shown to lessen chorea's severity. Dopamine receptor antagonists have been considered as possible useful symptomatic treatments [99]. Even though the current evidence-based guidelines point out the lack of supportive data [100], they remain a first-line treatment option together with tetrabenazine [11] as per expert opinion [101]. Unfortunately, these drugs sometimes fail in managing chorea. For neurodegenerative chorea, the high doses of these drugs needed for achieving more motor control can worsen other motor and neuropsychiatric symptoms of the condition [1]. Low-frequency repetitive transcranial magnetic stimulation has also shown some benefit in HD chorea [102].

In recent years, different surgical procedures have been tested to control motor symptoms of HD and eventually to learn if they can also provide a neuroprotective effect.

Ablative Neurosurgery

Although pallidotomy ameliorated behavioral and motor impairments in animal models of HD [103–105], there is only one case report about bilateral pallidal lesioning in a genetically proven HD patient [106]. In this case, the indication for surgery was dystonia rather than chorea. This 13-year-old girl was affected by the Westphal variant of HD and presented with medically intractable, severe dystonia and parkinsonism. Unfortunately, at 3 months after pallidotomy, the patient showed only minor improvement in dystonic features (the Burke-Fahn-Marsden Dystonia Rating Scale motor score changed from 56 preoperatively to 49 postoperatively). She also went on to develop progressive spasticity, eventually dying 7 months after surgery due to worsening of her medical conditions. The authors attributed the lack of clinical improvement to the possibly misplaced lesions due to marked GPi atrophy and high extent of pallidal degeneration. For the appearance of progressive and severe spasticity, the authors explained this on the basis of the potential lesion affecting the internal capsule. They concluded that the only advantage of the procedure was the possibility to manage the patient without continuous intravenous muscle relaxants. Interestingly, the microelectrode intraoperative recordings showed a relatively low firing rate activity in the GPi (29 ± 14 Hertz (Hz), similar to what was previously described in primary dystonia patients (21 ± 48 Hz) [107].

Deep Brain Stimulation

DBS surgery was already an established treatment for PD, essential tremor, and dystonia when it was firstly applied to patients with HD by the Toronto group [108] (see Table 17.2). Taking account the beneficial effects of pallidal stimulation for

Table 17.2 Reported cases of Huntington's disease patients treated with DBS in patients

Reference	Patient info	Optimal settings (F; PW; A)	Clinical assessment	Effects of stimulation/clinical outcome	Follow-up time (m)
Moro et al. [108]	43-year-old male, 8-year history of HD	40 Hz 90–120 µs 2.5–3.5 V	UHDRS Double-blind fashion	Chorea improvement of 44–70 %; 9.5 % increase in UHDRS functional subscore and 23.1 % in independence subscore. Induction of bradykinesia with high-frequency (130 Hz) stimulation	8
Hebb et al. [112]	41-year-old male, 13-year history of HD, 47 CAG repeats	180 Hz	UHDRS	UHDRSc improvement of 39 %. Improvement in behavioral features and ADL. Worsening of chorea with low-frequency (40 Hz) stimulation. Development of dysphagia and progressive rigidity	12
Biolsi et al. [114]	60-year-old male, 10-year history of HD, 44 CAG repeats	130 Hz 450 µs 1.9 V	UHDRS	Chorea score improvement of 56 %. Improvement in ADL after surgery (from 1/13 to 10/13) and independence subscore (from 25/100 to 75/100). Bradykinesia with 130 Hz, improved with LD	48
Fasano et al. [115]	72-year-old male, 17-year history of HD	40 Hz 90 µs 2.0 V	UHDRS Double-blind fashion	Chorea improvement of 76.5 %. Major disability due to progressive cognitive decline and apathy. Bradykinesia with 130 Hz, improved with LD	12
Groiss et al. [118]	65-year-old female	NR	NR	NR	NR
Kang et al. [116]	57-year-old male, 10-year history of HD, 42 CAG repeats	160/40 Hz 180 µs 3.6 V	UHDRS Neuropsychological evaluation	Chorea subscore improvement of 50 % and ADL improvement of 45 %. Development of mild cognitive decline. No clinical impairment with any frequency of stimulation	24
	50-year-old male, 3-year history of HD, 41 CAG repeats	130 Hz 210 µs 3.6 V	UHDRS Neuropsychological evaluation	Chorea subscore initially improved of 59 % leading to amelioration in functional status and ADL. Chorea worsened with 40 Hz stimulation. Further progressive worsening of dysphagia, dysarthria, dystonia, and parkinsonism and resulted in impairing in ADL, but chorea remained controlled. Symptoms deteriorated when stimulator was switched off. Mild cognitive deterioration	24

Study	Patient	Parameters	Scale	Outcome	Follow-up (m)
Garcia-Ruiz et al. [121]	30-year-old female, 10-year history of juvenile HD, 50 CAG repeats	130 Hz 60 µs 3.2 V	UHDRS	Improvement in chorea and involuntary vocalizations. Total UHDRS improved of 48 % (from 63 to 33). Improvement in ADL. No side effects	12
Spielberger et al. [122]	30-year-old male, 9 years of juvenile HD, 58 CAG repeats	130 Hz 60 µs 2.0 V	UHDRS	Chorea improvement of 56 %. No objective improvement in functional capacity score but improvement in subjective quality of life. Bradykinesia worsened over follow-up time	48
Velez-Lago et al. [123]	34-year-old male, 7 years of juvenile HD, 60 CAG repeats	60 Hz 150 µs 2 V.	UHDRS	Total UHDRS score improved by 50 %, chorea's subscore by 74 %. Postoperative impairment of dysphagia, gait, and balance never returned to baseline. The patient presented chorea worsening at 40 Hz stimulation	12
	25-year-old female, 6 years of juvenile HD, 68 CAG repeats	135/10 Hz 120/60 µs 2.8 V	UHDRS UDRS	Patient without chorea. Despite marked and sustained improvement in dystonia, the patient decided to switch off stimulation because of worsening in parkinsonism, falling, and drooling	13

F frequency, *Hz* hertz, *PW* pulse width, *µs* microseconds, *A* amplitude, *V* volts, *m* months, *HD* Huntington's disease, *UHDRS* Unified Huntington's Disease Rating Scale, *UHDRSc* Unified Huntington's Disease Rating Scale chorea subscale, *UDRS* Unified Dystonia Rating Scale, *ADL* activities of daily living, *LD* levodopa, *NR* not reported

LD-induced dyskinesia [109] and generalized dystonia, Moro et al. performed bilateral GPi implantation in a 43-year-old male with genetically confirmed HD who suffered from severe and disabling chorea and dystonia refractory to medical treatments. The main motor features were assessed preoperatively and after surgery under stimulation with several different parameters of stimulation. At 8 months after surgery, GPi stimulation at 40 Hz significantly improved chorea and dystonia. There was also a minor improvement of bradykinesia. With stimulation at high frequency, the improvement of dystonia and chorea was even greater but at the expense of marked worsening in bradykinesia. Moreover, regional cerebral blood flow (rCBF) PET scans were performed in three conditions: without stimulation and at both low- and high-frequency stimulation (40 and 130 Hz, respectively). In the on-stimulation condition while performing a hand movement, there was a major activation of cortical areas involved in motor decision making and execution (which is typically underactive in HD [110]) as compared to the off-stimulation condition, in association with a significant improvement in movement errors. These findings have supported the use of GPi DBS to improve motor symptoms in HD and highlight the importance of frequency between the parameters of stimulation. A further publication regarding this patient showed amelioration of the oculomotor deficits with GPi DBS [111].

Subsequently, Hebb and colleagues [112] published their results of DBS in another HD patient. They also observed that at high-frequency stimulation, the patient showed a dramatic reduction of chorea and improvement in overall motor functioning. Unlike the previous report, chorea deteriorated at low-frequency stimulation. Intraoperative micro-recordings exhibited irregular GPi activity at low frequency (10–25 Hz) and at high amplitude, resembling more the situation reported by Cubo et al. [106] (and that of primary dystonia) and not the activity found by the Toronto group, characterized by high-frequency (81.8 ± 4.3 Hz) pallidal rates of discharge [113]. During follow-up, the patient increased body weight, normalized in mood, and gained autonomy in daily living activities. However, he later developed dysphagia and worsened parkinsonism that did not respond to DBS adjustments.

In 2008, Biolsi and coworkers [114] reported another HD patient who was bilaterally implanted in the posteroventral GPi. Stimulation at high frequency induced a reduction of choreic movements in the limbs and trunk. However, the patient developed bradykinesia. At 3 months, chorea was suppressed. This improvement was sustained for 4 years, and behavioral and cognitive features showed no change.

Another research group [115] observed the improvement in chorea with high-frequency bilateral GPi stimulation (130 Hz), but this was accompanied by a marked worsening of bradykinesia and severe freezing of gait, impairments that were not observed at low stimulation frequencies. This patient showed progression of bradykinesia and cognitive performance without response to adjustment of DBS parameters.

Recently, Kang et al. [116] described two other HD patients who achieved lessened chorea with high-frequency GPi stimulation. However, despite reduction of chorea, only one patient improved in functioning during activities of daily living, since the other patient progressively deteriorated in gait, dystonia, and bradykinesia (which was unresponsive to decrease in stimulation frequency). The electrophysiological findings were in keeping with reports previously mentioned [106, 112] and showed a lower rate of discharge in the GPi of HD patients [117] as compared to that in PD patients.

Subsequently, new electrophysiological evidence appeared from Groiss and coworkers [118], who investigated GPi local field potential oscillations in a HD patient undergoing bilateral DBS. Spectral analysis found power increase in the 4–12 Hz theta/alpha-band and 35–40 Hz low gamma-band in the GPi. The authors considered the former findings to be a constant feature associated with involuntary movements as it can also be found in PD or dystonia [119]. Since the 40 Hz Piper rhythm has been proposed to be the translation of the physiological cortical motor drive [120], the authors suggested that the presence of this rhythm in the GPi could be a feature of HD, reflecting hyperactivation of the motor drive to the cortex through pallidal projections.

Garcia-Ruiz and colleagues [121] recently published a case of a juvenile-HD patient who received DBS for the presence of continuous and refractory involuntary vocalizations that had severely compromised her feeding. The patient also suffered from generalized chorea and mild akinetic-rigid syndrome. Bilateral GPi stimulation at 130 Hz induced rapid improvement in both vocalizations and chorea persisting at 12 months.

Another case [122] showed marked chorea reduction with GPi stimulation at 130 Hz. The patient and his caregivers reported improvement in quality of life and easier maintenance of nursing care.

In the most recent paper on the topic, Velez-Lago and colleagues [123] compared the outcome of DBS used for 2 HD patients. One patient received this surgery due to severe chorea and the other due to generalized dystonia without chorea. The authors observed a greater improvement for choreiform movements as compared to dystonia. The choreic patient showed marked reduction in chorea but experienced worsening in swallowing, gait, and bradykinesia. In the second patient, dystonia improved, but there was concomitant worsening in akinesia and rigidity that did not improve after several trials of programming stimulation.

The GPe has also been suggested as a potential target for functional neurosurgery in HD [124]. In one study (in a transgenic rat model), high-frequency (130 Hz) stimulation of the GPe improved cognitive performance and reduced chorea in the treated group as compared with controls [125]. A more recent rCBF PET study showed that GPe DBS (at 130 Hz) modulates connectivity in the basal ganglia-thalamocortical circuits of HD patients leading to a connectivity pattern similar to that found in healthy subjects [126]. This "normalization" of the disrupted basal ganglia circuitry went beyond sensory-motor networks (primary sensory-motor cortices, supplementary motor area, and premotor cortex) and included modulation of cortical areas involved in the anterior cingulated cortex, precuneus, cingulate cortices, superior temporal gyri, and prefrontal regions. Given the widespread range of effects, these findings suggested the possibility of DBS effects on cognitive and behavioral features of HD.

Taking together the results mentioned above (and despite their limitations), pallidal DBS seems to be relatively safe in HD patients and capable of providing a clinically meaningful, sustained, symptomatic relief for chorea. Nevertheless, clinical results have varied greatly among treated patients, and the optimal stimulation parameters are difficult to determine. Furthermore, the beneficial stimulation effect has often been limited by a trade-off of high frequency-induced bradykinesia. The development of dysphagia in some cases and disease progression (with worsening of other motor and cognitive features) can potentially reduce the impact of DBS benefits for chorea and dystonia. There is no evidence that GPi DBS induces any disease-modifying effect. The recent results of the multicenter TRACK-HD observational study could help to select patients without clinical markers of rapid progression [127]. When performing larger, randomized and controlled trials for testing DBS safety and efficacy.

Cellular Transplantation

Two types of HD animal models have been developed so far: a neurotoxin model based on the striatal injections of excitotoxic amino acids that impart target medium spiny projection neurons [128–130] and a transgenic knock-in model [131, 132]. Both animal models have been used in several studies testing cellular transplantation therapies with promising results [133–136]. The most frequently used transplant material has been fetal brain tissue [137] and, more recently, embryonic stem cells [138–141] and bone marrow cells [142]. Striatal implantation of genetically modified cells enabled to secrete neurotrophic factors (NTFs) (that is ex vivo delivery of NTFs) could be a promising approach in the future [143, 144]. The major difference between these techniques is that, in the latter method, the implanted cells do not need to integrate in the injured brain circuitry.

Restorative Cellular Transplantation

In recent years, several cellular transplantation trials have been conducted in HD patients [145–160]. A pilot study in France headed by Bachoud-Lévi and Peschanski found motor and cognitive improvements in three out of five patients, as compared to a control group, at 2 years after bilateral striatal implantation of fetal neural grafts. The improvements correlated with the increase of metabolic activity in grafted striatal areas and cortical connected regions using ^{18}F-labelled fluorodeoxy-glucose PET [149]. However, most of this clinical benefit was lost at 6 years follow-up in the three patients. The chorea, however, remained improved as compared to the presurgical state. Moreover, although functional neuroimaging continued to show a relative preservation of the metabolism in some striatal areas and in the frontal cortex, these hypothetical benefits were limited by the degeneration observed in other brain areas [151]. The recently reported results of the NEST-UK multicenter study (initiated in 1998) [159] showed no sustained clinical benefit in up to 10 years of follow-up. The findings of this study, in keeping with long-term pathological findings [160], showed no evidence for graft survival as judged by raclopride PET scans. However, the NEST-UK study demonstrated the feasibility and safety of cell transplant in HD.

Despite the overall findings of limited benefit, it may be premature to exclude restorative cell therapy as a possible treatment for HD. Methodological issues with the transplant techniques can have an important impact on graft survival [161] and, consequently, clinical results (although there is no demonstration of any direct correlation between a patient's functional status and pathological findings [162]). Large-scale multicentric trials designed to prove clinical efficacy as a primary endpoint are needed for making further progress with this form of therapy [162]. Finally, considering the widespread brain pathology beyond striatum in HD, implants in extra-striatal areas might enhance the benefit. The possibility of this awaits exploration in the transgenic mice models [163].

Ex Vivo Delivery of Neurotrophic Factors

NTFs are proteins secreted by neurons and glial cells whose physiological actions play a key role in neuronal functioning and survival [164]. In neurodegenerative diseases, deficiency of NTFs appears to be a consequence rather than a cause of the pathological process. However, their ability to control cell physiology gives NTFs a role for potentially modifying the pathological evolution of neurodegeneration. NTFs have been shown to enhance cell survival in vitro [165], to influence neuronal growth [166], and to prevent cell death in neurodegenerative diseases [167–169]. As NTFs are unable to cross the blood-brain barrier, different techniques have been developed for the delivery of NTFs into the central nervous system: implantation of cells genetically enabled to secrete NTFs, intraputaminal or intraventricular infusion of NTFs, and stereotactic striatal injections of inactivated viral vectors [170]. Studies with ex vivo NTFs delivery in animals have shown generally that the procedure can be carried out safely and that a protective effect can be achieved in striatal cells by implanting NTF-secreting cells both in encapsulated [168, 171, 172] and in grafted experiments [144, 173, 174]. In some instances, these protective actions were evident in behavioral effects [173, 175, 176].

To date, one completed Phase I clinical trial in 6 HD patients has demonstrated safety and feasibility from the intraventricular implantation of encapsulated baby hamster kidney cells engineered to secrete ciliary neutrophic factor (CNTF) [177, 178]. The capsules were composed of a semipermeable polymer membrane that allowed the entrance of oxygen and nutrients for the cells and the exit of the synthesized NTF. The device was changed every 6 months. The technique proved to be safe and well tolerated. However, 13 of the 24 capsules did not successfully release CNTF, suggesting the need for improved technical aspects for this promising therapeutic approach.

In addition to ex vivo delivery, several studies in animal models have directly injected NTFs in the striatum or intraventricularly with both a single-injection and a continuous-delivery device. Safety and limited benefit of this approach have been showed in preclinical assays [179–181]. To the best of our knowledge, there have been no published human trials using these methods.

Stereotactic Intrastriatal Injection of Viral Vectors

Viral-vector injection techniques are based on the modification of lentivirus, adenovirus, or adeno-associated viruses in order to enable them for safely delivering genetic material into a host cell. This viral integration can induce the cell to express a gene of interest and synthesize an encoding protein [182] or to silence cellular mRNA so that expression of an undesirable protein can be prevented (i.e., RNA interference (RNAi)) [170]. Like other macromolecules, viruses cannot cross the blood-brain barrier; thus, they need to be directly injected to the targeted brain region. Because viral vectors are inactivated and cannot replicate, they are reliably innocuous for the host.

Stereotactical injections of viral vectors encoding for NTFs have showed efficacy in a number of experiments using HD animal models (mostly models created by excitotoxicity, but some created by transgenic methods). The results have shown reduction in striatal cell loss and improvement of motor performance [183–185] or delay in motor phenotype onset [186]. Some contrary results have also been encountered, including worsening of clinical features, severe weight loss, and shorter survival [187]. In other instances, the neuroprotective effects of NTFs were dose-dependent [188]. As an example, in a recent study by Ellison and colleagues [189], striatal lentiviral vector-delivery injections of mutated *Htt* protein were used to generate a rat model of HD. While co-injection of a low dose of vectors of vascular endothelial growth factor165 exerted a protective effect over neuronal death and *Htt* aggregation, higher doses paradoxically led to a more extensive cell loss and early mortality for the animals.

Viral vectors have been used in RNAi procedures. Their experimental plan has been to introduce small, artificially synthesized interfering RNAs (siRNAs) into cells. The siRNAs bind to natural RNA-induced silencing complexes and subsequently hybridize to a target RNA for repressing RNA expression. RNA silencing has been attempted in several reported HD animal model investigations, which have implemented this technique, demonstrating improvement in pathological and behavioral abnormalities and prolonging animal survival [190–193].

Finally, viral vector injections have been also used to deliver specific anti-mutant *Htt* antibodies in HD animal models, increasing *Htt* turnover and thus reducing its neurotoxicity [194]. This method has led to behavioral improvement [195].

Antisense Oligonucleotides Injection

ASOs are synthetic DNA oligomers able to degrade a target mRNA inside the cell nucleus [170]. By this means, an ASO can suppress *Htt* expression. This was shown in *in vitro* studies [196] but not in intrastriatal infusion in HD animal model [197].

Neuroacanthocytosis

The relationship between neurological symptoms and the presence of acanthocytes (red blood cells with distorted spiculated cell membranes) in the peripheral blood was first reported by Bassen and Kornzweig in 1950 [198]. They described a patient who presented with loss of visual acuity at age 6; several years later, she developed ataxia and intentional tremor in her upper limbs. She was diagnosed with cerebellar ataxia and atypical retinitis pigmentosa. The abnormally shaped red cells were found in her blood smear. Subsequently, many authors have reported and documented the association between a high percentage of peripherally circulating acanthocytes and neurological syndromes. Levine and Critchley were the first to report the connection of these findings specifically with chorea [199, 200]. Even though the term neuroacanthocytosis (NA) was used years before [201], it was not until 1991 that Hardie and coworkers summarized the current understanding of this condition by their description of a series of 19 NA patients all affected with chorea [202]. Currently, the term NA is used to encompass a heterogeneous group of rare neurological syndromes that are known to be associated with red cell acanthocytosis. Between them, three syndromes can present with chorea: chorea-acanthocytosis (ChAc), McLeod syndrome (MLS), and Huntington's disease-like 2 (HDL2) [203]. This association can be explained by the pathological substrate of these diseases that shares with HD striatal cell loss as a major pathological hallmark, especially in the caudate nucleus [204], and resulting in reduced striatal metabolism [205].

ChAc is a rare autosomal recessive neurodegenerative disorder (autosomal dominant inheritance having been also reported [206]) caused by mutations in the VPS13A gene on chromosome 9q that encodes for the *chorein* protein [207, 208]. The clinical onset is usually in the third decade and rarely after 50 years of age [209]. ChAc often starts with subtle cognitive and behavioral symptoms. Later on during the third decade, orofacial chorea and dystonia start to develop, with the typical "feeding dystonia" characterized by a marked involuntary tongue protrusion [210]. With the disease progression, chorea tends to generalize, cognitive decline develops, and other movement disorders like tics and parkinsonism may appear. Seizures can also occur and may precede other symptoms by years [211]. Prognosis is variable, with 15–30 years of survival from diagnosis and with death usually related to bulbar or autonomic impairment [212].

MLS is due to X-linked mutations of the XK gene. It typically onsets in males between 40 and 60 years of age [213], but isolated cases of female mutation carriers who developed the disease have been also reported [202]. Like ChAc, neuropsychiatric symptoms can precede the appearance of movement disorders by years and seizures can also develop [213]. Generalized chorea is the predominant clinical feature occurring at any time during the course of the disease for the majority of patients. Cognitive decline can also develop in later stages. Survival is usually longer than for ChAc, with patients typically surviving for up to 30 years. Death is mostly related to cardiac disorders [212].

HDL2 age of onset is variable and, as in HD, correlates with the extent of trinu-cleotide repeat expansion [214]. The clinical picture resembles that of classic HD. Acanthocytes are less frequently present (10 % of cases) than in the two previous diseases.

No specific therapies for NA are available, and the treatment remains exclusively symptomatic. Treatment options include botulinum toxin injections, dopaminergic antagonists, or dopamine-depleting therapy (tetrabenazine) for the hyperkinetic movements. Antiepileptic drugs are used for seizures and as treatment for mood or other psychiatric disturbances [212].

Ablative Neurosurgery

As mentioned above, surgical approaches initially used for chorea were mostly ablative surgeries [28, 31] in a heterogeneous group of chorea of different unclear etiology [26]. The first (and to our knowledge, unique) case of ChAc treated with brain surgery was reported in 1997 [215]. Posteroventral pallidotomy improved choreoballistic movements in a 41-year-old ChAc patient. Surgery was initially per-formed unilaterally in the left pallidum with great improvement of involuntary limb, neck, face, and tongue movements such that the patient was able to feed himself. However, due to appearance of ballism on the left hemibody, right pallidotomy was needed 6 months later; this also achieved a positive effect. The response to bilateral pallidotomy was maintained through a 7-month follow-up period.

Deep Brain Stimulation

NA cases treated by DBS are summarized in Table 17.3. Wihl et al. [216] first pub-lished in 2001 a ChAc patient presenting with tongue protrusion, dysphagia, and lower limbs hyperkinesia inducing gait impairment. The symptoms progressed leading to a severe loss of weight. For this reason, bilateral GPi DBS surgery was performed. The absence of benefit led to the removal of electrodes 3 weeks later. The authors argued that too small a volume of neural tissue acted on by the electri-cal stimulation might have accounted for failure.

However, 1 year later, Burbaud and colleagues [217] reported on results of treating a ChAc patient who improved in severe trunk spasms with high-frequency bilateral stimulation of the posteroventral oral (Vop) thalamic nucleus. The patient died 1 year later for unknown reason in the long-term care facility where he was living [218]. The same group reported later the outcome of two other NA patients who underwent bilat-eral pallidal DBS [219]. The first was a ChAc patient who improved in chorea, belch-ing, and dysarthria with the use of 40 Hz stimulation. The second patient, diagnosed with MLS, had similar beneficial responses regarding chorea (improvement at low frequencies and worsening at high frequencies); dystonia improved regardless of stimulation frequency. The authors supported Wihl's hypothesis for the need to involve a larger field of stimulation to explain stimulation unresponsiveness.

Table 17.3 Reported cases of neuroacanthocytosis patients treated with DBS

Reference	Patient info	Target	Optimal settings (F; PW; A)	Clinical assessment	Effects of stimulation/clinical outcome	Follow-up time (m)
Wihl et al. [216]	38-year-old male, 3-year history of ChAc	Bilateral GPi	NA	NR	Worsening of chorea with high-frequency stimulation. No benefit and deterioration of speech and gait at 10–50 Hz frequency. Electrodes were removed	3 weeks
Burbaud et al. [218]	43-year-old male, 12-year history of ChAc	Bilateral Vop	160 Hz 90 μs 1.5/2.0 V	Marsden and Schachter choreic score Barthel index	Chorea improvement of 37 %. Improvement of trunk spasms frequency from 4.9 ± 0.1/min to 1.4 ± 1. 20 points increase in Barthel index (from 15 to 35). Improvement in ADL and gait. No side effects reported	9
Guehl et al. [219]	32-year-old female, 8-year history of ChAc	Bilateral GPi	40 Hz	UHDRS	Improvement in chorea subscore of 42 %. Improvement in belching and dysarthria. Slight improvement of dystonia but worsening in chorea and dysarthria with 130 Hz stimulation	3
	Age NR, MLS KX gene mutation	Bilateral GPi	40 HZ	UHDRS	Chorea subscore improvement of 48 %. Improvement of facial dystonia with 130 Hz but with worsening of chorea	3
Ruiz et al. [220]	35-year-old female, 10-year history of ChAc	Bilateral GPi	130 Hz 180 μs 4.2/4.4 V	UHDRS	Chorea-dystonia items improved of 41 % at 2 months enabling patient to walk. Sustained effect during follow-up but need to increase voltage due to chorea's relapse. Increase of trunk spasms at 40–50 Hz stimulation	24
Shin et al. [221]	39-year-old female, 4-year history of ChAc	Bilateral GPi	130 Hz 90 μs 2.9/3.0 V	UHDRS	Total UHDRS improvement of 72 %. Immediate benefit in chorea and bradykinesia and delayed in dystonia. Improvement in truncal bending. Increase in independence scale from 50 to 80 and in functional capacity from 1 to 9. Patient able to walk and perform ADL. No adverse effects	12

(continued)

Table 17.3 (continued)

Reference	Patient info	Target	Optimal settings (F; PW; A)	Clinical assessment	Effects of stimulation/clinical outcome	Follow-up time (m)
Li et al. [222]	39-year-old male, 22-year history of ChAc	Bilateral GPi	40 Hz 60 µs 3.5 V	UHDRS	Total UHDRS improvement of 72 % for patient 1 and of 49 % for patient 2. Both patients had further improvement in chorea and presented clinical worsening of both chorea and dystonia with high-frequency stimulation. Both patients improved dysphagia	9
	30-year-old male, 12-year history of ChAc	Bilateral GPi	40 Hz 60 µs 3.5 V	UHDRS		5
Kefalopoulou et al. [223]	54-year-old male	Bilateral GPi	130 Hz 60 µs 2.5 V	UHDRS AIMS	Chorea subscore reduction of 87.5 %, total UHDRS improvement of 42.4 %, and AIMS decrease of 75.6 %. Patient enabled to walk unaided. No side effects	5
	43-year-old male	Bilateral GPi	130 Hz 90 µs 2.5 V	UHDRS AIMS	Chorea subscore improvement of 32.4 % in, 66.7 % in total UHDRS, and 61.2 % in AIMS. IPG infection and subsequent intracerebral abscess	2
Lim et al. [224]	32-year-old male, 6-year history of ChAc	Bilateral GPi	60 Hz 60 µs 3.2/3.0 V	AIMS	AIMS improvement of 50 %. Improvement in chorea, dystonia, swallowing, posture, and balance. Initial and reversible worsening of dysarthria	8

F frequency, *Hz* hertz, *PW* pulse width, *µs* microseconds, *A* amplitude, *V* volts, *m* months, *ADL* activities of daily living, *GPi* internus globus pallidus, *Vop* posterior ventral oralis nucleus, *ChAc* chorea-acanthocytosis, *MLS* MacLeod syndrome, *UHDRS* Unified Huntington's Disease Rating Scale, *AIMS* Abnormal Involuntary Movement Scale, *NR* not reported, *NA* not applicable, *IPG* internal pulse generator

Ruiz et al. [220] also published their results in a ChAc patient who presented with progressive dystonia and chorea. Stimulation at 130 Hz first improved chorea and, next, dystonia. Of interest were the observations that stimulation with ventral contacts induced "freezing" of gait even if the patient was not affected with parkinsonism.

Up to six ChAc patients bilaterally implanted in the GPi have been reported in the last 2 years. Shin et al. [221] published their results in a 35-year-old patient who presented with choreic movements. High-frequency stimulation yielded progressive improvement of motor features and function. Li and colleagues [222] presented two ChAc patients whose problems started with lingual dyskinesia and tongue biting that subsequently progressed to medication-refractory generalized chorea, dystonia, and trunk spasms compromising gait. Low-frequency bilateral GPi stimulation provided marked improvement in chorea and mild amelioration of dystonic features 4 weeks after surgery. No metabolic basal ganglia changes in ^{18}F-fluorodeoxyglucose PET scans were postoperatively observed in any of the patients with respect to baseline. The National Hospital, Queen Square, group has recently added two new ChAc patients to the literature [223]: the first affected with orofacial dyskinesias, feeding dystonia, and trunk spasms; the second patient presenting with oromandibular dyskinesias, limb choreic movements, and axial spasms. Bilateral pallidal stimulation improved the symptoms in both patients without inducing side effects, although lower frequencies were not tested. The second patient developed a left intracerebral abscess that required removal of implants. The most recent case report by Lim et al. [224] described a patient who presented with rapid and marked improvement of chorea and dystonia with low-frequency stimulation. His cognitive status remained stable after surgery.

Overall, the results of DBS for NA appear to be quite variable. As already described above as outcomes in HD patients, different frequencies of GPi stimulation provided variable and sometimes led to opposite effects or no improvement at all. The reasons for these discrepancies are not well understood; they could be dependent on the different phenotypes and disease stage of the treated patients or on technical aspects (such as electrode position inside the target).

Hemichorea-Hemiballismus

Hemichorea refers to choreiform movements affecting one hemibody. If associated to fast and wide limb movements, the clinical picture accounts for the hemichorea-hemiballism syndrome (HcHb). It is a rare condition, estimated to occur in 0.7 % of patients seen in a movement disorder center [225].

When facing HcHb, contralateral lesions in the thalamus, basal ganglia, or associated pathways must always be searched for [226]. Neurovascular diseases are by far the first etiology, especially if the presentation is abrupt [56]. However, hemichorea can be triggered by several other conditions like nonketotic hyperglycemia (the second most frequent cause) [227] and other metabolic disturbances [228, 229], systemic diseases such as polycythemia vera [230] or lupus erythematosus [231], ventriculoperitoneal shunt insertion [232], oral contraceptive use [233], chorea gravidarum [234], and AIDS [235] (due to both infectious opportunistic brain lesions and the action of the virus itself). In addition, iatrogenic surgical brain lesions are another cause [236–238].

The prognosis for this disorder is quite variable, but in most cases, HcHb is transient and reverts spontaneously or else by resolution of its underlying cause. However, in rare occasions, the movements persist (the incidence of lingering HcHb after stroke is 0.04 % [239]) and can lead to severe disability. Treatments such as tetrabenazine [12], diazepam [240], neuroleptic therapy [241–244], antiepileptic drugs [245–247], and repetitive transcranial magnetic stimulation [248] have shown the capability for alleviating or suppressing HcHb.

Ablative Neurosurgery

Several reports have showed long-term efficacy of ablative surgery in refractory HcHb (mainly of vascular cause) by targeting the thalamic Vo-complex or the Vim [249–255], the zona incerta [253], the STN [249], the internal capsule [256], or the GPi [73, 76, 257–261]. In general, surgery has been highly effective, achieving remarkable reduction of the involuntary movements in the majority of patients as evident in the immediate postoperative period.

Krauss and Mundinger reported the clinical outcome of 14 patients affected by disabling hemiballism (7 of them also by hemichorea) who underwent stereotactic surgery in both zona incerta and Vo [253]. Patients showed excellent long-term response with sustained improvement. At their last follow-up (range: 3–27 years), seven patients were still free of symptoms whereas five presented minor residual hemichorea. They also exhibited a marked functional improvement with a 30 % mean reduction in the score of the HD Activities of Daily Living Scale.

Other reports cited intraoperative microrecording findings [73, 76, 257, 259, 261] that were in line with those published for HD (i.e., the presence of low-frequency GPi firing rates and an irregular bursting pattern [259]). They also found synchronous pauses following GPi bursts, a silence in activity that has been associated with the origin of the ballistic and choreic movements through the overactivation of thalamocortical projections [73].

Of interest are two patients with HcHb secondary to a cerebral angioma, one in the head of the caudate nucleus and the second in the putamen, and a third patient presented HcHb due to metastasis in the STN. The symptoms were completely abolished when the angioma was surgically removed [262, 263] in the first two cases and after Gamma Knife radiosurgery of the metastasis in the third case [264].

Deep Brain Stimulation

In 1995, Tsubokawa and coworkers [265] reported two patients with hemiballismus caused by vascular lesions in the STN who underwent Vim DBS (see Table 17.4). Both individuals exhibited attenuation of ballistic movements by stimulating at

Table 17.4 Reported cases of hemichorea-hemiballism patients treated with DBS

Reference	Patient info	Target	Optimal settings (F; PW; A)	Effects of stimulation/clinical outcome	Follow-up time (m)
Tsubokawa et al. [265]	57-year-old female, vascular HcHb (STN lesion)	VL and Vim	50 Hz 200–300 µs 4–7 V	For both patients, immediate improvement of ballism. Movement worsening at high intensities of stimulation	16
	62-year-old male, vascular HcHb (putaminal lesion)	VL and Vim	50 Hz 200–300 µs 4–7 V		16
Thompson et al. [266]	11-year-old girl, vascular monochorea (thalamic hemorrhage)	Unilateral Vim	180 Hz 150 µs 3.2 V	Amelioration without side effects. Setting adjustments needed because of partial lost effect	4
Nakano et al. [267]	65-year-old male, HcHb due to striatal lesion in context of severe DM	Unilateral Vo-c	130 Hz 90 µs 2 V	Immediate symptomatic alleviation without side effects	9
Hasegawa et al. [268]	56-year-old male, vascular HcHb (STN and midbrain hemorrhage)	Unilateral GPi	130 Hz 60 µs 4.5 V	Improvement of ballistic movements and gain of function. Persistence of wrist dystonic posture	15
Capelle et al. [269]	52-year-old male, HcHb secondary to postsurgical tumoral resection	Unilateral Vim and GPi	130 Hz 210 µs 0.8 V	HcHb completely disappeared with stimulation of both targets but less intensity was needed in the Vim thalamus (2 V vs. 5 V). The preoperative AIMS score was 15 and decreased to 1 at 9 months the patient remained stable	25

F frequency, *Hz* hertz, *PW* pulse width, *µs* microseconds, *A* amplitude, *V* volts, *m* months, *ADL* activities of daily living, *GPi* internum globus pallidus, *STN* subthalamic nucleus, *Vim* ventralis intermedius nucleus of the thalamus, *VL* ventrolateral part of the thalamus, *Vo-c* ventral oralis complex (anterior and posterior ventral oralis nucleus), *UHDRS* Unified Huntington's Disease Rating Scale, *AIMS* Abnormal Involuntary Movement Scale, *HcHb* hemichorea-hemiballism, *DM* diabetes mellitus, *NR* not reported

50 Hz (lower frequencies induced no improvement and stimulation with frequencies above 150 Hz resulted in exacerbation of ballistic movements). The effect was sustained during the 16-month follow-up.

Subsequently, Thompson et al. [266] reported a case of an 11-year-old girl with choreic movements in her right arm secondary to a left thalamic hemorrhage she suffered 4 years before. She was treated with left Vim DBS surgery with benefit, although the patient required several further setting adjustment because the benefit was partially lost over time.

Nakano et al. [267] reported a 65-year-old patient with severe diabetes who suffered from a striatal lesion that resulted in right intractable HcHb. The DBS electrode was placed in the thalamic Vo-complex and stimulation resulted in a rapid, complete, and persisting suppression of the movement disorder.

Another case was reported by Hasegawa and colleagues [268]. This was a 56-year-old male affected by HcHb due to a STN hemorrhage. High-frequency pallidal stimulation (130 Hz) resulted in cessation of ballistic movements with partial persistence of dystonia but allowing a better function up to 15 months.

Capelle et al. [269] described a patient with HcHb secondary to surgical resection of a craniopharyngioma who was targeted in both the Vim and the GPi. Intraoperative testing of each electrode within the two targets showed that, to achieve the same chorea control, it was necessary to raise up to 5.0 V the stimulation amplitude in the GPi whereas thalamic stimulation was already effective by 2.0 V. The authors concluded that, in HcHb, the GPi needed higher stimulation to exert the same beneficial effect as compared with the thalamus. Nevertheless, these findings are not enough to confirm superiority of one target over the other, and more data is needed achieve a more complete understanding.

Miscellaneous Results

Deep Brain Stimulation

Several other disorders have been treated by DBS in the last few years, as summarized in Table 17.5.

DBS has also been applied in children with a variety of movement disorders [270]. The primary indication has usually been dystonia, but DBS has also been used for chorea. Two cases of life-threatening dystonia-dyskinesias of unknown cause and refractory to medical treatment were successfully treated with bilateral GPi DBS that markedly diminished the hyperkinesia and enhanced the independent function of the patient [271, 272].

Another young male diagnosed with Cockayne syndrome (a premature aging disease that causes an early neurological degeneration and can present, among other symptoms, with movement disorders) responded favorably to bilateral thalamic DBS implanted to treat chorea and ballism [273].

Table 17.5 Miscellanea of patients presenting chorea from different causes treated with DBS

References	Indication for DBS	Target	Optimal settings (F; PW; A)	Effects of stimulation/clinical outcome
Angelini et al. [272] Sato et al. [271]	Child hyperkinesias of unknown cause	Bilateral GPi	130 Hz 60 µs 3.0 V	Almost complete control of chorea and dystonia. Gain of autonomy and walking in both patients
Thompson et al. [266] Vidailhet et al. [276] Apetauerova et al. [275]	Cerebral palsy chorea	Unilateral Vim Bilateral GPi	130/180 Hz 90–210 µs 2.0–4.0 V	Variable results (from no improvement to marked reduction of hyperkinetic movements and gain of autonomy)
Eltahawy et al. [278] Schrader et al. [279] Trottenberg et al. [283] Damier et al. [281] Kosel et al. [280] Kefalopoulou et al. [282] Spindler et al. [277]	Tardive dyskinesia	Bilateral and unilateral GPi	40–180 Hz 60–450 µs 2.5–6.5 V	Mean improvements in chorea of 35–78 %. Mean reduction of 56 % in AIMS in a 10-patient series. No worsening of underlying psychiatric disease
Hebb et al. [273]	Cockayne syndrome	Unilateral Vim	130 Hz 60 µs 4.0 V	Immediate bilateral reduction of chorea and myoclonus. No side effects. Spontaneous cessation of movements after 28 months probably due to disease progression
Yianni et al. [288]	Senile chorea	Unilateral Vop and GPi	185 Hz 90 µs 2.5 V (thalamic stimulation)	Initial suppression of movements with pallidal stimulation with relapse at 5 months. Thalamic stimulation induced 74 % reduction of AIMS score and effect was more sustained (18 months)
Loher et al. [289] Yamada et al. [290] Kaufman et al. [291]	Paroxysmal nonkinesigenic dyskinesia	Unilateral Vim Bilateral GPi	65 Hz 180 µs 5.4 V 130/150 Hz 90/120 µs 2.8/3.7 V	Reduction in frequency, duration, and intensity of paroxystic dystonic episodes with thalamic stimulation. A better effect with pallidal stimulation with complete suppression of nonkinesigenic dyskinesia

F frequency, *Hz* hertz, *PW* pulse width, *µs* microseconds, *A* amplitude, *V* volts, *ADL* activities of daily living, *GPi* internum globus pallidus, *Vim* ventralis intermedius nucleus of the thalamus, *Vop* posterior ventral oralis nucleus, *UHDRS* Unified Huntington's Disease Rating Scale, *AIMS* Abnormal Involuntary Movement Scale, *NR* not reported

Cerebral palsy-related movement disorders have also been treated with DBS with variable results [274]. In cases presenting with chorea (usually associated with dystonic phenomenology), both thalamic [266] and pallidal stimulation [275, 276] have shown to improve the symptoms. A Class III French study has demonstrated an overall clinical improvement of 24.4 % in the movement score of the Burke-Fahn-Marsden Dystonia Rating Scale and significant gain in independent functioning in 4 up to 13 patients with CP operated with bilateral GPi DBS.

Chronic antipsychotic drug treatment can induce hyperkinetic disorders (tardive dyskinesia and dystonia (TD)). Medical management is often limited and, for this reason, in recent years DBS surgery has been used. GPi DBS has been very successful in improving TD, with results comparable to those observed in primary dystonia patients [277]. Less frequently, the effect of pallidal stimulation has been assessed in TD presenting with choreiform movements [277–283]. Chorea improved significantly, usually in the immediate postoperative. The only controlled study with a 10-patient sample [281] showed improvement in dyskinetic movements to the same extent as observed with dystonic features. Different parameters of stimulation have been tested, finding major improvements with high frequencies (>100 Hz) rather than with low frequencies, although not in all cases [279]. In addition to these studies on DBS, efficacy of ablative surgery in TD has also been reported [284–287].

In a single case report, thalamic DBS was effective to reduce senile chorea [288]. Both the posteroventral GPi and the Vop were unilaterally targeted and stimulated at different times at high frequency (185 Hz). While GPi DBS improved contralateral limb chorea, but without effect on orofacial and shoulder movements, thalamic stimulation ameliorated the symptoms in the left arm, shoulder, and orofacial region, an effect that was sustained up to 18 months.

Up to three case reports of patients affected by refractory paroxysmal nonkinesigenic dyskinesia (PNKD) have been improved by both thalamic [289] and pallidal stimulation [290, 291]. In the case by Kaufman et al., a 26-year-old patient diagnosed with childhood-onset PNKD reported 90 % improvement in his global condition after surgery and an objective reduction in dyskinetic episodes from 20 per day before surgery to 0–3 after surgery, leading to a marked improvement of his independent functioning.

Conclusions

There is still no effective treatment to reverse or delay neurodegenerative chorea. However, several pharmacological and neurosurgical therapies have recently emerged to improve the disabling symptoms induced by choreic disorders. Among the surgical treatment options, pallidal DBS has shown to be the most effective therapy, although large randomized clinical trials are lacking. To this regard, a Phase I study from Düsseldorf University (comparing the efficacy of GPi and GPe stimulation in controlling motor symptoms in HD patients) has just been completed and will soon provide more information as to the clinical outcomes (NCT00902889).

In recent years, some experimental neurosurgical approaches have arisen, opening the door for some promising disease-modifying treatments in HD; as a result, new experimental trials are in the pipeline. The MIG-HD trial (Multicentric Intracerebral Grafting in Huntington's disease) headed by Bachoud-Levi and colleagues group (NCT00190450) is a Phase II study involving 60 HD patients treated with bilateral intrastriatal grafts of fetal neurons in a delayed-start clinical trial design (i.e., a first group of patients receiving the implantation at the beginning of the trial with the non-implanted patients serving as a control group and, later on, implantation of the control group with a final comparison of both groups after 18 months). This study will provide information about the possible clinical benefit of restorative cellular therapy in motor and non-motor features and interpretation as to a possible disease-modification effect.

Until obtaining further data with higher level of scientific evidence, the role of surgical approaches in the treatment of different types of chorea remains largely empirical.

References

1. Walker FO. Huntington's disease. Lancet. 2007;369(9557):218–28 [Review].
2. Goetz CG, Chmura TA, Lanska DJ. History of chorea: part 3 of the MDS-sponsored history of movement disorders exhibit, Barcelona, June 2000. Mov Disord. 2001;16(2):331–8 [Biography Historical Article Portraits].
3. Lanska DJ. Chapter 33: The history of movement disorders. Handb Clin Neurol. 2010;95:501–46 [Historical Article Review].
4. Huntington G. On chorea. Med Surg Reporter. 1872;26:320–1.
5. The Huntington's Disease Collaborative Research Group. A novel gene containing a trinucleotide repeat that is expanded and unstable on Huntington's disease chromosomes. Cell. 1993;72(6):971–83.
6. Ondo WG, Hanna PA, Jankovic J. Tetrabenazine treatment for tardive dyskinesia: assessment by randomized videotape protocol. Am J Psychiatry. 1999;156(8):1279–81 [Clinical Trial Randomized Controlled Trial].
7. Cardoso F, Seppi K, Mair KJ, Wenning GK, Poewe W. Seminar on choreas. Lancet Neurol. 2006;5(7):589–602 [Review].
8. Venuto CS, McGarry A, Ma Q, Kieburtz K. Pharmacologic approaches to the treatment of Huntington's disease. Mov Disord. 2012;27(1):31–41 [Review].
9. Reiner P, Galanaud D, Leroux G, Vidailhet M, Haroche J, du Huong LT, et al. Long-term outcome of 32 patients with chorea and systemic lupus erythematosus or antiphospholipid antibodies. Mov Disord. 2011;26(13):2422–7 [Comparative Study].
10. Kleinsasser BJ, Misra LK, Bhatara VS, Sanchez JD. Risperidone in the treatment of choreiform movements and aggressiveness in a child with "PANDAS". S D J Med. 1999;52(9):345–7 [Case Reports].
11. Huntington Study Group. Tetrabenazine as antichorea therapy in Huntington disease: a randomized controlled trial. Neurology. 2006;66(3):366–72 [Multicenter Study Randomized Controlled Trial Research Support, Non-U.S. Gov't].
12. Sitburana O, Ondo WG. Tetrabenazine for hyperglycemic-induced hemichorea-hemiballismus. Mov Disord. 2006;21(11):2023–5 [Case Reports].

13. Gras D, Jonard L, Roze E, Chantot-Bastaraud S, Koht J, Motte J, et al. Benign hereditary chorea: phenotype, prognosis, therapeutic outcome and long term follow-up in a large series with new mutations in the TITF1/NKX2-1 gene. J Neurol Neurosurg Psychiatry. 2012;83(10):956–62 [Case Reports Multicenter Study].

14. Calabro RS, Polimeni G, Gervasi G, Bramanti P. Postthalamic stroke dystonic choreoathetosis responsive to tetrabenazine. Ann Pharmacother. 2011;45(12):e65 [Case Reports].

15. Chatterjee A, Frucht SJ. Tetrabenazine in the treatment of severe pediatric chorea. Mov Disord. 2003;18(6):703–6 [Case Reports].

16. Genel F, Arslanoglu S, Uran N, Saylan B. Sydenham's chorea: clinical findings and comparison of the efficacies of sodium valproate and carbamazepine regimens. Brain Dev. 2002;24(2):73–6 [Clinical Trial Comparative Study Controlled Clinical Trial].

17. Chandra V, Spunt AL, Rusinowitz MS. Treatment of post-traumatic choreo-athetosis with sodium valproate. J Neurol Neurosurg Psychiatry. 1983;46(10):963 [Case Reports Letter].

18. Giroud M, Dumas R. Valproate sodium in postanoxic choreoathetosis. J Child Neurol. 1986;1(1):80 [Letter].

19. Lenton RJ, Copti M, Smith RG. Hemiballismus treated with sodium valproate. Br Med J. 1981;283(6283):17–8 [Case Reports].

20. Green LN. Corticosteroids in the treatment of Sydenham's chorea. Arch Neurol. 1978;35(1): 53–4.

21. Van Horn G, Arnett FC, Dimachkie MM. Reversible dementia and chorea in a young woman with the lupus anticoagulant. Neurology. 1996;46(6):1599–603 [Case Reports].

22. Min JH, Youn YC. Bilateral basal ganglia lesions of primary Sjogren syndrome presenting with generalized chorea. Parkinsonism Relat Disord. 2009;15(5):398–9 [Case Reports Letter].

23. Walter BL, Vitek JL. Surgical treatment for Parkinson's disease. Lancet Neurol. 2004;3(12):719–28 [Review].

24. Bucy PC, Case T. Tremor: physiologic mechanism and abolition by surgical means. Arch Neurol Psychiatry. 1939;41:721–46.

25. Spiegel EA, Wycis HT, Marks M, Lee AJ. Stereotaxic apparatus for operations on the human brain. Science. 1947;106(2754):349–50.

26. Hankinson J. Surgery of the dyskinesias. Proc R Soc Med. 1973;66(9):876–7.

27. Spiegel EA, Wycis HT. Pallidothalamotomy in chorea. Arch Neurol Psychiatry. 1950;64(2):295–6.

28. Wycis HT, Spiegel EA. Treatment of certain types of chorea, athetosis and tremor by stereoencephalotomy. J Int Coll Surg. 1956;25(2 Pt 1):202–7.

29. Blavier J, Blavier L. A case of Huntington's chorea ameliorated by electrocoagulation of the globus pallidus. Rev Med Liege. 1962;17:218–23.

30. Gioino GG, Dierssen G, Cooper IS. The effect os subcortical lesions on production and alleviation of hemiballic or hemichoreic movements. J Neurol Sci. 1966;3(1):10–36.

31. Mundinger F, Riechert T, Disselhoff J. Long term results of stereotaxic operations on extrapyramidal hyperkinesia (excluding parkinsonism). Confin Neurol. 1970;32(2):71–8.

32. Andrew J, Edwards JM, Rudolf Nde M. The placement of stereotaxic lesions for involuntary movements other than in Parkinson's disease. Acta Neurochir (Wien). 1974;Suppl 21:39–47 [Comparative Study].

33. Spiegel EA, Wycis HT. Thalamotomy and pallidotomy for treatment of choreic movements. Acta Neurochir (Wien). 1952;2(3–4):417–22.

34. Boixados JR. Pyramidotomy in the cerebral peduncle in treatment of choreo-athetosis. Rev Clin Esp. 1953;49(1):57–61.

35. Benabid AL, Pollak P, Louveau A, Henry S, de Rougemont J. Combined (thalamotomy and stimulation) stereotactic surgery of the VIM thalamic nucleus for bilateral Parkinson disease. Appl Neurophysiol. 1987;50(1–6):344–6.

36. Limousin P, Pollak P, Benazzouz A, Hoffmann D, Le Bas JF, Broussolle E, et al. Effect of parkinsonian signs and symptoms of bilateral subthalamic nucleus stimulation. Lancet. 1995;345(8942):91–5 [Research Support, Non-U.S. Gov't].

37. Limousin P, Krack P, Pollak P, Benazzouz A, Ardouin C, Hoffmann D, et al. Electrical stimulation of the subthalamic nucleus in advanced Parkinson's disease. N Engl J Med. 1998;339(16):1105–11 [Research Support, Non-U.S. Gov't].
38. Pahwa R, Wilkinson S, Smith D, Lyons K, Miyawaki E, Koller WC. High-frequency stimulation of the globus pallidus for the treatment of Parkinson's disease. Neurology. 1997;49(1):249–53 [Research Support, Non-U.S. Gov't].
39. Kumar R, Dagher A, Hutchison WD, Lang AE, Lozano AM. Globus pallidus deep brain stimulation for generalized dystonia: clinical and PET investigation. Neurology. 1999;53(4):871–4 [Case Reports Research Support, Non-U.S. Gov't].
40. Krack P, Hariz MI, Baunez C, Guridi J, Obeso JA. Deep brain stimulation: from neurology to psychiatry? Trends Neurosci. 2010;33(10):474–84 [Research Support, Non-U.S. Gov't Review].
41. Peschanski M, Cesaro P, Hantraye P. Rationale for intrastriatal grafting of striatal neuroblasts in patients with Huntington's disease. Neuroscience. 1995;68(2):273–85 [Research Support, Non-U.S. Gov't Review].
42. Harper SQ. Progress and challenges in RNA interference therapy for Huntington disease. Arch Neurol. 2009;66(8):933–8 [Research Support, Non-U.S. Gov't Review].
43. Cicchetti F, Soulet D, Freeman TB. Neuronal degeneration in striatal transplants and Huntington's disease: potential mechanisms and clinical implications. Brain. 2011;134(Pt 3):641–52 [Research Support, Non-U.S. Gov't Review].
44. Govert F, Schneider SA. Huntington's disease and Huntington's disease-like syndromes: an overview. Curr Opin Neurol. 2013;26(4):420–7.
45. Naito H, Oyanagi S. Familial myoclonus epilepsy and choreoathetosis: hereditary dentatorubral-pallidoluysian atrophy. Neurology. 1982;32(8):798–807 [Case Reports].
46. Danek A, Walker RH. Neuroacanthocytosis. Curr Opin Neurol. 2005;18(4):386–92 [Review].
47. Sempere AP, Aparicio S, Mola S, Perez-Tur J. Benign hereditary chorea: clinical features and long-term follow-up in a Spanish family. Parkinsonism Relat Disord. 2013;19(3):394–6 [Letter Research Support, Non-U.S. Gov't].
48. Miyajima H. Aceruloplasminemia, an iron metabolic disorder. Neuropathology. 2003;23(4):345–50.
49. Kubota A, Hida A, Ichikawa Y, Momose Y, Goto J, Igeta Y, et al. A novel ferritin light chain gene mutation in a Japanese family with neuroferritinopathy: description of clinical features and implications for genotype-phenotype correlations. Mov Disord. 2009;24(3):441–5 [Case Reports Research Support, Non-U.S. Gov't].
50. Walker RH. Update on the Non-Huntington's Disease Choreas with Comments on the Current Nomenclature. Tremor Other Hyperkinet Mov (NY). 2012;2:1–7.
51. Wild EJ, Tabrizi SJ. The differential diagnosis of chorea. Pract Neurol. 2007;7(6):360–73 [Review].
52. Edwards TC, Zrinzo L, Limousin P, Foltynie T. Deep brain stimulation in the treatment of chorea. Mov Disord. 2012;27(3):357–63 [Review].
53. Zesiewicz TA, Sullivan KL. Drug-induced hyperkinetic movement disorders by nonneuroleptic agents. Handb Clin Neurol. 2011;100:347–63 [Review].
54. Correll CU, Schenk EM. Tardive dyskinesia and new antipsychotics. Curr Opin Psychiatry. 2008;21(2):151–6 [Research Support, N.I.H., Extramural Research Support, Non-U.S. Gov't Review].
55. Piccolo I, Defanti CA, Soliveri P, Volonte MA, Cislaghi G, Girotti F. Cause and course in a series of patients with sporadic chorea. J Neurol. 2003;250(4):429–35.
56. Vidakovic A, Dragasevic N, Kostic VS. Hemiballism: report of 25 cases. J Neurol Neurosurg Psychiatry. 1994;57(8):945–9.
57. Oh SH, Lee KY, Im JH, Lee MS. Chorea associated with non-ketotic hyperglycemia and hyperintensity basal ganglia lesion on T1-weighted brain MRI study: a meta-analysis of 53 cases including four present cases. J Neurol Sci. 2002;200(1–2):57–62 [Meta-Analysis Research Support, Non-U.S. Gov't].

58. Hart DB. A Clinical Lecture on Two Cases of Chorea Gravidarum: delivered at the Extramural Class of Clinical Medicine in the Edinburgh Royal Infirmary. Br Med J. 1903;1(2194):126.
59. O'Toole O, Lennon VA, Ahlskog JE, Matsumoto JY, Pittock SJ, Bower J, et al. Autoimmune chorea in adults. Neurology. 2013;80(12):1133–44.
60. Gelosa G, Tremolizzo L, Galbussera A, Perego R, Capra M, Frigo M, et al. Narrowing the window for 'senile chorea': a case with primary antiphospholipid syndrome. J Neurol Sci. 2009;284(1–2):211–3 [Case Reports].
61. Tumas V, Caldas CT, Santos AC, Nobre A, Fernandes RM. Sydenham's chorea: clinical observations from a Brazilian movement disorder clinic. Parkinsonism Relat Disord. 2007; 13(5):276–83.
62. Fekete R, Jankovic J. Psychogenic chorea associated with family history of Huntington disease. Mov Disord. 2010;25(4):503–4 [Case Reports Letter].
63. Alexander GE, Crutcher MD. Functional architecture of basal ganglia circuits: neural substrates of parallel processing. Trends Neurosci. 1990;13(7):266–71 [Research Support, U.S. Gov't, P.H.S. Review].
64. Albin RL, Young AB, Penney JB. The functional anatomy of basal ganglia disorders. Trends Neurosci. 1989;12(10):366–75 [Research Support, Non-U.S. Gov't Research Support, U.S. Gov't, P.H.S. Review].
65. Berardelli A, Noth J, Thompson PD, Bollen EL, Curra A, Deuschl G, et al. Pathophysiology of chorea and bradykinesia in Huntington's disease. Mov Disord. 1999;14(3):398–403 [Research Support, Non-U.S. Gov't Review].
66. Reiner A, Albin RL, Anderson KD, D'Amato CJ, Penney JB, Young AB. Differential loss of striatal projection neurons in Huntington disease. Proc Natl Acad Sci U S A. 1988;85(15):5733–7 [Research Support, Non-U.S. Gov't Research Support, U.S. Gov't, P.H.S.].
67. Albin RL, Reiner A, Anderson KD, Penney JB, Young AB. Striatal and nigral neuron subpopulations in rigid Huntington's disease: implications for the functional anatomy of chorea and rigidity-akinesia. Ann Neurol. 1990;27(4):357–65 [Research Support, Non-U.S. Gov't Research Support, U.S. Gov't, P.H.S.].
68. Marsden CD, Obeso JA. The functions of the basal ganglia and the paradox of stereotaxic surgery in Parkinson's disease. Brain. 1994;117(Pt 4):877–97 [Review].
69. Thompson PD, Dick JP, Day BL, Rothwell JC, Berardelli A, Kachi T, et al. Electrophysiology of the corticomotoneurone pathways in patients with movement disorders. Mov Disord. 1986;1(2):113–7.
70. Miller BR, Walker AG, Shah AS, Barton SJ, Rebec GV. Dysregulated information processing by medium spiny neurons in striatum of freely behaving mouse models of Huntington's disease. J Neurophysiol. 2008;100(4):2205–16 [Research Support, N.I.H., Extramural Research Support, Non-U.S. Gov't].
71. Miller BR, Walker AG, Fowler SC, von Horsten S, Riess O, Johnson MA, et al. Dysregulation of coordinated neuronal firing patterns in striatum of freely behaving transgenic rats that model Huntington's disease. Neurobiol Dis. 2010;37(1):106–13 [Research Support, N.I.H., Extramural Research Support, Non-U.S. Gov't].
72. Obeso JA, Rodriguez-Oroz MC, Rodriguez M, DeLong MR, Olanow CW. Pathophysiology of levodopa-induced dyskinesias in Parkinson's disease: problems with the current model. Ann Neurol. 2000;47(4 Suppl 1):S22–32; discussion S-4.
73. Vitek JL, Chockkan V, Zhang JY, Kaneoke Y, Evatt M, DeLong MR, et al. Neuronal activity in the basal ganglia in patients with generalized dystonia and hemiballismus. Ann Neurol. 1999;46(1):22–35 [Case Reports Research Support, Non-U.S. Gov't].
74. Matsumura M, Tremblay L, Richard H, Filion M. Activity of pallidal neurons in the monkey during dyskinesia induced by injection of bicuculline in the external pallidum. Neuroscience. 1995;65(1):59–70 [Research Support, Non-U.S. Gov't].
75. Guridi J, Obeso JA. The subthalamic nucleus, hemiballismus and Parkinson's disease: reappraisal of a neurosurgical dogma. Brain. 2001;124(Pt 1):5–19 [Review].
76. Slavin KV, Baumann TK, Burchiel KJ. Treatment of hemiballismus with stereotactic pallidotomy. Case report and review of the literature. Neurosurg Focus. 2004;17(1):E7 [Case Reports Review].

77. Vlamings R, Zeef DH, Janssen ML, Oosterloo M, Schaper F, Jahanshahi A, et al. Lessons learned from the transgenic Huntington's disease rats. Neural Plast. 2012;2012:682712 [Research Support, Non-U.S. Gov't Review].
78. Trottier Y, Biancalana V, Mandel JL. Instability of CAG repeats in Huntington's disease: relation to parental transmission and age of onset. J Med Genet. 1994;31(5):377–82 [Research Support, Non-U.S. Gov't].
79. Hendricks AE, Latourelle JC, Lunetta KL, Cupples LA, Wheeler V, MacDonald ME, et al. Estimating the probability of de novo HD cases from transmissions of expanded penetrant CAG alleles in the Huntington disease gene from male carriers of high normal alleles (27-35 CAG). Am J Med Genet A. 2009;149A(7):1375–81 [Research Support, N.I.H., Extramural Research Support, Non-U.S. Gov't].
80. Kieburtz K, MacDonald M, Shih C, Feigin A, Steinberg K, Bordwell K, et al. Trinucleotide repeat length and progression of illness in Huntington's disease. J Med Genet. 1994;31(11):872–4 [Research Support, Non-U.S. Gov't Research Support, U.S. Gov't, P.H.S.].
81. MacDonald ME, Vonsattel JP, Shrinidhi J, Couropmitree NN, Cupples LA, Bird ED, et al. Evidence for the GluR6 gene associated with younger onset age of Huntington's disease. Neurology. 1999;53(6):1330–2 [Research Support, U.S. Gov't, P.H.S.].
82. Simonin C, Duru C, Salleron J, Hincker P, Charles P, Delval A, et al. Association between caffeine intake and age at onset in Huntington's disease. Neurobiol Dis. 2013;58C:179–82.
83. Pringsheim T, Wiltshire K, Day L, Dykeman J, Steeves T, Jette N. The incidence and prevalence of Huntington's disease: a systematic review and meta-analysis. Mov Disord. 2012;27(9):1083–91 [Meta-Analysis Review].
84. Quinn N, Schrag A. Huntington's disease and other choreas. J Neurol. 1998;245(11):709–16 [Review].
85. Craufurd D, Snowden J. Neuropsychological and neuropsychiatric aspects of Huntington's disease. In: Bates G, Harper P, Jones L, editors. Huntington's disease. New York: Oxford University Press; 2002. p. 62–94.
86. Di Maio L, Squitieri F, Napolitano G, Campanella G, Trofatter JA, Conneally PM. Suicide risk in Huntington's disease. J Med Genet. 1993;30(4):293–5 [Research Support, Non-U.S. Gov't].
87. Robins Wahlin TB, Backman L, Lundin A, Haegermark A, Winblad B, Anvret M. High suicidal ideation in persons testing for Huntington's disease. Acta Neurol Scand. 2000;102(3):150–61 [Comparative Study Research Support, Non-U.S. Gov't].
88. Farrer LA. Suicide and attempted suicide in Huntington disease: implications for preclinical testing of persons at risk. Am J Med Genet. 1986;24(2):305–11 [Research Support, Non-U.S. Gov't Research Support, U.S. Gov't, P.H.S.].
89. Roos RA. Huntington's disease: a clinical review. Orphanet J Rare Dis. 2010;5(1):40 [Review].
90. Mestre T, Ferreira J, Coelho MM, Rosa M, Sampaio C. Therapeutic interventions for disease progression in Huntington's disease. Cochrane Database Syst Rev. 2009(3):CD006455. [Meta-Analysis Review].
91. Kumar A, Sharma N, Mishra J, Kalonia H. Synergistical neuroprotection of rofecoxib and statins against malonic acid induced Huntington's disease like symptoms and related cognitive dysfunction in rats. Eur J Pharmacol. 2013;709(1–3):1–12 [Research Support, Non-U.S. Gov't].
92. Sarantos MR, Papanikolaou T, Ellerby LM, Hughes RE. Pizotifen activates ERK and provides neuroprotection in vitro and in vivo in models of Huntington's disease. J Huntingtons Dis. 2012;1(2):195–210.
93. Sagredo O, Pazos MR, Satta V, Ramos JA, Pertwee RG, Fernandez-Ruiz J. Neuroprotective effects of phytocannabinoid-based medicines in experimental models of Huntington's disease. J Neurosci Res. 2011;89(9):1509–18 [Research Support, Non-U.S. Gov't].
94. Wu J, Li Q, Bezprozvanny I. Evaluation of Dimebon in cellular model of Huntington's disease. Mol Neurodegener. 2008;3:15.
95. HORIZON Investigators of the Huntington Study Group and European Huntington's Disease Network. A randomized, double-blind, placebo-controlled study of latrepirdine in patients with mild to moderate Huntington disease. JAMA Neurol. 2013;70(1):25–33 [Multicenter Study Randomized Controlled Trial Research Support, Non-U.S. Gov't].

96. O'Suilleabhain P, Dewey Jr RB. A randomized trial of amantadine in Huntington disease. Arch Neurol. 2003;60(7):996–8 [Clinical Trial Randomized Controlled Trial].

97. Huntington Study Group. Dosage effects of riluzole in Huntington's disease: a multicenter placebo-controlled study. Neurology. 2003;61(11):1551–6 [Clinical Trial Multicenter Study Randomized Controlled Trial Research Support, Non-U.S. Gov't Research Support, U.S. Gov't, P.H.S.].

98. Curtis A, Mitchell I, Patel S, Ives N, Rickards H. A pilot study using nabilone for symptomatic treatment in Huntington's disease. Mov Disord. 2009;24(15):2254–9 [Clinical Trial Randomized Controlled Trial Research Support, Non-U.S. Gov't].

99. Bonelli RM, Wenning GK. Pharmacological management of Huntington's disease: an evidence-based review. Curr Pharm Des. 2006;12(21):2701–20 [Review].

100. Armstrong MJ, Miyasaki JM. Evidence-based guideline: pharmacologic treatment of chorea in Huntington disease: report of the guideline development subcommittee of the American Academy of Neurology. Neurology. 2012;79(6):597–603 [Practice Guideline Research Support, N.I.H., Extramural Research Support, Non-U.S. Gov't].

101. Reilmann R. Pharmacological treatment of chorea in Huntington's disease-good clinical practice versus evidence-based guideline. Mov Disord. 28(8):1030–3.

102. Brusa L, Versace V, Koch G, Bernardi G, Iani C, Stanzione P, et al. Improvement of choreic movements by 1 Hz repetitive transcranial magnetic stimulation in Huntington's disease patients. Ann Neurol. 2005;58(4):655–6 [Clinical Trial Comparative Study Letter].

103. Joel D, Ayalon L, Tarrasch R, Veenman L, Feldon J, Weiner I. Electrolytic lesion of globus pallidus ameliorates the behavioral and neurodegenerative effects of quinolinic acid lesion of the striatum: a potential novel treatment in a rat model of Huntington's disease. Brain Res. 1998;787(1):143–8 [Research Support, Non-U.S. Gov't].

104. Joel D, Ayalon L, Tarrasch R, Weiner I. Deficits induced by quinolinic acid lesion to the striatum in a position discrimination and reversal task are ameliorated by permanent and temporary lesion to the globus pallidus: a potential novel treatment in a rat model of Huntington's disease. Mov Disord. 2003;18(12):1499–507 [Evaluation Studies].

105. Ayalon L, Doron R, Weiner I, Joel D. Amelioration of behavioral deficits in a rat model of Huntington's disease by an excitotoxic lesion to the globus pallidus. Exp Neurol. 2004;186((1):46–58 [Comparative Study].

106. Cubo E, Shannon KM, Penn RD, Kroin JS. Internal globus pallidotomy in dystonia secondary to Huntington's disease. Mov Disord. 2000;15(6):1248–51 [Case Reports].

107. Lozano AM, Kumar R, Gross RE, Giladi N, Hutchison WD, Dostrovsky JO, et al. Globus pallidus internus pallidotomy for generalized dystonia. Mov Disord. 1997;12(6):865–70 [Case Reports Research Support, Non-U.S. Gov't].

108. Moro E, Lang AE, Strafella AP, Poon YY, Arango PM, Dagher A, et al. Bilateral globus pallidus stimulation for Huntington's disease. Ann Neurol. 2004;56(2):290–4 [Comparative Study].

109. Lang AE, Lozano AM, Montgomery E, Duff J, Tasker R, Hutchinson W. Posteroventral medial pallidotomy in advanced Parkinson's disease. N Engl J Med. 1997;337(15):1036–42 [Research Support, Non-U.S. Gov't].

110. Weeks RA, Ceballos-Baumann A, Piccini P, Boecker H, Harding AE, Brooks DJ. Cortical control of movement in Huntington's disease. A PET activation study. Brain. 1997;120(Pt 9):1569–78 [Research Support, Non-U.S. Gov't].

111. Fawcett AP, Moro E, Lang AE, Lozano AM, Hutchison WD. Pallidal deep brain stimulation influences both reflexive and voluntary saccades in Huntington's disease. Mov Disord. 2005;20(3):371–7 [Case Reports Research Support, Non-U.S. Gov't].

112. Hebb MO, Garcia R, Gaudet P, Mendez IM. Bilateral stimulation of the globus pallidus internus to treat choreathetosis in Huntington's disease: technical case report. Neurosurgery. 2006;58(2):E383; discussion E.

113. Tang JK, Moro E, Lozano AM, Lang AE, Hutchison WD, Mahant N, et al. Firing rates of pallidal neurons are similar in Huntington's and Parkinson's disease patients. Exp Brain Res.

2005;166(2):230–6 [Comparative Study Research Support, N.I.H., Extramural Research Support, Non-U.S. Gov't].

114. Biolsi B, Cif L, Fertit HE, Robles SG, Coubes P. Long-term follow-up of Huntington disease treated by bilateral deep brain stimulation of the internal globus pallidus. J Neurosurg. 2008;109(1):130–2 [Case Reports].

115. Fasano A, Mazzone P, Piano C, Quaranta D, Soleti F, Bentivoglio AR. GPi-DBS in Huntington's disease: results on motor function and cognition in a 72-year-old case. Mov Disord. 2008;23(9):1289–92 [Case Reports Research Support, Non-U.S. Gov't].

116. Kang GA, Heath S, Rothlind J, Starr PA. Long-term follow-up of pallidal deep brain stimulation in two cases of Huntington's disease. J Neurol Neurosurg Psychiatry. 2011;82(3):272–7 [Case Reports Research Support, Non-U.S. Gov't].

117. Starr PA, Kang GA, Heath S, Shimamoto S, Turner RS. Pallidal neuronal discharge in Huntington's disease: support for selective loss of striatal cells originating the indirect pathway. Exp Neurol. 2008;211(1):227–33 [Clinical Trial Comparative Study Research Support, N.I.H., Extramural Research Support, U.S. Gov't, Non-P.H.S.].

118. Groiss SJ, Elben S, Reck C, Voges J, Wojtecki L, Schnitzler A. Local field potential oscillations of the globus pallidus in Huntington's disease. Mov Disord. 2011;26(14):2577–8 [Case Reports Letter Research Support, Non-U.S. Gov't].

119. Silberstein P, Kuhn AA, Kupsch A, Trottenberg T, Krauss JK, Wohrle JC, et al. Patterning of globus pallidus local field potentials differs between Parkinson's disease and dystonia. Brain. 2003;126(Pt 12):2597–608 [Research Support, Non-U.S. Gov't].

120. Brown P. Cortical drives to human muscle: the Piper and related rhythms. Prog Neurobiol. 2000;60(1):97–108 [Research Support, Non-U.S. Gov't Review].

121. Garcia-Ruiz PJ, Ayerbe J, del Val J, Herranz A. Deep brain stimulation in disabling involuntary vocalization associated with Huntington's disease. Parkinsonism Relat Disord. 2012;18(6):803–4 [Case Reports Letter Research Support, Non-U.S. Gov't].

122. Spielberger S, Hotter A, Wolf E, Eisner W, Muller J, Poewe W, et al. Deep brain stimulation in Huntington's disease: a 4-year follow-up case report. Mov Disord. 2012;27(6):806–7; author reply 7–8 [Comment Letter].

123. Velez-Lago FM, Thompson A, Oyama G, Hardwick A, Sporrer JM, Zeilman P, et al. Differential and better response to deep brain stimulation of chorea compared to dystonia in Huntington's disease. Stereotact Funct Neurosurg. 2013;91(2):129–33.

124. Reiner A. Can lesions of GPe correct HD deficits? Exp Neurol. 2004;186(1):1–5 [Letter].

125. Temel Y, Cao C, Vlamings R, Blokland A, Ozen H, Steinbusch HW, et al. Motor and cognitive improvement by deep brain stimulation in a transgenic rat model of Huntington's disease. Neurosci Lett. 2006;406(1–2):138–41 [Research Support, Non-U.S. Gov't].

126. Ligot N, Krystkowiak P, Simonin C, Goldman S, Peigneux P, Van Naemen J, et al. External globus pallidus stimulation modulates brain connectivity in Huntington's disease. J Cereb Blood Flow Metab. 2011;31(1):41–6 [Research Support, Non-U.S. Gov't].

127. Tabrizi SJ, Scahill RI, Owen G, Durr A, Leavitt BR, Roos RA, et al. Predictors of phenotypic progression and disease onset in premanifest and early-stage Huntington's disease in the TRACK-HD study: analysis of 36-month observational data. Lancet Neurol. 2013;12(7):637–49.

128. Rieke GK, Scarfe AD, Hunter JF. L-pyroglutamate: an alternate neurotoxin for a rodent model of Huntington's disease. Brain Res Bull. 1984;13(3):443–56 [Comparative Study].

129. Coyle JT, Schwarcz R. Lesion of striatal neurones with kainic acid provides a model for Huntington's chorea. Nature. 1976;263(5574):244–6 [Research Support, U.S. Gov't, Non-P.H.S.].

130. Coyle JT, Schwarcz R, Bennett JP, Campochiaro P. Clinical, neuropathologic and pharmacologic aspects of Huntington's disease: correlates with a new animal model. Prog Neuropsychopharmacol. 1977;1(1–2):13–30 [Research Support, U.S. Gov't, P.H.S.].

131. Mangiarini L, Sathasivam K, Seller M, Cozens B, Harper A, Hetherington C, et al. Exon 1 of the HD gene with an expanded CAG repeat is sufficient to cause a progressive neurological phenotype in transgenic mice. Cell. 1996;87(3):493–506 [Research Support, Non-U.S. Gov't].

132. von Horsten S, Schmitt I, Nguyen HP, Holzmann C, Schmidt T, Walther T, et al. Transgenic rat model of Huntington's disease. Hum Mol Genet. 2003;12(6):617–24 [Research Support, Non-U.S. Gov't Research Support, U.S. Gov't, P.H.S.].

133. Deckel AW, Robinson RG, Coyle JT, Sanberg PR. Reversal of long-term locomotor abnormalities in the kainic acid model of Huntington's disease by day 18 fetal striatal implants. Eur J Pharmacol. 1983;93(3–4):287–8 [Research Support, Non-U.S. Gov't Research Support, U.S. Gov't, P.H.S.].

134. Isacson O, Brundin P, Gage FH, Bjorklund A. Neural grafting in a rat model of Huntington's disease: progressive neurochemical changes after neostriatal ibotenate lesions and striatal tissue grafting. Neuroscience. 1985;16(4):799–817 [Research Support, Non-U.S. Gov't Research Support, U.S. Gov't, P.H.S.].

135. Isacson O, Riche D, Hantraye P, Sofroniew MV, Maziere M. A primate model of Huntington's disease: cross-species implantation of striatal precursor cells to the excitotoxically lesioned baboon caudate-putamen. Exp Brain Res. 1989;75(1):213–20 [Research Support, Non-U.S. Gov't].

136. Hantraye P, Riche D, Maziere M, Isacson O. Intrastriatal transplantation of cross-species fetal striatal cells reduces abnormal movements in a primate model of Huntington disease. Proc Natl Acad Sci U S A. 1992;89(9):4187–91 [Research Support, Non-U.S. Gov't Research Support, U.S. Gov't, P.H.S.].

137. Dunnett SB, Carter RJ, Watts C, Torres EM, Mahal A, Mangiarini L, et al. Striatal transplantation in a transgenic mouse model of Huntington's disease. Exp Neurol. 1998;154(1):31–40 [Research Support, Non-U.S. Gov't].

138. Sadan O, Shemesh N, Barzilay R, Dadon-Nahum M, Blumenfeld-Katzir T, Assaf Y, et al. Mesenchymal stem cells induced to secrete neurotrophic factors attenuate quinolinic acid toxicity: a potential therapy for Huntington's disease. Exp Neurol. 2012;234(2):417–27 [Research Support, Non- .S. Gov't].

139. Yang CR, Yu RK. Intracerebral transplantation of neural stem cells combined with trehalose ingestion alleviates pathology in a mouse model of Huntington's disease. J Neurosci Res. 2009;87(1):26–33 [Research Support, N.I.H., Extramural Research Support, Non-U.S. Gov't].

140. Lee ST, Chu K, Jung KH, Im WS, Park JE, Lim HC, et al. Slowed progression in models of Huntington disease by adipose stem cell transplantation. Ann Neurol. 2009;66(5):671–81 [Comparative Study Research Support, Non-U.S. Gov't].

141. Snyder BR, Chiu AM, Prockop DJ, Chan AW. Human multipotent stromal cells (MSCs) increase neurogenesis and decrease atrophy of the striatum in a transgenic mouse model for Huntington's disease. PLoS One. 2010;5(2):e9347 [Research Support, N.I.H., Extramural].

142. Kwan W, Magnusson A, Chou A, Adame A, Carson MJ, Kohsaka S, et al. Bone marrow transplantation confers modest benefits in mouse models of Huntington's disease. J Neurosci. 2012;32(1):133–42 [Research Support, N.I.H., Extramural Research Support, Non-U.S. Gov't].

143. Emerich DF, Winn SR. Neuroprotective effects of encapsulated CNTF-producing cells in a rodent model of Huntington's disease are dependent on the proximity of the implant to the lesioned striatum. Cell Transplant. 2004;13(3):253–9.

144. Giralt A, Friedman HC, Caneda-Ferron B, Urban N, Moreno E, Rubio N, et al. BDNF regulation under GFAP promoter provides engineered astrocytes as a new approach for long-term protection in Huntington's disease. Gene Ther. 2010;17(10):1294–308 [Research Support, Non-U.S. Gov't].

145. Madrazo I, Franco-Bourland RE, Castrejon H, Cuevas C, Ostrosky-Solis F. Fetal striatal homotransplantation for Huntington's disease: first two case reports. Neurol Res. 1995;17(4):312–5 [Case Reports Research Support, Non-U.S. Gov't].

146. Philpott LM, Kopyov OV, Lee AJ, Jacques S, Duma CM, Caine S, et al. Neuropsychological functioning following fetal striatal transplantation in Huntington's chorea: three case presentations. Cell Transplant. 1997;6(3):203–12 [Case Reports].

147. Kopyov OV, Jacques S, Lieberman A, Duma CM, Eagle KS. Safety of intrastriatal neurotransplantation for Huntington's disease patients. Exp Neurol. 1998;149(1):97–108 [Case Reports Research Support, Non-U.S. Gov't].

148. Fink JS, Schumacher JM, Ellias SL, Palmer EP, Saint-Hilaire M, Shannon K, et al. Porcine xenografts in Parkinson's disease and Huntington's disease patients: preliminary results. Cell Transplant. 2000;9(2):273–8 [Clinical Trial Clinical Trial, Phase I].

149. Bachoud-Levi A, Bourdet C, Brugieres P, Nguyen JP, Grandmougin T, Haddad B, et al. Safety and tolerability assessment of intrastriatal neural allografts in five patients with Huntington's disease. Exp Neurol. 2000;161(1):194–202 [Clinical Trial Research Support, Non-U.S. Gov't].

150. Bachoud-Levi AC, Remy P, Nguyen JP, Brugieres P, Lefaucheur JP, Bourdet C, et al. Motor and cognitive improvements in patients with Huntington's disease after neural transplantation. Lancet. 2000;356(9246):1975–9 [Clinical Trial Research Support, Non-U.S. Gov't].

151. Bachoud-Levi AC, Gaura V, Brugieres P, Lefaucheur JP, Boisse MF, Maison P, et al. Effect of fetal neural transplants in patients with Huntington's disease 6 years after surgery: a long-term follow-up study. Lancet Neurol. 2006;5(4):303–9 [Clinical Trial Comparative Study Research Support, Non-U.S. Gov't].

152. Hauser RA, Sandberg PR, Freeman TB, Stoessl AJ. Bilateral human fetal striatal transplantation in Huntington's disease. Neurology. 2002;58(11):1704; author reply.

153. Rosser AE, Barker RA, Harrower T, Watts C, Farrington M, Ho AK, et al. Unilateral transplantation of human primary fetal tissue in four patients with Huntington's disease: NEST-UK safety report ISRCTN no 36485475. J Neurol Neurosurg Psychiatry. 2002;73(6):678–85 [Research Support, Non-U.S. Gov't].

154. Keene CD, Sonnen JA, Swanson PD, Kopyov O, Leverenz JB, Bird TD, et al. Neural transplantation in Huntington disease: long-term grafts in two patients. Neurology. 2007;68(24):2093–8 [Case Reports Research Support, N.I.H., Extramural Research Support, Non-U.S. Gov't Research Support, U.S. Gov't, Non-P.H.S.].

155. Reuter I, Tai YF, Pavese N, Chaudhuri KR, Mason S, Polkey CE, et al. Long-term clinical and positron emission tomography outcome of fetal striatal transplantation in Huntington's disease. J Neurol Neurosurg Psychiatry. 2008;79(8):948–51 [Controlled Clinical Trial Research Support, Non-U.S. Gov't].

156. Gallina P, Paganini M, Lombardini L, Saccardi R, Marini M, De Cristofaro MT, et al. Development of human striatal anlagen after transplantation in a patient with Huntington's disease. Exp Neurol. 2008;213(1):241–4 [Clinical Trial].

157. Cicchetti F, Saporta S, Hauser RA, Parent M, Saint-Pierre M, Sanberg PR, et al. Neural transplants in patients with Huntington's disease undergo disease-like neuronal degeneration. Proc Natl Acad Sci U S A. 2009;106(30):12483–8 [Research Support, Non-U.S. Gov't].

158. Gallina P, Paganini M, Lombardini L, Mascalchi M, Porfirio B, Gadda D, et al. Human striatal neuroblasts develop and build a striatal-like structure into the brain of Huntington's disease patients after transplantation. Exp Neurol. 2010;222(1):30–41 [Clinical Trial].

159. Barker RA, Mason SL, Harrower TP, Swain RA, Ho AK, Sahakian BJ, et al. The long-term safety and efficacy of bilateral transplantation of human fetal striatal tissue in patients with mild to moderate Huntington's disease. J Neurol Neurosurg Psychiatry. 2013;84(6):657–65 [Clinical Trial Research Support, Non-U.S. Gov't].

160. Cisbani G, Freeman TB, Soulet D, Saint-Pierre M, Gagnon D, Parent M, et al. Striatal allografts in patients with Huntington's disease: impact of diminished astrocytes and vascularization on graft viability. Brain. 2013;136(Pt 2):433–43 [Research Support, Non-U.S. Gov't].

161. Freeman TB, Cicchetti F, Bachoud-Levi AC, Dunnett SB. Technical factors that influence neural transplant safety in Huntington's disease. Exp Neurol. 2011;227(1):1–9.

162. Bachoud-Levi AC. Neural grafts in Huntington's disease: viability after 10 years. Lancet Neurol. 2009;8(11):979–81 [Letter].

163. van Dellen A, Deacon R, York D, Blakemore C, Hannan AJ. Anterior cingulate cortical transplantation in transgenic Huntington's disease mice. Brain Res Bull. 2001;56(3–4):313–8 [Research Support, Non-U.S. Gov't].

164. Alberch J, Perez-Navarro E, Canals JM. Neurotrophic factors in Huntington's disease. Prog Brain Res. 2004;146:195–229 [Research Support, Non-U.S. Gov't Review].

165. Alderson RF, Alterman AL, Barde YA, Lindsay RM. Brain-derived neurotrophic factor increases survival and differentiated functions of rat septal cholinergic neurons in culture. Neuron. 1990;5(3):297–306.
166. Snider WD. Functions of the neurotrophins during nervous system development: what the knockouts are teaching us. Cell. 1994;77(5):627–38 [Review].
167. Fischer W, Wictorin K, Bjorklund A, Williams LR, Varon S, Gage FH. Amelioration of cholinergic neuron atrophy and spatial memory impairment in aged rats by nerve growth factor. Nature. 1987;329(6134):65–8 [Research Support, Non-U.S. Gov't Research Support, U.S. Gov't, Non-P.H.S. Research Support, U.S. Gov't, P.H.S.].
168. Emerich DF, Winn SR, Hantraye PM, Peschanski M, Chen EY, Chu Y, et al. Protective effect of encapsulated cells producing neurotrophic factor CNTF in a monkey model of Huntington's disease. Nature. 1997;386(6623):395–9 [Research Support, Non-U.S. Gov't].
169. Bjorklund A, Kirik D, Rosenblad C, Georgievska B, Lundberg C, Mandel RJ. Towards a neuroprotective gene therapy for Parkinson's disease: use of adenovirus, AAV and lentivirus vectors for gene transfer of GDNF to the nigrostriatal system in the rat Parkinson model. Brain Res. 2000;886(1–2):82–98 [Research Support, Non-U.S. Gov't Review].
170. Demeestere J, Vandenberghe W. Experimental surgical therapies for Huntington's disease. CNS Neurosci Ther. 2011;17(6):705–13 [Research Support, Non-U.S. Gov't Review].
171. Tornoe J, Torp M, Jorgensen JR, Emerich DF, Thanos C, Bintz B, et al. Encapsulated cell-based biodelivery of meteorin is neuroprotective in the quinolinic acid rat model of neurodegenerative disease. Restor Neurol Neurosci. 2012;30(3):225–36 [Research Support, Non-U.S. Gov't].
172. Emerich DF, Cain CK, Greco C, Saydoff JA, Hu ZY, Liu H, et al. Cellular delivery of human CNTF prevents motor and cognitive dysfunction in a rodent model of Huntington's disease. Cell Transplant. 1997;6(3):249–66.
173. Dey ND, Bombard MC, Roland BP, Davidson S, Lu M, Rossignol J, et al. Genetically engineered mesenchymal stem cells reduce behavioral deficits in the YAC 128 mouse model of Huntington's disease. Behav Brain Res. 2010;214(2):193–200 [Research Support, Non-U.S. Gov't].
174. Schumacher JM, Short MP, Hyman BT, Breakefield XO, Isacson O. Intracerebral implantation of nerve growth factor-producing fibroblasts protects striatum against neurotoxic levels of excitatory amino acids. Neuroscience. 1991;45(3):561–70 [Research Support, Non-U.S. Gov't Research Support, U.S. Gov't, P.H.S.].
175. Ebert AD, Barber AE, Heins BM, Svendsen CN. Ex vivo delivery of GDNF maintains motor function and prevents neuronal loss in a transgenic mouse model of Huntington's disease. Exp Neurol. 2010;224(1):155–62 [Research Support, Non-U.S. Gov't].
176. Mittoux V, Joseph JM, Conde F, Palfi S, Dautry C, Poyot T, et al. Restoration of cognitive and motor functions by ciliary neurotrophic factor in a primate model of Huntington's disease. Hum Gene Ther. 2000;11(8):1177–87 [Research Support, Non-U.S. Gov't Research Support, U.S. Gov't, P.H.S.].
177. Bachoud-Levi AC, Deglon N, Nguyen JP, Bloch J, Bourdet C, Winkel L, et al. Neuroprotective gene therapy for Huntington's disease using a polymer encapsulated BHK cell line engineered to secrete human CNTF. Hum Gene Ther. 2000;11(12):1723–9 [Research Support, Non-U.S. Gov't].
178. Bloch J, Bachoud-Levi AC, Deglon N, Lefaucheur JP, Winkel L, Palfi S, et al. Neuroprotective gene therapy for Huntington's disease, using polymer-encapsulated cells engineered to secrete human ciliary neurotrophic factor: results of a phase I study. Hum Gene Ther. 2004;15(10):968–75 [Clinical Trial Clinical Trial, Phase I Research Support, Non-U.S. Gov't].
179. Altar CA, Armanini M, Dugich-Djordjevic M, Bennett GL, Williams R, Feinglass S, et al. Recovery of cholinergic phenotype in the injured rat neostriatum: roles for endogenous and exogenous nerve growth factor. J Neurochem. 1992;59(6):2167–77.
180. Davies SW, Beardsall K. Nerve growth factor selectively prevents excitotoxin induced degeneration of striatal cholinergic neurones. Neurosci Lett. 1992;140(2):161–4 [Research Support, Non-U.S. Gov't].
181. Araujo DM, Hilt DC. Glial cell line-derived neurotrophic factor attenuates the excitotoxin-induced behavioral and neurochemical deficits in a rodent model of Huntington's disease. Neuroscience. 1997;81(4):1099–110.

182. Ramaswamy S, Kordower JH. Gene therapy for Huntington's disease. Neurobiol Dis. 2012;48(2):243–54 [Review].
183. de Almeida LP, Zala D, Aebischer P, Deglon N. Neuroprotective effect of a CNTF-expressing lentiviral vector in the quinolinic acid rat model of Huntington's disease. Neurobiol Dis. 2001;8(3):433–46 [Research Support, Non-U.S. Gov't].
184. Ramaswamy S, McBride JL, Han I, Berry-Kravis EM, Zhou L, Herzog CD, et al. Intrastriatal CERE-120 (AAV-Neurturin) protects striatal and cortical neurons and delays motor deficits in a transgenic mouse model of Huntington's disease. Neurobiol Dis. 2009;34(1):40–50 [Research Support, Non-U.S. Gov't].
185. Jorgensen JR, Emerich DF, Thanos C, Thompson LH, Torp M, Bintz B, et al. Lentiviral delivery of meteorin protects striatal neurons against excitotoxicity and reverses motor deficits in the quinolinic acid rat model. Neurobiol Dis. 2011;41(1):160–8 [Research Support, Non-U.S. Gov't].
186. Arregui L, Benitez JA, Razgado LF, Vergara P, Segovia J. Adenoviral astrocyte-specific expression of BDNF in the striata of mice transgenic for Huntington's disease delays the onset of the motor phenotype. Cell Mol Neurobiol. 2011;31(8):1229–43 [Research Support, Non-U.S. Gov't].
187. Denovan-Wright EM, Attis M, Rodriguez-Lebron E, Mandel RJ. Sustained striatal ciliary neurotrophic factor expression negatively affects behavior and gene expression in normal and R6/1 mice. J Neurosci Res. 2008;86(8):1748–57 [Comparative Study Research Support, N.I.H., Extramural Research Support, Non-U.S. Gov't].
188. Regulier E, Pereira-de-Almeida L, Sommer B, Aebischer P, Deglon N. Dose-dependent neuroprotective effect of ciliary neurotrophic factor delivered via tetracycline-regulated lentiviral vectors in the quinolinic acid rat model of Huntington's disease. Hum Gene Ther. 2002;13(16):1981–90 [Research Support, Non-U.S. Gov't].
189. Ellison SM, Trabalza A, Tisato V, Pazarentzos E, Lee S, Papadaki V, et al. Dose-dependent Neuroprotection of VEGF in Huntington's Disease Striatum. Mol Ther. 2013;21(10):1862–75.
190. Harper SQ, Staber PD, He X, Eliason SL, Martins IH, Mao Q, et al. RNA interference improves motor and neuropathological abnormalities in a Huntington's disease mouse model. Proc Natl Acad Sci U S A. 2005;102(16):5820–5 [Research Support, Non-U.S. Gov't Research Support, U.S. Gov't, P.H.S.].
191. Rodriguez-Lebron E, Denovan-Wright EM, Nash K, Lewin AS, Mandel RJ. Intrastriatal rAAV-mediated delivery of anti-huntingtin shRNAs induces partial reversal of disease progression in R6/1 Huntington's disease transgenic mice. Mol Ther. 2005;12(4):618–33 [Research Support, N.I.H., Extramural Research Support, Non-U.S. Gov't Research Support, U.S. Gov't, P.H.S.].
192. Zuleta A, Vidal RL, Armentano D, Parsons G, Hetz C. AAV-mediated delivery of the transcription factor XBP1s into the striatum reduces mutant Huntingtin aggregation in a mouse model of Huntington's disease. Biochem Biophys Res Commun. 2012;420(3):558–63 [Research Support, Non-U.S. Gov't].
193. Wang YL, Liu W, Wada E, Murata M, Wada K, Kanazawa I. Clinico-pathological rescue of a model mouse of Huntington's disease by siRNA. Neurosci Res. 2005;53(3):241–9 [Research Support, Non-U.S. Gov't].
194. Southwell AL, Khoshnan A, Dunn DE, Bugg CW, Lo DC, Patterson PH. Intrabodies binding the proline-rich domains of mutant huntingtin increase its turnover and reduce neurotoxicity. J Neurosci. 2008;28(36):9013–20 [In Vitro Research Support, Non-U.S. Gov't].
195. Wang CE, Zhou H, McGuire JR, Cerullo V, Lee B, Li SH, et al. Suppression of neuropil aggregates and neurological symptoms by an intracellular antibody implicates the cytoplasmic toxicity of mutant huntingtin. J Cell Biol. 2008;181(5):803–16 [Research Support, N.I.H., Extramural].
196. Boado RJ, Kazantsev A, Apostol BL, Thompson LM, Pardridge WM. Antisense-mediated down-regulation of the human huntingtin gene. J Pharmacol Exp Ther. 2000;295(1):239–43 [Research Support, Non-U.S. Gov't].
197. Haque N, Isacson O. Antisense gene therapy for neurodegenerative disease? Exp Neurol. 1997;144(1):139–46 [Research Support, Non-U.S. Gov't Research Support, U.S. Gov't, Non-P.H.S. Research Support, U.S. Gov't, P.H.S.].

198. Bassen FA, Kornzweig AL. Malformation of the erythrocytes in a case of atypical retinitis pigmentosa. Blood. 1950;5(4):381–7.
199. Critchley EM, Clark DB, Wikler A. Acanthocytosis and neurological disorder without betalipoproteinemia. Arch Neurol. 1968;18(2):134–40.
200. Levine IM, Estes JW, Looney JM. Hereditary neurological disease with acanthocytosis. A new syndrome. Arch Neurol. 1968;19(4):403–9.
201. Yamamoto T, Hirose G, Shimazaki K, Takado S, Kosoegawa H, Saeki M. Movement disorders of familial neuroacanthocytosis syndrome. Arch Neurol. 1982;39(5):298–301 [Case Reports].
202. Hardie RJ, Pullon HW, Harding AE, Owen JS, Pires M, Daniels GL, et al. Neuroacanthocytosis. A clinical, haematological and pathological study of 19 cases. Brain. 1991;114(Pt 1A):13–49 [Case Reports Research Support, Non-U.S. Gov't].
203. Edwards M, Quinn N, Bhatia K, editors. Parkinson's disease and other movement disorders. New York: Oxford University Press; 2008.
204. Vital A, Bouillot S, Burbaud P, Ferrer X, Vital C. Chorea-acanthocytosis: neuropathology of brain and peripheral nerve. Clin Neuropathol. 2002;21(2):77–81 [Case Reports].
205. Muller-Vahl KR, Berding G, Emrich HM, Peschel T. Chorea-acanthocytosis in monozygotic twins: clinical findings and neuropathological changes as detected by diffusion tensor imaging, FDG-PET and (123)I-beta-CIT-SPECT. J Neurol. 2007;254(8):1081–8 [Case Reports].
206. Saiki S, Sakai K, Kitagawa Y, Saiki M, Kataoka S, Hirose G. Mutation in the CHAC gene in a family of autosomal dominant chorea-acanthocytosis. Neurology. 2003;61(11):1614–6 [Research Support, Non-U.S. Gov't].
207. Rubio JP, Danek A, Stone C, Chalmers R, Wood N, Verellen C, et al. Chorea-acanthocytosis: genetic linkage to chromosome 9q21. Am J Hum Genet. 1997;61(4):899–908 [Research Support, Non-U.S. Gov't Research Support, U.S. Gov't, P.H.S.].
208. Ueno S, Maruki Y, Nakamura M, Tomemori Y, Kamae K, Tanabe H, et al. The gene encoding a newly discovered protein, chorein, is mutated in chorea-acanthocytosis. Nat Genet. 2001;28(2):121–2 [Research Support, Non-U.S. Gov't].
209. Walker RH, Jung HH, Dobson-Stone C, Rampoldi L, Sano A, Tison F, et al. Neurologic phenotypes associated with acanthocytosis. Neurology. 2007;68(2):92–8 [Meta-Analysis Review].
210. Bader B, Walker RH, Vogel M, Prosiegel M, McIntosh J, Danek A. Tongue protrusion and feeding dystonia: a hallmark of chorea-acanthocytosis. Mov Disord. 2010;25(1):127–9 [Case Reports Letter].
211. Al-Asmi A, Jansen AC, Badhwar A, Dubeau F, Tampieri D, Shustik C, et al. Familial temporal lobe epilepsy as a presenting feature of choreoacanthocytosis. Epilepsia. 2005;46(8):1256–63 [Case Reports Comparative Study Research Support, Non-U.S. Gov't].
212. Jung HH, Danek A, Walker RH. Neuroacanthocytosis syndromes. Orphanet J Rare Dis. 2011;6:68 [Review].
213. Rampoldi L, Danek A, Monaco AP. Clinical features and molecular bases of neuroacanthocytosis. J Mol Med (Berl). 2002;80(8):475–91 [Research Support, Non-U.S. Gov't Review].
214. Margolis RL, Holmes SE, Rosenblatt A, Gourley L, O'Hearn E, Ross CA, et al. Huntington's disease-like 2 (HDL2) in North America and Japan. Ann Neurol. 2004;56(5):670–4 [Comparative Study Research Support, Non-U.S. Gov't Research Support, U.S. Gov't, Non--P.H.S. Research Support, U.S. Gov't, P.H.S.].
215. Fujimoto Y, Isozaki E, Yokochi F, Yamakawa K, Takahashi H, Hirai S. A case of chorea-acanthocytosis successfully treated with posteroventral pallidotomy. Rinsho Shinkeigaku. 1997;37(10):891–4 [Case Reports].
216. Wihl G, Volkmann J, Allert N, Lehrke R, Sturm V, Freund HJ. Deep brain stimulation of the internal pallidum did not improve chorea in a patient with neuro-acanthocytosis. Mov Disord. 2001;16(3):572–5 [Case Reports].
217. Burbaud P, Rougier A, Ferrer X, Guehl D, Cuny E, Arne P, et al. Improvement of severe trunk spasms by bilateral high-frequency stimulation of the motor thalamus in a patient with chorea-acanthocytosis. Mov Disord. 2002;17(1):204–7 [Case Reports].

218. Burbaud P, Vital A, Rougier A, Bouillot S, Guehl D, Cuny E, et al. Minimal tissue damage after stimulation of the motor thalamus in a case of chorea-acanthocytosis. Neurology. 2002;59(12):1982–4 [Case Reports].

219. Guehl D, Cuny E, Tison F, Benazzouz A, Bardinet E, Sibon Y, et al. Deep brain pallidal stimulation for movement disorders in neuroacanthocytosis. Neurology. 2007;68(2):160–1 [Case Reports].

220. Ruiz PJ, Ayerbe J, Bader B, Danek A, Sainz MJ, Cabo I, et al. Deep brain stimulation in chorea acanthocytosis. Mov Disord. 2009;24(10):1546–7 [Letter Research Support, Non-- U.S. Gov't].

221. Shin H, Ki CS, Cho AR, Lee JI, Ahn JY, Lee JH, et al. Globus pallidus interna deep brain stimulation improves chorea and functional status in a patient with chorea-acanthocytosis. Stereotact Funct Neurosurg. 2012;90(4):273–7 [Case Reports].

222. Li P, Huang R, Song W, Ji J, Burgunder JM, Wang X, et al. Deep brain stimulation of the globus pallidus internal improves symptoms of chorea-acanthocytosis. Neurol Sci. 2012;33(2):269–74 [Case Reports].

223. Kefalopoulou Z, Zrinzo L, Aviles-Olmos I, Bhatia K, Jarman P, Jahanshahi M, et al. Deep brain stimulation as a treatment for chorea-acanthocytosis. J Neurol. 2013;260(1):303–5 [Case Reports Letter].

224. Lim TT, Fernandez HH, Cooper S, Wilson KM, Machado AG. Successful deep brain stimulation surgery with intraoperative magnetic resonance imaging on a difficult neuroacanthocytosis case: case report. Neurosurgery. 2013;73(1):E184–8.

225. Shannon KR. Ballism. In: Jankovic J, Tolosa E, editors. Parkinson's disease and movement disorders. 3rd ed. Baltimore: Williams and Wilkins; 1998. p. 365–75.

226. Lee MS, Marsden CD. Movement disorders following lesions of the thalamus or subthalamic region. Mov Disord. 1994;9(5):493–507 [Review].

227. Lin JJ, Chang MK. Hemiballism-hemichorea and non-ketotic hyperglycaemia. J Neurol Neurosurg Psychiatry. 1994;57(6):748–50 [Case Reports Review].

228. Miao J, Liu R, Li J, Du Y, Zhang W, Li Z. Meige's syndrome and hemichorea associated with hyperthyroidism. J Neurol Sci. 2010;288(1–2):175–7 [Case Reports].

229. el Maghraoui A, Birouk N, Zaim A, Slassi I, Yahyaoui M, Chkili T. Fahr syndrome and dysparathyroidism. 3 cases. Presse Med. 1995;24(28):1301–4 [Case Reports].

230. Morre HH, van Woerkom TC, Endtz LJ. A case of chorea due to polycythaemia vera. Clin Neurol Neurosurg. 1982;84(2):125–30 [Case Reports].

231. Parikh S, Swaiman KF, Kim Y. Neurologic characteristics of childhood lupus erythematosus. Pediatr Neurol. 1995;13(3):198–201.

232. Alakandy LM, Iyer RV, Golash A. Hemichorea, an unusual complication of ventriculoperitoneal shunt. J Clin Neurosci. 2008;15(5):599–601 [Case Reports Review].

233. Buge A, Vincent D, Rancurel G, Cheron F. Hemichorea and oral contraceptives. Rev Neurol (Paris). 1985;141(10):663–5 [Case Reports].

234. Dike GL. Chorea gravidarum: a case report and review. Md Med J. 1997;46(8):436–9 [Case Reports Review].

235. Gastaut JL, Nicoli F, Somma-Mauvais H, Bartolomei F, Dalecky A, Bruzzo M, et al. Hemichorea-hemiballismus and toxoplasmosis in AIDS. Rev Neurol (Paris). 1992;148(12):785–8 [Case Reports].

236. Dewey Jr RB, Jankovic J. Hemiballism-hemichorea. Clinical and pharmacologic findings in 21 patients. Arch Neurol. 1989;46(8):862–7.

237. Borremans JJ, Krauss JK, Fanardjian RV, Seeger W. Hemichorea-hemiballism associated with an ipsilateral intraventricular cyst after resection of a meningioma. Parkinsonism Relat Disord. 1996;2(3):155–9.

238. Krauss JK, Borremans JJ, Nobbe F, Mundinger F. Ballism not related to vascular disease: a report of 16 patients and review of the literature. Parkinsonism Relat Disord. 1996;2(1):35–45.

239. Ghika-Schmid F, Ghika J, Regli F, Bogousslavsky J. Hyperkinetic movement disorders during and after acute stroke: the Lausanne Stroke Registry. J Neurol Sci. 1997;146(2):109–16.

240. Becker RE, Lal H. Pharmacological approaches to treatment of hemiballism and hemichorea. Brain Res Bull. 1983;11(2):187–9 [Review].
241. Evidente VG, Gwinn-Hardy K, Caviness JN, Alder CH. Risperidone is effective in severe hemichorea/hemiballismus. Mov Disord. 1999;14(2):377–9 [Case Reports].
242. Safirstein B, Shulman LM, Weiner WJ. Successful treatment of hemichorea with olanzapine. Mov Disord. 1999;14(3):532–3 [Case Reports].
243. Emre M, Landis T. Haloperidol in hemichorea-hemiballismus. J Neurol. 1984;231(5):280 [Case Reports Letter].
244. Bashir K, Manyam BV. Clozapine for the control of hemiballismus. Clin Neuropharmacol. 1994;17(5):477–80 [Case Reports].
245. Hernandez-Latorre MA, Roig-Quilis M. The efficiency of carbamazepine in a case of post-streptococcal hemichorea. Rev Neurol. 2003;37(4):322–6 [Case Reports].
246. Kothare SV, Pollack P, Kulberg AG, Ravin PD. Gabapentin treatment in a child with delayed-onset hemichorea/hemiballismus. Pediatr Neurol. 2000;22(1):68–71 [Case Reports].
247. Gatto EM, Uribe Roca C, Raina G, Gorja M, Folgar S, Micheli FE. Vascular hemichorea/hemiballism and topiramate. Mov Disord. 2004;19(7):836–8 [Case Reports].
248. Di Lazzaro V, Dileone M, Pilato F, Contarino MF, Musumeci G, Bentivoglio AR, et al. Repetitive transcranial magnetic stimulation of the motor cortex for hemichorea. J Neurol Neurosurg Psychiatry. 2006;77(9):1095–7 [Case Reports Letter].
249. Grimm E. Therapy of ballistic hyperkinesia (case report). Psychiatr Neurol Med Psychol (Leipz). 1980;32(6):369–72 [Case Reports].
250. Kawashima Y, Takahashi A, Hirato M, Ohye C. Stereotactic Vim-Vo-thalamotomy for choreatic movement disorder. Acta Neurochir Suppl (Wien). 1991;52:103–6 [Case Reports].
251. Siegfried J, Lippitz B. Chronic electrical stimulation of the VL-VPL complex and of the pallidum in the treatment of movement disorders: personal experience since 1982. Stereotact Funct Neurosurg. 1994;62(1–4):71–5 [Case Reports].
252. Cardoso F, Jankovic J, Grossman RG, Hamilton WJ. Outcome after stereotactic thalamotomy for dystonia and hemiballismus. Neurosurgery. 1995;36(3):501–7; discussion 7–8 [Comparative Study Review].
253. Krauss JK, Mundinger F. Functional stereotactic surgery for hemiballism. J Neurosurg. 1996;85(2):278–86.
254. Astradsson A, Schweder P, Joint C, Forrow B, Thevathasan W, Pereira EA, et al. Thalamotomy for postapoplectic hemiballistic chorea in older adults. J Am Geriatr Soc. 2010;58(11):2240–1 [Case Reports Letter].
255. Goto S, Kunitoku N, Hamasaki T, Nishikawa S, Ushio Y. Abolition of postapoplectic hemichorea by Vo-complex thalamotomy: long-term follow-up study. Mov Disord. 2001;16(4):771–4 [Case Reports Research Support, Non-U.S. Gov't].
256. Yasargil MG. The results of stereotactic operations in hyperkinesia. Schweiz Med Wochenschr. 1962;92:1550–5.
257. Suarez JI, Metman LV, Reich SG, Dougherty PM, Hallett M, Lenz FA. Pallidotomy for hemiballismus: efficacy and characteristics of neuronal activity. Ann Neurol. 1997;42(5):807–11 [Case Reports Research Support, U.S. Gov't, P.H.S.].
258. Choi SJ, Lee SW, Kim MC, Kwon JY, Park CK, Sung JH, et al. Posteroventral pallidotomy in medically intractable postapoplectic monochorea: case report. Surg Neurol. 2003;59(6):486–90; discussion 90 [Case Reports Research Support, Non-U.S. Gov't].
259. Hashimoto T, Morita H, Tada T, Maruyama T, Yamada Y, Ikeda S. Neuronal activity in the globus pallidus in chorea caused by striatal lacunar infarction. Ann Neurol. 2001;50(4):528–31 [Case Reports].
260. Tseng KY, Tang CT, Chang CF, Chen KY. Treatment of delayed-onset post-stroke monochorea with stereotactic pallidotomy. J Clin Neurosci. 2010;17(6):779–81 [Case Reports].
261. Goto T, Hashimoto T, Hirayama S, Kitazawa K. Pallidal neuronal activity in diabetic hemichorea-hemiballism. Mov Disord. 2010;25(9):1295–7 [Case Reports Letter].
262. Carpay HA, Arts WF, Kloet A, Hoogland PH, Van Duinen SG. Hemichorea reversible after operation in a boy with cavernous angioma in the head of the caudate nucleus. J Neurol Neurosurg Psychiatry. 1994;57(12):1547–8 [Case Reports Letter].

263. Zabek M, Sobstyl M, Dzierzecki S, Gorecki W, Jakucinski M. Right hemichorea treated successfully by surgical removal of a left putaminal cavernous angioma. Clin Neurol Neurosurg. 2013;115(6):844–6.
264. Karampelas I, Podgorsak MB, Plunkett RJ, Fenstermaker RA. Subthalamic nucleus metastasis causing hemichorea-hemiballism treated by gamma knife stereotactic radiosurgery. Acta Neurochir (Wien). 2008;150(4):395–6; discussion 7.
265. Tsubokawa T, Katayama Y, Yamamoto T. Control of persistent hemiballismus by chronic thalamic stimulation. Report of two cases. J Neurosurg. 1995;82(3):501–5 [Case Reports Research Support, Non-U.S. Gov't].
266. Thompson TP, Kondziolka D, Albright AL. Thalamic stimulation for choreiform movement disorders in children. Report of two cases. J Neurosurg. 2000;92(4):718–21 [Case Reports].
267. Nakano N, Uchiyama T, Okuda T, Kitano M, Taneda M. Successful long-term deep brain stimulation for hemichorea-hemiballism in a patient with diabetes. Case report. J Neurosurg. 2005;102(6):1137–41 [Case Reports].
268. Hasegawa H, Mundil N, Samuel M, Jarosz J, Ashkan K. The treatment of persistent vascular hemidystonia-hemiballismus with unilateral GPi deep brain stimulation. Mov Disord. 2009;24(11):1697–8 [Case Reports Letter].
269. Capelle HH, Kinfe TM, Krauss JK. Deep brain stimulation for treatment of hemichorea-hemiballism after craniopharyngioma resection: long-term follow-up. J Neurosurg. 2011;115(5):966–70 [Case Reports].
270. Air EL, Ostrem JL, Sanger TD, Starr PA. Deep brain stimulation in children: experience and technical pearls. J Neurosurg Pediatr. 2011;8(6):566–74 [Research Support, Non-U.S. Gov't].
271. Sato K, Nakagawa E, Saito Y, Komaki H, Sakuma H, Sugai K, et al. Hyperkinetic movement disorder in a child treated by globus pallidus stimulation. Brain Dev. 2009;31(6):452–5 [Case Reports].
272. Angelini L, Nardocci N, Estienne M, Conti C, Dones I, Broggi G. Life-threatening dystonia-dyskinesias in a child: successful treatment with bilateral pallidal stimulation. Mov Disord. 2000;15(5):1010–2 [Case Reports].
273. Hebb MO, Gaudet P, Mendez I. Deep brain stimulation to treat hyperkinetic symptoms of Cockayne syndrome. Mov Disord. 2006;21(1):112–5 [Case Reports].
274. Koy A, Hellmich M, Pauls KA, Marks W, Lin JP, Fricke O, et al. Effects of deep brain stimulation in dyskinetic cerebral palsy: a meta-analysis. Mov Disord. 2013;28(5):647–54 [Research Support, Non-U.S. Gov't].
275. Apetauerova D, Schirmer CM, Shils JL, Zani J, Arle JE. Successful bilateral deep brain stimulation of the globus pallidus internus for persistent status dystonicus and generalized chorea. J Neurosurg. 2010;113(3):634–8 [Case Reports].
276. Vidailhet M, Yelnik J, Lagrange C, Fraix V, Grabli D, Thobois S, et al. Bilateral pallidal deep brain stimulation for the treatment of patients with dystonia-choreoathetosis cerebral palsy: a prospective pilot study. Lancet Neurol. 2009;8(8):709–17 [Clinical Tria Multicenter Study Research Support, Non-U.S. Gov't].
277. Spindler MA, Galifianakis NB, Wilkinson JR, Duda JE. Globus pallidus interna deep brain stimulation for tardive dyskinesia: case report and review of the literature. Parkinsonism Relat Disord. 2013;19(2):141–7 [Case Reports Review].
278. Eltahawy HA, Feinstein A, Khan F, Saint-Cyr J, Lang AE, Lozano AM. Bilateral globus pallidus internus deep brain stimulation in tardive dyskinesia: a case report. Mov Disord. 2004;19(8):969–72 [Case Reports Comparative Study].
279. Schrader C, Peschel T, Petermeyer M, Dengler R, Hellwig D. Unilateral deep brain stimulation of the internal globus pallidus alleviates tardive dyskinesia. Mov Disord. 2004;19(5):583–5 [Case Reports].
280. Kosel M, Sturm V, Frick C, Lenartz D, Zeidler G, Brodesser D, et al. Mood improvement after deep brain stimulation of the internal globus pallidus for tardive dyskinesia in a patient suffering from major depression. J Psychiatr Res. 2007;41(9):801–3 [Case Reports].
281. Damier P, Thobois S, Witjas T, Cuny E, Derost P, Raoul S, et al. Bilateral deep brain stimulation of the globus pallidus to treat tardive dyskinesia. Arch Gen Psychiatry. 2007;64(2):170–6 [Clinical Trial Comparative Study Multicenter Study Research Support, Non-U.S. Gov't].

282. Kefalopoulou Z, Paschali A, Markaki E, Vassilakos P, Ellul J, Constantoyannis C. A double-blind study on a patient with tardive dyskinesia treated with pallidal deep brain stimulation. Acta Neurol Scand. 2009;119(4):269–73 [Case Reports].
283. Trottenberg T, Volkmann J, Deuschl G, Kuhn AA, Schneider GH, Muller J, et al. Treatment of severe tardive dystonia with pallidal deep brain stimulation. Neurology. 2005;64(2):344–6.
284. Wang Y, Turnbull I, Calne S, Stoessl AJ, Calne DB. Pallidotomy for tardive dyskinesia. Lancet. 1997;349(9054):777–8 [Case Reports Letter Research Support, Non-U.S. Gov't].
285. Weetman J, Anderson IM, Gregory RP, Gill SS. Bilateral posteroventral pallidotomy for severe antipsychotic induced tardive dyskinesia and dystonia. J Neurol Neurosurg Psychiatry. 1997;63(4):554–6 [Case Reports Letter].
286. Hillier CE, Wiles CM, Simpson BA. Thalamotomy for severe antipsychotic induced tardive dyskinesia and dystonia. J Neurol Neurosurg Psychiatry. 1999;66(2):250–1 [Case Reports Letter].
287. Lenders MW, Buschman HP, Vergouwen MD, Steur EN, Kolling P, Hariz M. Long term results of unilateral posteroventral pallidotomy for antipsychotic drug induced tardive dyskinesia. J Neurol Neurosurg Psychiatry. 2005;76(7):1039 [Case Reports Letter Research Support, Non-U.S. Gov't].
288. Yianni J, Nandi D, Bradley K, Soper N, Gregory R, Joint C, et al. Senile chorea treated by deep brain stimulation: a clinical, neurophysiological and functional imaging study. Mov Disord. 2004;19(5):597–602 [Case Reports Research Support, Non-U.S. Gov't].
289. Loher TJ, Krauss JK, Burgunder JM, Taub E, Siegfried J. Chronic thalamic stimulation for treatment of dystonic paroxysmal nonkinesigenic dyskinesia. Neurology. 2001;56(2):268–70 [Case Reports].
290. Yamada K, Goto S, Soyama N, Shimoda O, Kudo M, Kuratsu J, et al. Complete suppression of paroxysmal nonkinesigenic dyskinesia by globus pallidus internus pallidal stimulation. Mov Disord. 2006;21(4):576–9 [Case Reports Research Support, Non-U.S. Gov't].
291. Kaufman CB, Mink JW, Schwalb JM. Bilateral deep brain stimulation for treatment of medically refractory paroxysmal nonkinesigenic dyskinesia. J Neurosurg. 2010;112(4):847–50 [Case Reports].

Chapter 18
Chorea in Childhood

Emilio Fernández-Alvarez

Abstract Chorea is a rare presenting symptom starting in childhood. Chorea can be seen as a monosymptomatic movement disorder or as an associated symptom in a wide number of diseases such as immunomediated, neurodegenerative, paroxysmal, and metabolic diseases. Its onset may be acute or subtle. The most common causes of acute chorea in children are autoimmune, especially Sydenham chorea. In chronic encephalopathies, chorea usually occurs in conjunction with other neurologic and extraneurologic symptoms. In these cases, family and clinical history, physical examination, and a specific set of ancillary studies such as genetic, neuroimaging, and biochemical studies will often enable a correct diagnosis.

Keywords Pediatric chorea • Athetosis • Choreoathetotic cerebral palsy • Glut1 deficiency

Introduction

From a series of 922 pediatric patients with movement disorders (tics excluded), 6 % had chorea as their predominant abnormal movement [1]. In childhood, chorea can be seen in a wide array of acquired, neurodegenerative, metabolic, autoimmune, or iatrogenic disorders. This chapter will focus on clinical diagnosis and management of some acute and chronic chorea in childhood that are not extensively discussed in other chapters of this book.

E. Fernández-Alvarez
Neuropediatric Department, Hospital Sant Joan de Déu-Barcelona,
Rua Paxarinos 4, 2A, 27002 Lugo, Spain
e-mail: efernandez@hsjdbcn.org

F.E. Micheli, P.A. LeWitt (eds.), *Chorea*,
DOI 10.1007/978-1-4471-6455-5_18, © Springer-Verlag London 2014

History

Familial history is important in suspected genetic (including metabolic and neurodegenerative) disorders. The history should identify in chronological order the onset and setting of the symptoms. It is essential to consider that intellectually normal children beyond the age of 3 years can contribute to the history and particularly to the current symptomatology. The events surrounding the delivery and developmental assessment must be carefully documented. When chorea is monosymptomatic and starts acutely or subacutely in childhood (such as in Sydenham chorea [SC]), parents can describe the onset and the way in which the functional impairment in child's purposeful movements and/or speech have occurred. In subtle onset cases, particularly in young children, a parent's report that coordination or speech has changed and has functional impairment must be relied upon.

In two scenarios, the report of the parents can be misleading. The first situation is when chorea appears in infancy. In this instance, parents most often report that their child is floppy because generalized hypotonia can be the predominant feature and the choreic movements are subtle and frequently observed only obvious in perioral region. Second, when chorea occurs as a late or minor feature of a chronic neurologic disease, parents might not accurately report chorea onset or neurologic impairment. However, from the diagnostic point of view, in both situations the presence of chorea may be a useful diagnostic clue.

The Exam of Chorea in Children

Observation of the child begins during the history taking. It is best to observe an infant in the parent's lap, where the child will be much more comfortable. If the child is crying, important information mainly concerning abnormal movements can be misinterpreted or lost.

Chorea may result from simple restlessness with mild intermittent exaggeration of gestures or expressions to produce a continuous flow of violent and incapacitating movements (ballismus). Because the motor activity in typical children is greater than that in adults, in the mildest cases childhood chorea may be difficult to differentiate from exaggerated normal motor activity. A helpful clue is that normal exaggerated motor activity is a habitual motor pattern without associated functional impairment. It is necessary to consider that, in children, there is a continuum from exaggerated normal motor activity to chorea and to ballism (and chorea and ballism differ in their phenomenology only by the intensity and amplitude of the movements). Moreover, the degree of this continuum may be state dependent. For example, typical chorea may evolve into ballism when the same child is agitated. However, chorea and ballism are characterized by some salient features such as their abrupt character (that reach a continuous flow of violent movements in ballism) and, moreover, both interferes the normal voluntary movements. So, a child with suspected chorea should be observed carefully in spontaneous and structured activities rather than required to engage in a "formal" neurologic examination. In practice, many children with this kind of hyperkinetic movement disorders may have a combination of chorea, athetosis, and ballism. Speech may be impaired to the extent that a child might be anarthria. A rare and difficult diagnostic situation is when chorea in children is associated with severe hypotonia ("chorea mollis"). In this instance, the choreic movements may be so subtle that it may be necessary to be an experienced clinician to detect it.

The adventitious movements of chorea involve predominantly the upper limbs (usually asymmetrically). The arms should be investigated not only from the point of view of spontaneous activity but also in imposed attitudes (such as extension of both arms) and in "classical" neurologic testing such as finger-to-nose maneuver. In older children, drawing and writing are especially sensitive for testing.

The patient may often try to mask involuntary movements by incorporating or prolonging them with seemingly purposeful activity such as touching the face. Speech may be slurred or slow because of involvement of the tongue and facial muscles.

When chorea involves the legs, walking may take on motions like those of a dancer, or else a child may tilt intermittently. The latter situation may make it easy to confound with ataxia, but, in chorea, there is not the consistent broad-based gait seen in ataxia.

Choreic movements can be exacerbated by action, stress, or mental concentration.

Motor impersistence (a particularly frequent feature of SC) is the inability to maintain a posture or stable motor command. Asking the patient to gently press the extended hands on those of the examiner while the attention is distracted, maintain a steady grip force (the "milkmaid's sign"), and keep the tongue protruded (looking for the "darting tongue") are useful maneuvers to detect motor impersistence. Requesting the patient to engage into fine activities such as threading pearls unmasks or increases choreic movements.

Some degree of hypotonia is often apparent with chorea. With the patient sitting and leg unsupported, the quadriceps reflex may elicit a swing of the leg or a sustained leg extension ("hung up" sign). This is best seen with the patient sitting and leg unsupported.

Ballismus is easy to recognize. Their impressive, high-amplitude, flinging movements typically originate in limbs proximally. It can occur in children with severe static encephalopathies.

Finally, it can be useful to consider two scenarios regarding these involuntary movements: one, that the immature movements of typically developing infants may, in some instances, mimic mild chorea ("choreiform movements"). These do not indicate neurologic disease. Second, children with neurodevelopmental disorders (such as attention-deficit/hyperactivity disorder) are significantly more likely to manifest some mild choreiform movements.

Diagnostic Approach of Chorea in Childhood

Chorea is rare in children but may be a consequence of a large number of diseases, so the differential diagnosis of chorea is extensive. Several aspects must be taken into consideration in the diagnostic approach of chorea in childhood: (1) the age of onset, (2) if the onset is insidious, subacute, or acute, (3) if it is "pure" or associated to other neurologic (such as intellectual impairment, convulsions, or other problems) or extraneurologic signs or symptoms (such as fever, rash, or other manifestations), and (4) if it appears as a new symptom in the course of a chronic encephalopathy or following an encephalopathic crisis.

Tables 18.1, 18.2, 18.3, 18.4, 18.5, and 18.6 show some diseases and syndromes featuring chorea.

In otherwise healthy children who acquire chorea, poststreptococcal SC is most common, but chorea associated with systemic lupus erythematosus, antiphospholipid syndrome, iatrogenic, or acute metabolic and vascular injuries may need to be considered. In chronic encephalopathies, chorea may also occur. The presence of acute or chronic chorea suggests disordered neural transmission or structural pathology in the basal ganglia or occasionally in other structures. Athetosis and ballism have similar etiologies. Athetosis is more likely to occur in chronic neurologic conditions, and ballism may occur in SC or as hemiballism in acute vascular basal ganglia injury.

Complementary Studies

Peripheral and central neurophysiologic studies are rarely helpful in diagnosis, prognosis, or treatment decisions for chorea. Neuroimaging will help to rule out structural causes of chorea and should be obtained in all cases of acute and subacute hemichorea or hemiballism (particularly if there is no history supportive of Sydenham chorea). MRI should be done to exclude vascular, neoplastic, or inflammatory pathology in the basal ganglia or adjacent structures. In addition, MRI can also reveal evidence of pathologies of the basal ganglia in immunological or metabolic choreas. MRI can detect pathology in genetic choreas like frontal and caudate atrophy in Huntington's disease or striatal hyperintensities in choreoacanthocytosis. Specific studies must be performed according to the diagnostic suspicion of the causal disease mainly if chorea is suspected to be an associated symptom of metabolic disease.

Chorea in Cerebral Palsy (CP)

CP is an umbrella term that includes a group of disorders with a chronic disturbance of movement or posture due to nonprogressive defects or lesions in an immature brain [43]. Classical forms of CP are spastic (about 50 %), dyskinetic (about 20 %),

Table 18.1 Some diseases or syndromes featuring chorea with onset in infancy

Disease	Hallmark features other than chorea	References
Cerebral palsy	This chapter	
Kernicterus	This chapter	
HIV infection	Vertical transmission	
Infantile bilateral striatal necrosis	Autosomal recessive. Mutation of nup62. Choreoathetosis, developmental regression, intellectual impairment, optic atrophy, dystonia, spasticity, and severe bilateral striatal atrophy	[2]
Infantile bronchopulmonary dysplasia		[3]
Moyamoya disease	Progressive occlusion of the arteries of the circle of Willis	[4–6]
Lesch-Nyhan disease	Athetoid movements are often the early symptom	[7]
Dopamine synthesis defects	Several diseases both autosomal dominant and recessive, with variable age of onset and response to l-dopa treatment	[8, 9]
Benign familial chorea	See Chap. 3	
TSEN54-related pontocerebellar hypoplasia (PCH)	Autosomal recessive. Gen TSEN54. Three types (PCH2, 4, 5). Brainstem and cerebellar hypoplasia. Microcephaly, severe cognitive impairment, seizures, poor feeding	[10, 11]
Leigh syndrome	Autosomal recessive. Mutation in the E1-alpha polypeptide 1 of the pyruvate dehydrogenase complex 1 gene. Also parkinsonism, dystonia, and/or myoclonus. High T2 signal in basal ganglia. Elevated CSF lactate/pyruvate	[12, 13]
3-methylglutaconic aciduria type III	Autosomal recessive. Mutations in the *OPA3* gene. Early-onset optic atrophy and choreiform movement, later spasticity, ataxia, dementia	[14]
Isolated sulfite oxidase deficiency	Autosomal recessive. Increased urinary excretion of sulfite, thiosulfate, taurine, and sulfocysteine. Early-onset neonatal form and late-onset form have been described. Choreoathetosis is a frequent sign. Neuroimaging may show diffuse brain edema and multicystic leukoencephalopathy	[15, 16]
Molybdenum cofactor deficiency	Autosomal recessive. Xanthine and hypoxanthine increased secretion. Clinical similar to isolated sulfite oxidase deficiency	[17]
Nonketotic hyperglycinemia	Autosomal recessive. Mutations in several genes in mitochondrial glycine cleavage syndrome. Neonatal onset. Hypotonia, severe myoclonic epilepsy, vertical gaze palsy, profound intellectual impairment	[18, 19]
Alternating hemiplegia of childhood	Onset before age of 18 months. Mutations de novo in *ATP1A* gene. Episodes of hemiplegia, dystonia, or chorea affecting either side and disappearing with sleep	[20]
Idiopathic basal ganglia calcification (IBGC)	Autosomal recessive. Basal ganglia calcification. Tetraplegia, severe intellectual impairment, microcephaly	[21, 22]

Table 18.2 Some diseases with acute "pure" chorea with onset in childhood

Disease	Hallmark features other than chorea	References
Sydenham chorea	See Chap. 5	
Systemic lupus erythematosus-SLE/ antiphospholipid antibody syndrome APS	See Chap. 5	
Borrelia infection	Occurs commonly in the 5–14-year age group	[23]
Glut1 deficiency	This chapter	
Drug induced	See Chap. 16	
Psychogenic chorea	See Chap. 22. About 3 % of movement disorders in children have functional origin. Female and prepuberal age predominates	[24, 25]

Table 18.3 Some diseases with acute/subacute associated chorea with onset in childhood

Disease	Hallmark features other than chorea	References
GM1 and GM2 gangliosidosis	Also dystonia, tremor, rigidity	[26]
	Frequency dystonia: GM1 95 %, GM2 33 %, GM2 – signs of anterior horn	
	Cell involvement, high T2 signal in basal ganglia, atypical cherry red	
	Spot (GM1)	
Brain creatine deficiency	This chapter	
Cavernous hemangioma	Heterogeneous, popcorn-like hyperintense lesion on MRI	[27, 28]
Niemann-Pick type C		[29]
Biotinidase deficiency	Also parkinsonism, dystonia, myoclonus	[30]
Dentatorubral-pallidoluysian atrophy	Autosomal dominant. Mainly onset in adults. Expanded trinucleotide CAG repeat in the gene encoding atrophin-1. Ataxia, tics, dementia, seizures	[31, 32]
Infectious mononucleosis		[33]
Early Huntington disease	See Chaps. 6 and 7	
Huntington-like disease 3	See Chap. 8	
Wilson disease	Autosomal recessive treatable disease. Mutations in ATP7B gene. Onset neurologic manifestations after 8 years of age. Great variability of symptoms. Insidious onset. A choreic form is infrequent in children. Dystonia, dysarthria, facial grimacing, and ataxia are frequent symptoms. Kayser-Fleischer ring is a helpful sign	[34]
Chorea-acanthocytosis	See Chap. 4	
Ataxia telangiectasia	Autosomal recessive. Mutations in the ATM gene	[35]
	Onset early childhood. Ataxia. Conjunctival telangiectasias	
Nonketotic hyperglycemia	Mainly in diabetic adults. MRI shows increased signal in putamen on T1 imaging	[36]

Table 18.4 Some diseases with chorea following encephalopathic crises		
Sequel to status epilepticus	[37]
Glutaric aciduria	[38]

Table 18.5 Other choreas

Postpump chorea	This chapter

Table 18.6 Rare X-linked genetic conditions involving choreoathetosis

	Inheritance	Hallmark features	References
Allan- Herndon-Dudley syndrome	X-linked	Athetosis, severe mental retardation, dysarthria spastic paraplegia	[39, 40]
Schimke X-linked mental retardation syndrome	X-linked	Childhood-onset choreoathetosis, spasticity, acquired microcephaly, growth and mental retardation, external ophthalmoplegia, deafness	[41]
Pettigrew syndrome	Xq26	Choreoathetosis, seizures, severe mental retardation, hypotonia with progression to spasticity, Dandy-Walker malformation, and iron accumulation in the basal ganglia	[42]

ataxic (about 10 %), and mixed (about 20 %), but most cerebral palsy patients are mixed in type to some degree; for example, mild dyskinetic signs are often present in spastic CP. It is common to use dystonic and choreoathetotic (the combination of chorea and athetosis) subtypes for subgrouping dyskinetic CP [44–46].

The incidence of the choreoathetotic type in Western countries is about 0.15 per 1,000 newborn infants [46].

Clinical Features

Choreoathetotic subtype of dyskinetic CP is characterized by the predominance of choreic and distal athetotic movements and varying muscle tone. Some degree of spasticity is often present. The essential defect is a disturbed capacity to organize and execute voluntary movements. It is important to realize that in choreoathetotic CP the pure forms are rare. More commonly, an individual child will coexist in some degree with spasticity and dystonia.

The typical presentation is a delay in attaining motor milestones and abnormal persistence of archaic motor reactions. In the first months of life, diagnosis may be difficult as the clinical manifestations are nonspecific with a predominance of hypotonia, brisk primary reflexes, (especially tonic asymmetrical neck reflexes), exaggerated startle responses, and a tendency to neck and back hyperextension on stimulation.

The full-blown picture generally begins to be progressively suggestive between the first and second birthday and may progress slowly for several years. Choreic and athetotic movements usually involve all four extremities with a predominance in the upper. They show marked variability depending on the state of the individual; they are decreased during relaxation and sleep and increased by crying and stress. Oral dyskinesia is a common symptom and causes swallowing difficulties. Drooling may be a major problem [47]. Tendon reflexes are normal or increased in the lower limbs. In very rare cases febrile illness due to nonstreptococcal infection can be a precipitating factor of severe, long-lasting episodes of continuous intense chorea or ballismus.

Intellectual impairment and/or learning difficulties are present in 30 % of patients [45], but often children are considered as retarded because of motor and speech difficulties despite a normal cognitive level. Epilepsy, often relatively benign, is present in 20–25 % of cases [45, 48]. Hypoacusia and visual problems, especially strabismus and abnormalities of refraction [49], are common.

The resulting variable disability is a lifelong one, and some of the patients are unable to walk independently.

Neuroimaging

In the diagnostic assessment of the child with choreoathetotic CP, a neuroimaging study is recommended, preferably magnetic resonance imaging (MRI). CT scan usually shows only nonspecific abnormalities such as ventricular dilatation and increased width of sulci suggestive of cerebral atrophy [50]. MRI in series of dyskinetic CP patients of probable perinatal origin showed bilateral lesions of increased T2 signal in the ventrolateral thalamus, putamen, and/or hemispherical white matter [51–53].

In the study of Krägeloh-Mann et al. on [54] MRI in CP, athetoid CP cases had periventricular MRI lesions (8 % of the total CP cases). Basal ganglia/thalamus or bilateral cortico-subcortical lesions were found in 12 athetoid cases from 62 term-born (18 %) children with CP.

Etiology

Choreoathetotic form of CP tends to occur in term infants with severe perinatal asphyxia.

Diagnosis

When there is parental consanguinity, another similarly affected sibling, or when no clear abnormality was observed in the perinatal, it is important to follow the patient for progressive changes and to consider a broader diagnostic evaluation. Metabolic encephalopathies of early infantile onset can be misdiagnosed as choreoathetotic CP. Glutaric aciduria, Lesch-Nyhan disease, Pelizaeus-Merzbacher disease (the latter two when the patient is a boy), and mitochondrial diseases are the metabolic diseases most frequently erroneously diagnosed as choreoathetotic CP. Special attention has to be given to defects of the dopamine synthesis (such as tyrosine hydroxylase deficiency) because there exists effective treatment.

Management

The management of choreoathetotic CP requires a comprehensive multidisciplinary team approach that can deal with the numerous physical and behavioral needs of the child and family. Clear information to the family through the process is essential.

Physiotherapy, speech therapy, prevention of contractures, and orthopedic help to avoid secondary are essential in the long-term management of these children.

Drug therapy has only a minor place in most patients and is largely empirical. Anticholinergic medication in patients with chorea or choreoathetosis does not benefit or often even worsen [55]. Benzodiazepines, sodium valproate [56], carbamazepine, levetiracetam [57], and neuroleptics have been reported useful.

Botulinum toxin is useful in drooling problems [58].

Deep brain stimulation has been tried in small series [59] with inconclusive results, and further studies in children are ongoing.

Kernicterus

Kernicterus or chronic bilirubin encephalopathy is a permanent sequel of brain deposition (generally in the globus pallidus, subthalamic nuclei, hippocampus, substantia nigra, and cranial nerve nuclei), in the neonatal period, of high serum levels (total serum bilirubin >25 mg/mL) of unconjugated bilirubin. A role of prenatal factors is suggested by the relative frequency of preterm birth or low birthweight for term [60], and moreover, in some sick children kernicterus may occur in the absence of marked hyperbilirubinemia [60]. Kernicterus is now very rare in industrialized countries [61, 62]. Generalized choreoathetosis is the predominant symptom. Hearing loss and impaired upgaze are frequent. Intellectual disability is variable. Paroxysmal dyskinesia has been reported [63].

In about 2/3 of these cases, MRI findings shift from bilateral T1 hypersignal in the globus pallidus in the acute situation to permanent T2 hypersignal in sequel status [60, 64, 65]. But sequel lesions may not be visible even on MRI.

Postmortem studies show bilirubin staining and neuronal damage in globus pallidi, subthalamic nuclei, hippocampus, substantia nigra pars reticulate, and brainstem and cerebellar nuclei.

Management

The general considerations previously related concerning treatment of choreoathetotic CP are applied to kernicterus. In kernicterus, special attention should be given to the hearing loss that often necessitates the use of sign language. New developments in computer science open the hope of better substitute for verbal language.

Postpump Chorea

Generalized chorea, mainly involving the limbs, facial musculature, and tongue, occurs as a complication of *cardiac surgery* (postpump chorea) in about 1 % of children with cardiopulmonary bypass and deep hypothermia [66–70]. Associated neurologic manifestations include hypotonia, oral-facial dyskinesia, and pseudobulbar signs. Symptoms appear between 2 days and 2 weeks postoperatively. MRI shows no lesions. About 50 percent of the cases have a complete regression of the abnormal movements 1–4 weeks after onset, but others can have a severe irreversible chorea associated with dysphagia and/or dysarthria. The cause of the syndrome is unclear. Curless et al. [70] suggested that chorea results from cerebral vasoconstriction caused by hypocapnia and that close monitoring of pCO_2 reduces the incidence of this syndrome.

Glut1 Deficiency

Glut1 is the major transporter of glucose across the blood-brain barrier and into the glia. As glucose is the main source of energy for the brain. Glut1 deficiency results in cerebral energy failure. Glut1 protein is encoded by the SLC2A1 gene. SLC2A1 mutations usually have an autosomal mode of inheritance. Autosomal recessive mutations have been reported [71].

Clinical Features

Glut1 deficiency is characterized by a wide range of clinical manifestations and various degrees of severity. Globally, chorea is present in 75 % of all these cases [72]. Classical phenotype presents with infantile epilepsy, developmental delay, acquired microcephaly, intellectual impairment, and varying degrees of spasticity, ataxia, and movement disorders [72, 73]. Another phenotype is characterized by chronic choreoathetosis and dystonia and paroxysmal episodes of abnormal head and eye movements [74]. According to the review of Pons [73], chorea was noted in

43 patients from 57. In 12 patients chorea was noted during walking and resembled the gait of patients with SC. Cases with chorea as the main symptom have been reported [75].

Paroxysmal exercise-induced dyskinesia (PED) is another phenotype [76]. PED is triggered by prolonged exercise or fasting. The episodes last from minutes to hours, with variable frequency. Dystonia (mainly as postural limb dystonia and dystonic gait) and chorea (most commonly involving the limbs and face) are the main types of abnormal movements during the episodes.

Biochemical Investigations

The most useful diagnostic test for Glut1 deficiency is hypoglycorrhachia with fasting CSF glucose <40 mg/dL or a fasting CSF/plasma glucose ratio less than 45 %. The diagnosis can be confirmed by mutation analysis of the SLC2A1 gene. Normal CSF analysis does not exclude the diagnosis, and genetic screening is warranted if clinical suspicion is strong [72].

Neuroimaging

Brain MRI in patients with Glut1 deficiency is usually normal, and PET scan typically shows decreased glucose in the cortex, especially the mesial temporal regions and the thalami [77].

In childhood, EEG might show generalized discharges of spikes or polyspike waves. In most patients, a significant decrease of EEG abnormalities can be observed in postprandial EEG recordings [72, 78].

Treatment

Ketogenic diet (KD) provides ketone bodies that bypass the Glut1 defect and represent an alternative energy source to the brain. KD is very effective in controlling seizures and movement disorders.

Anti-NMDA-R Encephalitis

Anti-N-methyl-D-aspartate receptor (anti-NMDA-R) encephalitis is a severe autoimmune disease with autoantibodies binding to the NR1/NR2 subunits of the NMDA-R [79]. Other terms as "immune-mediated chorea encephalopathy syndrome" [80] and "encephalopathy with abnormal movements and cognitive disturbance" [81] very probably correspond to this disease. Although the initial cases were described in adults as paraneoplastic, a quarter of children with this disease do not have associated tumors. This may represent an important difference between pediatric and adult disease.

This encephalitis affects children of all ages (1–16 years) with a female predominance [82]. Previously well children often first present with an upper respiratory tract infection, headache, vomiting, sudden change in behavior, and seizures [82, 83]. Although paroxysmal events may appear as epileptic, EEG often does not show epileptic discharges and proves not to be epileptic [83, 84]. This prodromal phase is described in about a half of the children. It may persist for 5 days to 2 weeks [83, 85–87]. This may progress to full-blown catatonia, self-injurious behaviors, aggression, mutism, and restlessness. Sleep disorders including insomnia and sleep inversion are common. In this phase, many of the children had abnormal movements [82] including chorea, oculogyric crises, and characteristic stereotyped motor automatisms (sustained jaw movements, grimacing, oro-buccal "rabbit" movements) [88, 89].

Despite the dramatic clinical features, brain MRI is often normal. CSF often shows slight pleocytosis. Oligoclonal bands appear to be a useful marker.

Diagnosis is made by finding positive autoantibodies in CSF and/or serum against NR1/NR2 subunits of the NMDA-R. Due to the association with ovarian teratoma in adults, screening for ovarian tumors is recommended but as say before the most young children do not have a paraneoplastic process [82]. The encephalitis episode typically lasts for 2–4 months, requiring hospital prolonged inpatient management. Recovery may take 3 years or longer [90].

Acyclovir, steroids, intravenous immunoglobulin, anti-inflammatory agents, immunosuppressants (cyclophosphamide and rituximab), plasma exchange, and monoclonal antibodies directed against CD20 lymphocytes have all been trialed with variable response [83].

The outcome is variable from complete or near-complete recovery (75 % in patients with prompt aggressive treatment, to death, including sever sequelae [87, 89].

Ballismus

Ballismus is usually considered as an extreme form of chorea. It is rare in infancy and childhood. At this age, some cases of rheumatic chorea with ballismus have been reported. Other cases have an unclear cause. Ballismus can also be unilateral (hemiballismus) due to vascular lesions, infection [91], or tumors of the subthalamic body or of the caudate nuclei [92] which is the common lesion in adult cases.

Febrile illness due to nonstreptococcal infection can be a precipitating factor of severe, long-lasting episodes of continuous intense chorea or ballismus. The movements stop while the patient is asleep but promptly recur on awakening.

Such patients had usually preexisting dyskinetic cerebral palsy often of undetermined etiology [93–96]. Transaminases, creatine kinase, and other enzymes are markedly elevated coinciding with the maximal ballistic activity, but return to normal range once the movement disorder improves. The marked muscle activity is also responsible for a further rise in body temperature. EEG does not show paroxysmal activity.

At the first episode these patients cause diagnostic difficulties. Status epilepticus (in a child with a preexisting encephalopathy), metabolic disorder (mainly mitochondrial encephalopathy triggered by the increased catabolism produced by the fever), severe rheumatic chorea, and encephalitis are the main differential diagnoses. The severe prolonged muscular activity in these patients may be dangerous. Tachycardia, poor peripheral perfusion, and dehydration must be strictly controlled. Lowering of the temperature is essential. Haloperidol, levetiracetam, benzodiazepines, carbamazepine, and phenytoin are the most effective drugs.

Brain Creatine Deficiency

Patients with brain creatine (alpha-methyl-guanidinoacetic acid) deficiency show a complex syndrome starting at infancy with psychomotor developmental arrest, autistic features, epilepsy, and later dystonic and/or choreic movements [97]. Creatine and creatine phosphate play essential roles in the storage and transmission of phosphate-bound energy and in other brain activities such as neurotransmitter release. Three causes of deficit of creatine have been, until now, reported: guanidinoacetate methyltransferase (GAMT) deficiency, arginine glycine amidinotransferase (AGAT) deficiency (both autosomal recessive disorders), and an X-linked creatine transport defect. In all these creatine deficiencies, the brain MR spectroscopy shows absence or marked reduction of brain creatine. MRI shows pallidal hypointensities in T1-weighted images and hyperintensity in T2-weighted images [97] or myelination delay of white matter. Guanidinoacetic acid, the immediate precursor of creatine, is elevated in urine.

Oral administration of creatine monohydrate in GAMT and AGAT deficiencies increases cerebral creatine and improves the abnormal movements. Dietary treatment (arginine restriction and ornithine supplementation) has been reported to be beneficial [98].

Symptomatic Treatment of Chorea in Children

Only a few cases of chorea are due to a disease with a treatable cause (Sydenham chorea, hyperthyroidism, drugs, infections). Treatment differs between acute and chronic choreas.

In autoimmune-mediated forms of acute choreas, mainly Sydenham chorea, corticosteroids [99, 100] and immune-modulating treatments such as intravenous immunoglobulins have been used [101].

Because both chorea and ballismus are considered to be due to increased dopaminergic activity, symptomatic treatment with dopamine receptor-blocking agents has been used for many years. Phenothiazines such as chlorpromazine or butyrophenones (haloperidol), pimozide [28], and tetrabenazine [102, 103] are the most frequently used. Anticonvulsants such as carbamazepine and valproate may also be helpful [104, 105].

Concerning chronic choreas, the response to treatment is usually poor. Haloperidol, pimozide, and carbamazepine may be tried.

References

1. Fernández-Alvarez E. Prevalence of paediatric movement disorders in a Spanish tertiary Neuropediatric Department. In: Fernández-Alvarez E, Arzimanoglou A, Tolosa E, editors. Paediatric movement disorders. Progress in understanding. Paris: John Libbey Eurotext; 2005. p. 1–18.
2. Basel-Vanagaite L, Muncher L, Straussberg R, et al. Mutated nup62 causes autosomal recessive infantile bilateral striatal necrosis. Ann Neurol. 2006;60:214–22.
3. Hadders-Algra M, Boss AF, Martijn A, Prechtl HFR. Infantile chorea in an infant with severe bronchopulmonary dysplasia: an EMG study. Dev Med Child Neurol. 1994;36:177–82.
4. Hong YH, Ahn TB, Oh CW, Jeon BS. Hemichorea as an initial manifestation of moyamoya disease: reversible striatal hypoperfusion demonstrated on single photon emission computed tomography. Mov Disord. 2002;17:1380–3.
5. Shanahan P, Hutchinson M, Bohan A, O'Donohue D, Sheahan K, Owens A. Hemichorea, moya-moya, and ulcerative colitis. Mov Disord. 2001;16:570–2.
6. Wantanabe K, Negoro T, Maehara H, Takahashi I, Nomura K, Miura K. Moyamoya disease presenting with chorea. Pediatr Neurol. 1990;6:40–2.
7. Jinnah HA, De Gregorio L, Harris JC, et al. The spectrum of inherited mutations causing HPRT deficiency: 75 new cases and a review of 196 previously reported cases. Mutat Res. 2000;463:309–26.
8. Fernandez-Alvarez E. Dystonia. The paediatric perspective. Eur J Neurol. 2010;17 Suppl 1:46–51.
9. Pons R. The phenotypic spectrum of paediatric neurotransmitter diseases and infantile parkinsonism. J Inherit Metab Dis. 2009;32:321–32.
10. Barth PG, Blennow G, Lenard HG, et al. The syndrome of autosomal recessive pontocerebellar hypoplasia, microcephaly, and extrapyramidal dyskinesia (pontocerebellar hypoplasia type 2): compiled data from 10 pedigrees. Neurology. 1995;45(2):311–7.
11. Namavar Y, Barth PG, Poll-The BT, Baas F. Classification, diagnosis and potential mechanisms in pontocerebellar hipoplasia. Orphanet J Rare Dis. 2011;12:6–50.
12. Macaya A, Munell F, Burke RE, De Vivo DC. Disorders of movement in Leigh syndrome. Neuropediatrics. 1993;24:60–7.
13. Distelmaier F, Koopman WJ, van den Heuvel LP, Rodenburg RJ, Mayatepek E, Willems PH, et al. Mitochondrial complex I deficiency: from organelle dysfunction to clinical disease. Brain. 2009;132:833–42.
14. Anikster Y, Kleta R, Shaag A, Gahl WA, Elpeleg O. Type III 3-methylglutaconic aciduria (optic atrophy plus syndrome, or Costeff optic atrophy syndrome): identification of the OPA3 gene and its founder mutation in Iraqi Jews. Am J Hum Genet. 2001;69(6):1218–24.

15. Barbot C, Martins E, Vilarinho L, Dorche C, Cardoso ML. A mild form of infantile isolated sulphite oxidase deficiency. Neuropediatrics. 1995;26:322–4.
16. Lee HF, Mak BSC, Chi CS, Tsai CR, Chen CH, Shu SG. A novel mutation in neonatal isolated sulphite oxidase deficiency. Neuropediatrics. 2002;33:174–9.
17. Mize C, Johnson JL, Rajagopalan KV. Defective molybdopterin biosynthesis: clinical heterogeneity associated with molybdenum cofactor deficiency. J Inherit Metab Dis. 1995;18:283–90.
18. Branca D, Gervasio O, Le Piane E, Russo C, Aguglia U. Chorea induced by non-ketotic hyperglycaemia: a case report. Neurol Sci. 2005;26(4):275–7.
19. Hoover-Fong JE, Shah S, van Hove JLK, Applegarth D, Toone J, Hamosh A. Natural history of nonketotic hyperglycinemia in 65 patients. Neurology. 2004;63:1847–53.
20. Rosewich H, et al. Heterozygous *de novo* mutations in *ATP1A3* in patients with alternating hemiplegia of childhood: a whole exome sequencing gene-identification study. Lancet Neurol. 2012;11(9):764–73.
21. Billard C, Dulac O, Bouloche J, et al. Encephalopathy with calcifications of the basal ganglia in children. A reappraisal of Fahr's syndrome with respect to 14 new cases. Neuropediatrics. 1989;20(1):12–9.
22. Geschwind DH, Loginov M, Stern JM. Identification of a locus on chromosome 14q for idiopathic basal ganglia calcification (Fahr disease). Am J Hum Genet. 1999;65:764–72.
23. Bingham PM, Galetta SM, Athreya B, Sladky J. Neurologic manifestations in children with Lyme disease. Pediatrics. 1995;96:1053–6.
24. Schwingenschuh P, Pont-Sunyer C, Surtees R, Edwards MJ, Bhatia KP. Psychogenic movement disorders in children: a report of 15 cases and a review of the literature. Mov Disord. 2008;23:1882–8.
25. Ferrara J, Jankovic J. Psychogenic movement disorders in children. Mov Disord. 2008;23: 1875–81.
26. Guazzi GC, D'Amore I, Van Hoof F, Fruschelli C, Alessandrini C, Palmeri S, Federico A. Type 3 (chronic) GM1 gangliosidosis presenting as infanto-choreo-athetotic dementia, without epilepsy, in three children. Neurology. 1988;38:1124–7.
27. Carpay HA, Arts WF, Kloet A, et al. Hemichorea reversible after operation in a boy with cavernous angioma in the head of caudate nucleus. J Neurol Neurosurg Psychiatry. 1994;57:1547–8.
28. Yakinci C, Durmaz Y, Korkut M, Aladag A, Onal C, Aydinli M. Cavernous hemangioma in a child presenting with hemichorea: response to pimozide. J Child Neurol. 2001;16:685–8.
29. Uc EY, Wenger DA, Jankovic J. Niemann-Pick disease type C: two cases and an update. Mov Disord. 2000;15:1199–203.
30. Gascon GG, Ozand PT, Brismar J. Movement disorders in childhood organic acidurias. Clinical, neuroimaging, and biochemical correlations. Brain Dev. 1994;16(Suppl):94–103.
31. Becher MW, Rubinstein DC, Leggo J, Wagster MV, Stine OC, Ranen NG, et al. Dentatorubral and pallidoluysian atrophy (DRPLA) clinical and neuropathological findings in genetically confirmed North American and European pedigrees. Mov Disord. 1997;12:519–30.
32. Brunetti-Pierri N, Wilfong AA, Hunter JV, Craigen WJ. A severe case of dentatorubropallidoluysian atrophy (DRPLA) with microcephaly, very early onset of seizures, and cerebral white matter involvement. Neuropediatrics. 2006;37:308–11.
33. Connelly KP, De Witt LD. Neurologic complications of infectious mononucleosis. Pediatr Neurol. 1994;10:180–4.
34. Ferenci P. Pathophysiology and clinical features of Wilson disease. Metab Brain Dis. 2004;19:229–39.
35. Woods CG, Taylor AMR. Ataxia telangiectasia in the British Isles: the clinical and laboratory features of 70 affected individuals. Q J Med. 1992;298:169–79.
36. Oh SH, Lee KY, Im JH, Lee MS. Chorea associated with non-ketotic hyperglycemia and hyperintensity basal ganglia lesion on T1-weighted brain MRI study: a meta-analysis of 53 cases including four present cases. J Neurol Sci. 2002;200:57–62.

37. Fowler WE, Kriel RL, Krach LE. Movement disorders after status epilepticus and other brain injuries. Pediatr Neurol. 1992;8:281–4.
38. Gitiuax C, Roze E, Kinugawa K, Lamand-Rouviere C, Boddaert N, et al. Spectrum of movement disorders associated with glutaric aciduria type I. A study of 16 patients. Mov Disord. 2008;23:2392–7.
39. Stevenson RE, Goodman HO, Schwartz CE, et al. Allan-Herndon syndrome. I. Clinical studies. Am J Hum Genet. 1990;65:446–53.
40. Reyniers E, Van Bogaert P, Peeters N, et al. A new neurological syndrome with mental retardation, choreoathetosis, and abnormal behavior maps to chromosome Xp11. Am J Hum Genet. 1999;65:1406–12.
41. Schimke RN, Horton WA, Collins DL, Therou L. A new X-linked syndrome comprising progressive basal ganglia dysfunction, mental and growth retardation, external ophthalmoplegia, postnatal microcephaly and deafness. Am J Med Genet. 1984;17:323–32.
42. Pettigrew AL, Jackson LG, Ledbetter DH. New X-linked mental retardation disorder with Dandy-Walker malformation, basal ganglia disease, and seizures. Am J Med Genet. 1991; 38:200–7.
43. Bax M, Goldstein M, Rosenbaum P, Leviton A, Paneth M, et al. Proposed definition and classification of cerebral palsy. Dev Med Child Neurol. 2005;47:571–6.
44. Hagberg B, Hagberg G, Olow I. The changing panorama of cerebral palsy in Sweden 1954-1970. II. Analysis of the various syndromes. Acta Paediatr Scand. 1975;64:193–200.
45. Kyllerman M, Bager B, Bensch J, Bille B, Olow J, Voss H. Dyskinetic cerebral palsy. I. Clinical categories, associated neurological abnormalities and incidences. Acta Paediatr Scand. 1982;71:543–50.
46. SCPE Prevalence and characteristics of children with cerebral palsy in Europe. Dev Med Child Neurol. 2002;44:633–40.
47. Harris SR, Purdy AH. Drooling and its management in cerebral palsy. Dev Med Child Neurol. 1987;29:805–14.
48. Krageloh-Mann I, Bax M. Cerebral palsy. In: Aicardi J, editor. Diseases of the nervous system in childhood. 3rd ed. London: Mac Keith Press; 2009. p. 210–42.
49. Schenk-Rootlieb AJF, van Nieuwenhuizen O, van der Graaf Y, Wittebol-Post D, Willemse J. The prevalence of cerebral visual disturbance in children with cerebral palsy. Dev Med Child Neurol. 1992;34:473–80.
50. Kulakowski S, Larroche JC. Cranial computerized tomography in cerebral palsy. An attempt at anatomo-clinical and radiological correlation. Neuropediatrics. 1980;11:339–53.
51. Yokochi K, Aiba K, Kodama M, Fujimoto S. Magnetic resonance imaging in athetotic cerebral palsied children. Acta Paediatr Scand. 1991;80:818–23.
52. Rutherford MA, Pennock JM, Murdoch-Eaton DM, Cowan FM, Dubowitz LM. Athetoid cerebral palsy with cysts in the putamen after hypoxic-ischemic encephalopathy. Arch Dis Child. 1992;67:846–50.
53. Hoon AH, Reinhardt EM, Kelley RJ, Breiter SN, Morton H, Naidu SB, Johnston MV. Brain magnetic resonance imaging in suspected extrapyramidal cerebral palsy: observations in distinguishing genetic-metabolic from acquired causes. J Pediatr. 1997;131:240–5.
54. Krägeloh-Mann I, Helber A, Mader I, Staudt M, Wolff M, Groenendaal F, De Vries L. Bilateral lesions of thalamus and basal ganglia: origin and outcome. Dev Med Child Neurol. 2002;44:477–84.
55. Sanger TD, Bastian A, Brunstrom J, et al. Prospective open-label clinical trial of trihexyphenidyl in children with secondary dystonia due to cerebral palsy. J Child Neurol. 2007;22:530–7.
56. Kulkarni ML. Sodium valproate controls choreoathetoid movements of kernicterus. Indian Pediatr. 1992;29:1029–30.
57. Vles GF, Hendriksen JG, Visschers A, Speth L, Nicolai J, Vles JSH. Levetiracetam therapy for treatment of choreoathetosis in dyskinetic cerebral palsy. Dev Med Child Neurol. 2009;51:487–90.
58. Wilken B, Salami B, Backes H. Successful treatment of drooling in children with neurological disorders with Botulinum toxin A or B. Neuropediatrics. 2008;39:200–4.

59. Vidailhet M, Yelnik J, Lagrange C, et al. Bilateral pallidal deep brain stimulation for the treatment of patients with dystonia-choreoathetosis cerebral palsy: a prospective pilot study. Lancet Neurol. 2009;8:709–17.

60. Govaert P, Lequin M, Swarte R, Robben S, De Coo R, et al. Changes in globus pallidus with (pre)term kernicterus. Pediatrics. 2003;112:1256–63.

61. Johnson LH, Bhutani VK, Brown AK. System-based approach to management of neonatal jaundice and prevention of kernicterus. J Pediatr. 2002;140:396–403.

62. Manning D, Todd P, Maxwell M, Platt MJ. Perspective surveillance study of severe hyperbilirubinaemia in the newborn in the UK and Ireland. Arch Dis Child. 2007;92:F342–6.

63. Blakeley J, Jankovic J. Secondary paroxysmal dyskinesias. Mov Disord. 2002;17:726–34.

64. Martich-Kriss V, Kollias SS, Ball WS. MR findings in kernicterus. AJNR Am J Neuroradiol. 1995;16:819–21.

65. Coskun A, Yikilmaz A, Kumandas S, Karahan OI, Akcakus M, Manav A. Hyperintense globus pallidus on T1-weighted MR imaging in acute kernicterus: is it common or rare? Eur Radiol. 2005;15:1263–7.

66. Bjork VO, Hultquist G. Brain damage in children after deep hypothermia for open heart surgery. Thorax. 1960;15:284–91.

67. Brunberg JA, Doty DB, Reilly EL. Choreoathetosis in infants following cardiac surgery with deep hypothermia and circulatory arrest. J Pediatr. 1974;84:232–5.

68. Barratt-Boyes BG. Choreoathetosis as a complication of cardiopulmonary bypass. Ann Thorac Surg. 1990;50:693–4.

69. De Leon S, Ilbawi M, Archilla R, Cutilletta A, Egel R, Wong A, Quiñones J, Gerpelli JD, Azaka E, Riso A, Atik E, Ebaid M, Barbero-Marcial M. Choreoathetosis after cardiac surgery with hypothermia and extracorporeal circulation. Pediatr Neurol. 1998;19:113–8.

70. Curless RG, Katz DA, Perryman RA, Ferrer PL, Gelblum J, Weiner WJ. Choreoathetosis after surgery for congenital heart disease. J Pediatr. 1994;124:737–9.

71. Brockmann K. The expanding phenotype of GLUT1-deficiency syndrome. Brain Dev. 2009;31(7):545–52.

72. Pons R, Collins A, Rotstein M, Engelstad K, De Vivo DC. The spectrum of movement disorders in Glut-1 deficiency. Mov Disord. 2010;25(3):275–81.

73. Wang D, Pascual JM, Yang H. Glut-1 deficient syndrome: clinical, genetic, and therapeutic aspects. Ann Neurol. 2005;57:111–8.

74. Friedman JR, Thiele EA, Wang D, et al. Atypical GLUT1 deficiency with prominent movement disorder responsive to ketogenic diet. Mov Disord. 2006;21:241–5.

75. Perez-Duenas B, Prior C, Fernandez-Alvarez E, et al. Childhood chorea with cerebral hypotrophy. A treatable GLUT1 energy failure syndrome. Arch Neurol. 2009;66(11):1410–4.

76. Suls A, Dedeken P, Goffin K, et al. Paroxysmal exercise-induced dyskinesia and epilepsy is due to mutations in SLC2A1, encoding the glucose transporter GLUT1. Brain. 2008;131:1831–44.

77. Pascual JM, Van Heertum RL, Wang D, Engelstad K, De Vivo DC. Imaging the metabolic footprint of GLUT1 deficiency on the brain. Ann Neurol. 2002;52:458e64.

78. Verrotti A, D'Egidio C, Agostinelli S, Gobbi G. GLUT1 deficiency: when to suspect and how to diagnose? Eur J Paediatr Neurol. 2012;16:3–9.

79. Dalmau J, Tuzun E, Wu HY, et al. Paraneoplastic anti-N-methyl-D-aspartate receptor encephalitis associated with ovarian teratoma. Ann Neurol. 2007;61:25–36.

80. Hartley LM, Ng SY, Dale RC, et al. Immune mediated chorea encephalopathy syndrome in childhood. Dev Med Child Neurol. 2002;44:273–7.

81. Sebire G, Devictor D, Huault G, et al. Coma associated with intense bursts of abnormal movements and longlasting cognitive disturbances: an acute encephalopathy of obscure origin. J Pediatr. 1992;121:845–51.

82. Florance NR, Davis RL, Lam C, et al. Anti-N-methyl-D-aspartate receptor (NMDAR) encephalitis in children and adolescents. Ann Neurol. 2009;66:11–8.

83. Gataullina S, Plouin P, Vincent A, et al. Paroxysmal EEG pattern in a child with N-methyl-D-aspartate receptor antibody encephalitis. Dev Med Child Neurol. 2011;53:764–7.

84. Johnson N, Henry C, Fessler AJ, et al. Anti-NMDA receptor encephalitis causing prolonged nonconvulsive status epilepticus. Neurology. 2010;75:1480–2.
85. Iizuka T, Yoshii S, Kan S, et al. Reversible brain atrophy in anti-NMDA receptor encephalitis: a long-term observational study. J Neurol. 2010;257:1686–91.
86. Irani SR, Bera K, Waters P, et al. N-methyl-D-aspartate antibody encephalitis: temporal progression of clinical and paraclinical observations in a predominantly nonparaneoplastic disorder of both sexes. Brain. 2010;133:6–67.
87. Dalmau J, Lancaster E, Martinez-Hernandez E, et al. Clinical experience and laboratory investigations in patients with anti-NMDAR encephalitis. Lancet Neurol. 2011;10:63–74.
88. Barry H, Hardiman O, Healy DG, et al. Anti-NMDA receptor encephalitis: an important differential diagnosis in psychosis. Br J Psychiatry. 2011;199:508–9.
89. Dale RC, Irani SR, Brilot F, et al. N-methyl-D-aspartate receptor antibodies in pediatric dyskinetic encephalitis lethargica. Ann Neurol. 2009;66:704–9.
90. Fawcett RG. Acute psychosis associated with anti-NMDA-receptor antibodies and bilateral ovarian teratomas: a case report. J Clin Psychiatry. 2010;71:504.
91. Yoshikawa H, Oda Y. Hemiballismus associated with influenza A infection. Brain Dev. 1999;21:132–4.
92. Bhatia KP, Marsden CD. The behavioural and motor consequences of focal lesions of the basal ganglia in man. Brain. 1994;117:859–76.
93. Erickson GR, Chun RWM. Acquired paroxysmal movement disorders. Pediatr Neurol. 1987;3:226–9.
94. Kakinuma H, Hori A, Itoh M, Nakamura T, Takahashi H. An inherited disorder characterized by repeated episodes of bilateral ballism: a case report. Mov Disord. 2007;22:2110–31.
95. Recio MV, Hauser RA, Louis ED, Radhashakar H, Sullivan KL, Zesiewicz TA. Chorea in a patient with cerebral palsy: treatment with levetiracetam. Mov Disord. 2005;20:762–4.
96. Beran-Koenh MA, Zupanc ML, Patterson MC, et al. Violent recurrent ballism associated with infections in two children with static encephalopathy. Mov Disord. 2000;15:570–4.
97. Stöckler S, Isbrandt D, Hanefeld F, Schmidt B, von Figura K. Guanidinoacetate methyltransferase deficiency: the first inborn error of creatine metabolism. Am J Hum Genet. 1996;58:914–22.
98. Stockler S, Hanefeld F, Frahm J. Creatine replacement therapy in guanidinoacetate methyltransferase deficiency, a novel inborn error of metabolism. Lancet. 1996;348:789–90.
99. Paz JA, Silva CA, Marques-Dias MJ. Randomized double-blind study with prednisone in Sydenham's chorea. Pediatr Neurol. 2006;34:264–9.
100. Walker AR, Tani LY, Thompson JA, et al. Rheumatic chorea: relationship to systemic manifestations and response to corticosteroids. J Pediatr. 2007;151:679–83.
101. Garvey MA, Snider LA, Leitman SF, Werden R, Swedo SE. Treatment of Sydenham's chorea with intravenous immunoglobulin, plasma exchange, or prednisone. J Child Neurol. 2005;20:424–9.
102. Jankovic J, Clarence-Smith K. Tetrabenazine for the treatment of chorea and other hyperkinetic movement disorders. Expert Rev Neurother. 2011;11:1509–23.
103. Chatterjee A, Frucht SJ. Tetrabenazine in the treatment of severe pediatric chorea. Mov Disord. 2003;18:703–6.
104. Harel L, Zecharia A, Straussberg R, et al. Successful treatment of rheumatic chorea with carbamazepine. Pediatr Neurol. 2000;23:147–51.
105. Genel F, Arsnaloglu S, Uran N, Saylan B. Sydenhan's chorea: clinical findings and comparison of the efficacies of sodium valproate and carbamazepine regimens. Brain Dev. 2002;24:73–6.

Chapter 19
Psychogenic Chorea

Juan Carlos Giugni, Daniel Martínez-Ramírez,
and Ramon L. Rodríguez-Cruz

Abstract *Objective*: To describe the clinical and epidemiological characteristics of psychogenic chorea and report current treatment recommendations for psychogenic movement disorders (PMDs).

Summary: Neurologists and movement disorder specialists alike are frequently called upon to evaluate patients with bizarre movement disorders and suspected PMDs. The clinical evaluation and treatment of these disorders are frequently complex and time consuming. Psychogenic chorea is not as common as other psychogenic movement disorders, with reports ranging from 0 to 12 % in specialized movement disorder clinics. A comprehensive history, paying particular attention to onset of the symptoms; progression, inconsistency, and incongruence of movements; and the presence of psychiatric disturbance and distractibility, along with other false neurological signs are helpful in making the proper diagnosis. Although there are no specific diagnostic criteria for psychogenic chorea, we can use the available criteria already published for PMD. Treatment should be implemented in an organized process where the diagnosis has to be informed clearly to the patient and the family; a treatment plan should be explained including a psychiatrist in the team and expectations should be clearly explained.

Conclusion: Psychogenic choreas are an uncommon presentation for PMD, but still represent a diagnostic challenge for physicians. It is of great importance to recognize the red flags that suggest a psychogenic etiology. The presence of psychiatric disorders can be related with poor prognosis. Treatment must be initiated immediately for better prognosis.

Keywords Psychogenic movement disorders • Chorea • Conversion disorders • History • Functional movement disorders

J.C. Giugni, MD (✉) • D. Martínez-Ramírez, MD • R.L. Rodríguez-Cruz, MD
Department of Neurology, University of Florida Center for Movement
Disorders and Neurorestoration, Gainesville, FL, USA
e-mail: juan.giugni@neurology.ufl.edu; daniel.martinez-ramirez@neurology.ufl.edu;
ramon.rodriguez@neurology.ufl.edu

F.E. Micheli, P.A. LeWitt (eds.), *Chorea*,
DOI 10.1007/978-1-4471-6455-5_19, © Springer-Verlag London 2014

Introduction

It is common for physicians, in particular neurologists, to confront patients with underlying psychogenic disorders. In movement disorder centers, psychogenic movement disorders (PMDs) have been reported in 2–4 % of visits [1, 2], highlighting the significance for specialists and nonspecialists to learn how to diagnose and approach PMD. In this chapter, we will describe the definitions for PMD and chorea, a brief history of chorea, current epidemiological and clinical characteristics of psychogenic chorea, and clinical criteria for PMD and give an update of treatment and prognosis.

Definitions

Psychogenic movement disorders (PMDs) are defined as hyperkinetic or hypokinetic movements that cannot be attributed to any known structural or neurochemical disease of the nervous system and are interpreted as a result from an underlying psychiatric illness or malingering [2–4]. Despite notable developments in diagnostic techniques, advances in pathophysiological knowledge and treatments in PMD, diagnosis and treatment are still challenging for neurologists and psychiatrists. A variety of movement disorders can present as psychogenic and are classified according to the dominant phenomenology. Tremor, dystonia, gait disorders, and myoclonus are the most often reported movement disorders in large different series, with chorea representing a small percentage of PMD [1, 2, 4, 5]. Chorea is defined as involuntary movements of the trunk, extremities, face, and neck, which rapidly flow from region to region in an irregular pattern. The term chorea is derived from the Greek word χορεία (= dance).

History of Chorea

At the beginning of the sixteenth century, Paracelsus, an alchemist and physician, introduced for the first time the term Chorea Sancti Viti, describing the dancing mania and strange choreiform disease with epidemic proportion during the dancing procession in Echternach, Luxembourg, as part of mass hysteria characterized by outbreaks of collective movement disorders (Fig. 19.1) [6]. Later, in 1686, Sydenham erroneously used the same name of Chorea Sancti Viti to describe a totally different disease in childhood, currently known as Sydenham's chorea [7]. The most recognized disease related to chorea is a progressive and hereditary form, known as Huntington's chorea, described in 1872 by George Summer Huntington [8].

Fig. 19.1 The Dancing Plague of 1518". Engraving of Hendrik Hondius. Originally drawn by Peter Brueghel the Elder (1564)

History of Psychogenic Chorea

In 1889, Smith [9] perhaps was the first one to describe patients with hysterical chorea, but was Osler 5 years later who reported patients in his monograph named *On Chorea and Choreiform Affections* [10], where he established different manifestations of chorea, such as Sydenham's and Huntington's, but also reported what could be the first definition of hysterical chorea: "under this term are now embraced both the dancing mania and the various forms of rhythmical or hysterical disorders of motion." Arthur Van Gehuchten, in his posthumously published work *Les Maladies nerveuses*, reported an outbreak of sudden involuntary movement in 13 adolescent girls residing in orphanage [11]. He described the movements as "chorée salutatoire" (saluting chorea) and defined this type of chorea as "a nervous disease without known anatomic-pathological lesions" and also distinguished four types of chorea, Sydenham's chorea, chorea of pregnancy, Huntington's chorea, and hysterical chorea [11].

Epidemiology

In contrast to the nineteenth-century literature, psychogenic chorea is notably rare in the current large series of PMD, and this could be explained by several factors. First, the accelerated development in diagnostic techniques and knowledge of more organic or structural etiologies of chorea make psychogenic diagnosis less probable. We just need to compare the different etiologies reported by Osler and Van Gehuchten (Sydenham's chorea, chorea of pregnancy, hysterical chorea, and Huntington's chorea) with current reported etiologies in this field [7]. Second, the phenomenology of movement disorders has evolved into a unified framework. Many patients with PMD present with multiple or complex movements. So even if chorea is part of PMD, movements are probably reported by its dominant phenomenology. A third issue to consider is the cultural development around the world. In 1900, Henri Meige visited Echternach to observe the annual dancing procession, and he was disappointed with the lack of hysteria and concluded that the outbreaks of mass hysteria with a background of religious fervor, pagan traditions, or superstition are the most likely explanation for the medieval dancing mania [8]. After Sirois publication in 1974 [12], few outbreaks of mass hysteria have been reported [10, 11] and were associated to odd behavior and anxiety. All this together could contribute to the low percentage of psychogenic chorea in all the modern series of PMD. One last point to consider is that in previous centuries, bizarre movement disorders were classified as chorea and some of the phenomenologies previously diagnosed as chorea might have been another movement disorder [13].

The overall prevalence of psychogenic chorea is low, ranging from 0 to 12 % [1, 4] in different series if we take chorea, ballism, and athetosis in the same group (Table 19.1). In a large study reported by Thomas et al. [14], of 517 patients with PMD, only 3 (0.6 %) had chorea and the rest of the patients had the following movement disorders: 40.8 % tremor, 40.2 % dystonia, 17 % myoclonus, 4.3 % tics, 3.9 % gait disorder, 3.1 % parkinsonism, and 1.4 % dyskinesia, and 7.3 % had more than one form of PMD. In another large study of PMD in children,

Table 19.1 Reports of psychogenic chorea in different series of PMD

Author	PMD (N)	Age	Female (%)	Psychogenic chorea (%)
Factor et al. [1]	28	50 (17–83)	61	0
Hinson et al. [26]	88	40.3 (14–71)	75	12[a]
Thomas et al. [14]	571	42.3 (28–57)	72.8	0.6
Anderson et al. [38]	66	49.6 (37–62)	65.2	6
Ferrara et al. [15]	54	14.2 (7.6–17.7)	77.8	0[b]
Ertan et al. [5]	49	39 (8–86)	69.4	6.1[c]

[a]Included psychogenic chorea (7 %), ballism (2 %), and athetosis (3 %)
[b]In children
[c]Included chorea and ballism

tremor and dystonia were the most common movement disorders, and no chorea was reported [15]. Ganos et al. [16] reported 26 out of 245 patients having psychogenic paroxysmal movement, and of these, 42.2 % had complex generalized movement established as hyperkinesias, 26.9 % had combination of movement disorders, 15.3 % had dystonia, and 7.6 % had tremors or jerks. None had chorea, but the 42.2 % had been classified as hyperkinesia and this in the same point could be interpreted as chorea.

Interestingly, PMD usually does not present sex predominance during childhood. However, in the adult, women are more frequently affected [15, 17], with most common onset between 35 and 50 years old [18, 19]. Fekete [20] described a 38-year-old woman with psychogenic chorea and family history of Huntington's disease that fits the above description.

Clinical Features

A Even though this amazing clinical description was done by Smith [9] many years before the diagnostic criteria were proposed by Fahn and Williams in 1988 [3], it covered the clinical manifestations we currently use. Many of the clinical characteristics proposed by Fahn and Williams in 1988 [3] for psychogenic chorea, were amazingly described by Smith many years before, when he reported a female patient presenting with abnormal movements after being frightened by a dog. A great description of psychogenic movement disorders was made by Gupta et al. [21], which describes that PMD may affect any topographical distribution and can spread from one region to another, not necessarily respecting anatomical distribution. Symptom onset is often abrupt with a static course and severity tends to be maximal since the beginning. Symptoms can spread to multiple sites and be mixed with paroxysmal symptoms. The inconsistency and incongruence of the symptoms are often the rule. Inconsistency is being defined as a movement increasing with attention, decreasing with distraction, entrainment of tremor to the frequency of repetitive movements, and variability of phenomenology, while incongruence is defined as a mixture of different movements, atypical stimulus sensitivity, and paroxysmal attacks. Distractibility is another important characteristic we should keep in mind [22]. In chorea we can see movement rapidly flit from region to region in an irregular pattern, but in psychogenic chorea this pattern is not respected, the pattern is more regular, and there is commonly a combination of different type of movements, producing a bizarre clinical picture. Additional neurological signs can also suggest a PMD or in this case psychogenic chorea, such as the lack of motor impersistence, false weakness, false sensory signs, deliberate slowness, and suggestibility; patients commonly are able to maintain tongue protrusion [20], and saccadic eye movements are usually normal. In addition, there are different historical and clinical signs proposed to recognize a PMD. The presence of other psychogenic signs such as nonanatomical sensory loss, unusual gait, selective disability, change in reflexes, or exaggerated response to the stimulus should make you think of a psychogenic origin.

Psychiatric Features

It is important for the clinician to recognize that significant psychiatric disorders could develop with time in psychogenic movement disorders because their presence has been related with poor outcome and brings out the importance of including a psychiatrist in the treatment plan. Feinstein et al. [17] described a sample of 88 patients that were followed up on an average of 3.2 years, diagnosing a mental illness in 95.3 % of the subjects according to Axis I mental disorders. Lifetime prevalence rates for Axis I diagnoses were as follows: major depression 42.9 %, anxiety disorders 61.9 %, comorbid major depression and anxiety disorders 28.6 %, somatoform disorder in about 5 %, and adjustment disorder in 10 %. Personality disorders were also present in 45 % of the sample. There have been reports of suicide in patients suffering from PMD [17].

Diagnosis

Although there is not a specific set of diagnostic criteria for psychogenic chorea, there are elements in the history and physical examination that can help in the diagnosis (similar to other PMDs). We should look for red flags from the history, clinical examination, and diagnostic testing. As previously mentioned, an abrupt onset, static course with early development of maximal severity, spontaneous remissions, history of litigation, presence of secondary gain, and presence of psychiatric disturbance are suggestive of PMD, particularly with a bizarre or difficult to recognize movement disorder [22]. However, we should never forget that organic movement disorders might have bizarre presentation as well [23]. In addition, the diagnosis of PMD can coexist with an organic disease. We recently evaluated a female patient with antiphospholipid syndrome who developed psychogenic chorea [24]. Another pearl to take home is that family history could be misleading, as in Fekete's case of a female with psychogenic chorea but with family history of Huntington's disease [20].

Three sets of diagnostic criteria have been proposed to assist in diagnosis of psychogenic movement disorders [21, 25, 26]. The Fahn and Williams criteria proposed in 1988 have been the most used worldwide. They specify four levels of diagnostic certainty for psychogenic movement disorder (documented, clinically established, probable, and possible). Later in 1995, Fahn and colleagues [25] proposed to combine categories 1 and 2 (documented and clinically diagnosis) under one category "clinically definite." The Fahn and Williams criteria are based on incongruence or inconsistency of the movements and use the response to suggestion, to placebo, or to psychotherapy to define the highest level of diagnostic certainty. Although electrophysiological tests can help distinguish organic versus psychogenic movements, they are helpful only for tremors or myoclonus. Currently, we do not have a test to differentiate between organic and psychogenic chorea. Therefore, the diagnosis of psychogenic chorea is still based on clinical examination and the experience of the examiner is critical in this step. Another tool we can utilize when faced with a suspected PMD is the Rating Scale for Psychogenic Movement Disorders, which has been shown to be useful measuring the response to therapeutic interventions [27].

Treatment and Prognosis

Treatment of psychogenic movement disorders is complex, requiring good communication skills and building exceptional rapport between patient and doctor. The most important step is to rule out an organic etiology of the bizarre presentation and begin treatment [28]. Effective presentation of the diagnosis is a very important step [22]. A treatment approach has been recommended when a suspected PMD is encountered in practice, which consists [29] informing the patient and family about the diagnosis, which has to be in a tactful and careful manner [30], and the treatment plan has to be carefully explained. It is important to explain the expectations and role of each member of the therapeutic team [31]. Unfortunately, one of the main obstacles for the treatment of these disorders is the lack of resources to take care of this type of patients. These visits tend to take longer and for some practitioners might not be rewarding. Overall, the prognosis is considered to be poor with large numbers of patients reporting inability to work [32]. One of the main challenges is to determine whether the patient sincerely wishes to improve; otherwise, psychotherapy and physiotherapy would be a waste of time and resources [25, 33]. It is also important to remember that factitious disorders and malingering are usually not benefited from therapy and are considered negative prognostic factors [34]. On the other hand, somatoform disorders can usually be treated. Younger age and shorter duration of symptoms seem to have the most favorable outcome [17]. A personality disorder is also a negative prognostic factor [35].

Unfortunately, there are no official guidelines for the treatment of psychogenic movement disorders, and there continues to be a lack of prospective, randomized controlled trials. Additional therapies, such as antidepressants [34], cognitive behavioral therapy [30], transcranial magnetic stimulation [36], and physical therapy [33], have been used to treat these disorders along with psychotherapy [25] providing good results in case reports or small series. The importance of undergoing psychotherapy, either early or late opposed to neurological observation and support, has been recently reported in a randomized study [37].

What should we expect in these patients after treatment? The grade of benefit varies according to published literature. Improvements of the psychogenic movements have been reported in 35–56 % of patients [1, 14, 25]. Relapses have been reported to occur in 21 % [25], symptoms could even be worse in 22 % [14], and in 12–21 % [14, 25] symptoms remained the same. Development of additional unexplained medical symptoms in almost 40 % of patients in long-term follow-up studies has also been reported [17].

Conclusions

Psychogenic chorea represents a diagnostic challenge for clinicians and movement disorder specialists, and clinical experience aids in proper diagnosis. It is of great importance to recognize red flags and when to suspect a psychogenic origin, as early initiation of treatment increases the rate of sustained remission. Psychogenic chorea is not as common as other PMDs, such as tremor, dystonia, or myoclonus. It is important to recognize psychiatric features associated with PMD. Early intervention along with psychiatric assessment can be effective, but, unfortunately, overall prognosis for full remission can be poor.

References

1. Factor SA, Podskalny GD, Molho ES. Psychogenic movement disorders: frequency, clinical profile, and characteristics. J Neurol Neurosurg Psychiatry. 1995;59:406–12.
2. Thomas M, Jankovic J. Psychogenic movement disorders: diagnosis and management. CNS Drugs. 2004;18:437–52.
3. Fahn S, Williams DT. Psychogenic dystonia. Adv Neurol. 1988;50:431–55.
4. Hinson VK, Haren WB. Psychogenic movement disorders. Lancet Neurol. 2006;5:695–700.
5. Ertan S, Uluduz D, Ozekmekci S, Kiziltan G, Ertan T, Yalcinkaya C, et al. Clinical characteristics of 49 patients with psychogenic movement disorders in a tertiary clinic in Turkey. Mov Disord. 2009;24:759–62.
6. Aubert G. Charcot revisited: the case of Bruegel's chorea. Arch Neurol. 2005;62:155–61.
7. Walker RH. Chorea. Continuum (Minneap Minn). 2013;19:1242–63.
8. Krack P. Relicts of dancing mania: the dancing procession of Echternach. Neurology. 1999;53:2169–72.
9. Smith PB. Hysterical chorea. Br Med J. 1889;2:11–2.
10. Ellis SJ. Hysteria over doctors with HIV. Lancet. 1993;341:764.
11. Moscrop A. Mass hysteria is seen as main threat from bioweapons. BMJ. 2001;323:1023.
12. Sirois F. Epidemic hysteria. Acta Psychiatr Scand Suppl. 1974;252:1–46.
13. Goetz CG, Chmura TA, Lanska DJ. History of chorea: part 3 of the MDS-sponsored history of movement disorders exhibit, Barcelona, June 2000. Mov Disord. 2001;16:331–8.
14. Thomas M, Vuong KD, Jankovic J. Long-term prognosis of patients with psychogenic movement disorders. Parkinsonism Relat Disord. 2006;12:382–7.
15. Ferrara J, Jankovic J. Psychogenic movement disorders in children. Mov Disord. 2008;23:1875–81.

16. Ganos C, Aguirregomozcorta M, Batla A, Stamelou M, Schwingenschuh P, Munchau A, et al. Psychogenic paroxysmal movement disorders – clinical features and diagnostic clues. Parkinsonism Relat Disord. 2014;20:41–6.
17. Feinstein A, Stergiopoulos V, Fine J, Lang AE. Psychiatric outcome in patients with a psychogenic movement disorder: a prospective study. Neuropsychiatry Neuropsychol Behav Neurol. 2001;14:169–76.
18. Lang AE. Psychogenic dystonia: a review of 18 cases. Can J Neurol Sci. 1995;22:136–43.
19. Kim YJ, Pakiam AS, Lang AE. Historical and clinical features of psychogenic tremor: a review of 70 cases. Can J Neurol Sci. 1999;26:190–5.
20. Fekete R, Jankovic J. Psychogenic chorea associated with family history of Huntington disease. Mov Disord. 2010;25:503–4.
21. Gupta A, Lang AE. Psychogenic movement disorders. Curr Opin Neurol. 2009;22:430–6.
22. Hallet M. Psychogenic movement disorders and other conversion disorders. Cambridge: Cambridge University Press; 2011. p. 324.
23. Simon DK, Nishino S, Scammell TE. Mistaken diagnosis of psychogenic gait disorder in a man with status cataplecticus ("limp man syndrome"). Mov Disord. 2004;19:838–40.
24. Antiphospholipid Antibody Syndrome presenting as Psychogenic Chorea. 18th Interantional Congress of Parkinson's disease and Movement disorders. Abstract: 551531.
25. Williams DT, Ford B, Fahn S. Phenomenology and psychopathology related to psychogenic movement disorders. Adv Neurol. 1995;65:231–57.
26. Shill H, Gerber P. Evaluation of clinical diagnostic criteria for psychogenic movement disorders. Mov Disord. 2006;21:1163–8.
27. Hinson VK, Cubo E, Comella CL, Goetz CG, Leurgans S. Rating scale for psychogenic movement disorders: scale development and clinimetric testing. Mov Disord. 2005;20:1592–7.
28. Ljungberg L. Hysteria; a clinical, prognostic and genetic study. Acta Psychiatr Neurol Scand Suppl. 1957;112:1–162.
29. Fahn S, Jankovic J, Hallett M. Principles and practice of movement disorders. Edinburgh/New York: Elsevier Saunders; 2011. p. 1, online resource (vii, 548 p.).
30. LaFrance Jr WC, Friedman JH. Cognitive behavioral therapy for psychogenic movement disorder. Mov Disord. 2009;24:1856–7.
31. Williams DT, Ford B, Fahn S. Treatment issues in psychogenic-neuropsychiatric movement disorders. Adv Neurol. 2005;96:350–63.
32. Crimlisk HL, Bhatia K, Cope H, David A, Marsden CD, Ron MA. Slater revisited: 6 year follow up study of patients with medically unexplained motor symptoms. BMJ. 1998;316:582–6.
33. Czarnecki K, Thompson JM, Seime R, Geda YE, Duffy JR, Ahlskog JE. Functional movement disorders: successful treatment with a physical therapy rehabilitation protocol. Parkinsonism Relat Disord. 2012;18:247–51.
34. Voon V, Lang AE. Antidepressant treatment outcomes of psychogenic movement disorder. J Clin Psychiatry. 2005;66:1529–34.
35. Binzer M, Kullgren G. Motor conversion disorder. A prospective 2- to 5-year follow-up study. Psychosomatics. 1998;39:519–27.
36. Garcin B, Roze E, Mesrati F, Cognat E, Fournier E, Vidailhet M, et al. Transcranial magnetic stimulation as an efficient treatment for psychogenic movement disorders. J Neurol Neurosurg Psychiatry. 2013;84:1043–6.
37. Kompoliti K, Wilson B, Stebbins G, Bernard B, Hinson V. Immediate vs. delayed treatment of psychogenic movement disorders with short term psychodynamic psychotherapy: randomized clinical trial. Parkinsonism Relat Disord. 2014;20:60–3.
38. Anderson KE, Gruber-Baldini AL, Vaughan CG, Reich SG, Fishman PS, Weiner WJ, et al. Impact of psychogenic movement disorders versus Parkinson's on disability, quality of life, and psychopathology. Mov Disord. 2007;22(15):2204–9.

Index

F.E. Micheli, P.A. LeWitt (eds.), *Chorea*,
DOI 10.1007/978-1-4471-6455-5, © Springer-Verlag London 2014

363

CPSIA information can be obtained at www.ICGtesting.com
Printed in the USA
LVOW02*1233220714

395480LV00002B/7/P